Third Edition

Classical
Charged
Particles

Third Edition

Classical
Charged
Particles

Fritz Rohrlich

Syracuse University, New York, USA

World Scientific

NEW JERSEY • LONDON • SINGAPORE • BEIJING • SHANGHAI • HONG KONG • TAIPEI • CHENNAI

Published by

World Scientific Publishing Co. Pte. Ltd.

5 Toh Tuck Link, Singapore 596224

USA office: 27 Warren Street, Suite 401-402, Hackensack, NJ 07601

UK office: 57 Shelton Street, Covent Garden, London WC2H 9HE

British Library Cataloguing-in-Publication Data
A catalogue record for this book is available from the British Library.

CLASSICAL CHARGED PARTICLES (3rd Edition)

ISBN-13 978-981-270-004-9
ISBN-10 981-270-004-8

Printed in Singapore by World Scientific Printers (S) Pte Ltd

To Emily and Paul,
in the hope that some day
they too will learn to appreciate
the great beauty that lies
in the mathematical description
of our natural world

Preface to the Third Edition

Why would an author prepare a new edition of a forty-odd year old physics book? The short answer is that only in the last few years, a century old problem has been solved. The classical dynamics of charged particles has now become a completed theory.

Consider this history: the brilliant work of Hendrick Lorentz and Max Abraham deduced the dynamics of charged particles from Maxwell's equations; that dynamics was relativistically correct although it was developed before 1905 when Einstein published the special theory of relativity. Later, Dirac rederived the Lorentz-Abraham equations in a manifestly relativistic way (1938). For a full century, the Lorentz-Abraham-Dirac (LAD) equations have thus been the generalization to charged particles of Newton's fundamental dynamics.

But during all that time, people knew of a serious flaw in the LAD equations: they violate Newton's fundamental "law of inertia" that a body is not accelerated unless a force acts on it. Despite numerous unsuccessful attempts throughout the twentieth century (supplementing or modifying the LAD equations), this problem remained unsolved. The previous editions of the present book were written under that handicap.

Would it be worth a new edition if, finally, that old problem got solved? This is exactly what happened at the dawn of our new century. The present edition includes the new results. I tell how they came about, and I show how they complete the theory.

But there is more. Of all the symmetries of classical physics, the most recent and most intriguing symmetry is invariance under time reversal. This is the invariance of the fundamental equations under the reversal of time. Such a symmetry has in the past been shown to hold for quantum mechanics and quantum electrodynamics. It was also shown to hold for Maxwell's equations, but it has caused problems for the classical dynamics of charged particles. These problems have now also been resolved. The corresponding sections in the previous edition have been rewritten.

Finally, the theory as a whole must be related to the other physical theories: there is only one real world to be described by science so that the various descriptions by different theories must be interrelated. The present book is framed by this concern. An introductory discussion on this interrelation (Chapter 1) is followed at the end by placing the classical dynamics of charged particles into the general web of theories (Section 9-5). I show how the foundations of the present theory are now complete within the theory's prescribed validity domain.

This concludes the long answer to the question for a justification of this new edition. There are of course also more conventional reasons for a new edition: better ways of explaining things, more recent and better references, and the correction of the inevitable misprints that plague all printed matter (without introducing new ones).

I want to thank the many people who have helped me over the years to improve my understanding of the subject matter. They have talked to me, written to me, and they published books and articles that were invaluable to me. I cannot possibly recall the names of all those to whom I am indebted for help. The references throughout the book do justice to only a fraction of the many sources I consulted.

Finally, I am grateful to World Scientific for producing such a well-printed book despite the complication due to the many mathematical symbols.

Fritz Rohrlich
Fall 2006

Contents

9   The Theory's Structure and Place in Physics

S   The Physically Correct Dynamics

A1   The Space-Time of Special Relativity

A2   The Space-Time of General Relativity

Classical
Charged Particles

CHAPTER 1

Philosophy and Logic
of Physical Theory

We have undertaken to describe and study a particular theory in physics. It is therefore necessary that we ask ourselves exactly what this entails. Is it simply a set of equations from which other equations are derived? And if it were so, where do the first equations come from? Are we doomed from the start, because we are thinking about a classical theory when we know that classical physics is "wrong" and quantum physics is "right"? To what extent does it make sense to talk about an electron, say, in classical terms? These and similar questions clearly indicate that ignoring philosophy in physics means not understanding physics. For there is no theoretical physics without some philosophy; not admitting this fact would be self-deception. Let us then start with a very brief excursion into the philosophy of theoretical physics.

1–1 THE NATURE AND AIM OF THEORY IN PHYSICS

It is commonplace to say that physics is an experimental science. It is usually admitted that theory is at least useful. It is not always appreciated that theory is in general necessary in order to carry out meaningful experiments.

One could attempt to define theoretical physics as an intellectual enterprise in which one tries to bring order into the unending variety of physical phenomena by means of a logical-mathematical scheme. But this "scheme" is obviously of a very special kind. While one often speaks of "theory" in a very vague fashion, something quite specific emerges when the study of a particular subject matter has sufficiently advanced to permit the statement, "We understand this now."

A physical theory, in the narrow sense of the word, is a logical structure based on assumptions and definitions which permits one to predict the outcome of a maximum number of different experiments on the basis of a minimum number of postulates. One usually desires the postulates (or axioms) to be as self-evident as possible; one demands simplicity, beauty, and even elegance.

The "maximum number of different experiments" is better expressed as "maximum domain of validity." This, of course, contains the admission that *the validity domain of a theory is limited,* as is obvious to the physicist who has become acquainted with such theories as thermodynamics, geometrical optics, Newtonian mechanics, or Maxwell's electrodynamics.

We shall return to the question of validity limits in Section 1-2. Now we shall assume that we stay within these limits and that we have such a theory. We can then *explain* all phenomena that fall within the validity limits. This means that we are able to account for any such particular happening on the basis of some principles or postulates which are basic to the theory. For example, when two containers of the same gas at different temperatures are suitably connected, a certain average temperature will establish itself; the time-reversed process does not take place. This is explained by means of the second law of thermo-dynamics. We are satisfied with this explanation and we do not question the second law since it is an axiom on which thermodynamics is built.

Two remarks are essential in this connection. First, the validity limits are not drawn *ad hoc*. When a particular phenomenon is not accounted for by the theory, we cannot gerrymander the validity limits of the theory in order to "save the theory." At some point in the development these limits are established once and for all and cannot be moved arbitrarily. This point of development is reached when the covering theory has been developed, i.e. when the next higher level of theory is invented which contains the first one as an approximation. We shall see this in detail in the next section.

The second remark is the observation that the explanation achieved by the theory is in a certain sense tautological. For how was the theory obtained in the first place? Let us try to reconstruct this at least in a very cursory fashion: Experimental data are studied and compared; regularities are found; new experiments are suggested. Eventually one discovers a "law of nature." This means that one has *invented a proposition* which is confirmed whenever relevant experiments are carried out and for which there does not exist even a single violation. This law was inferred from the results of observations. By its validity we mean that we can safely turn things around, put the law first and deduce from it how the physical system "should behave;" the *prediction* will then be confirmed. This process is possible when nature continues to behave in the same way as it did before (a silent assumption underlying all of science) and when the law is indeed valid.

As we progress in discovering laws of nature (Faraday's law, Ampere's law) we attempt to establish logical relationships between them (which is usually done by means of mathematics) and we hope to arrive eventually at a *deductive logical scheme* (Maxwell's theory of electricity and magnetism) where all the specific instances can be deduced from a small number of fundamental postulates (Maxwell's equations). In this way a theory can be found.

Having gone through all the trouble that is necessary to find the "right" laws and the "right" theory, no one should be surprised that the predictions of the theory indeed agree with experiment or, for that matter, that a given phenomenon can be fully accounted for by the theory. Nevertheless, we say that the theory *explains* the phenomenon. In this sense *scientific explanation* is circular. The emphasis should really be on the existence and correctness of the deductive system, not on the explanation, for the theory is *not derived* or derivable from experiments. It is a mental step into the abstract and general, postulating validity in the future and for all experiments including those that have never been carried out before. The *existence* of a valid theory is therefore the nontrivial and indeed the very remarkable facet of scientific explanation.

We can now define the aim of theoretical physics not as an attempt at explaining the phenomena, but as a striving for the construction of more and more inclusive physical theories and the exploration of their ramifications.

The axiomatization of a theory is usually left to the philosophers of science. Few physicists take interest in this process beyond the purely mathematical aspects of it. Nevertheless, it should be realized that from the logical point of view the final product is an *axiomatic, deductive, logical-mathematical system*.

1-2 THE HIERARCHY OF THEORIES

Every theory is phenomenological to some extent. This means that every theory contains certain quantities which are not determinable by it and which must be fed into it on the basis of appropriate measurements. In optics the index of refraction must be given. In electrodynamics the relationship between **D** and **E** and between **B** and **H** must be known; if these relations are linear, the dielectric constant ϵ and the permeability μ must be given. In the theory of charged particles, which we shall discuss in the following chapters, the mass and charge of each particle must be given.

As our knowledge of the physical world increases, we are able to construct theories which are more and more sophisticated and which require fewer and fewer given data. The theories become less phenomenological. For example, solid state physics enables us to *compute* the dielectric constant on the basis of the chemical constituents and the physical conditions of a solid. Progress in theoretical physics, then, clearly points toward the eventual construction of an all-inclusive theory which is phenomenologically minimal. This minimal property is given by the minimum number of units (dimensions) that are necessary: the units of length, time, and mass, or any other three independent ones constructed from them. For a theory can give only dimensionless numbers; and these must be constructed from the quantities *with* dimensions which are fed into it.

A physical theory seems to have only a finite lifetime. Newtonian mechanics was replaced by relativistic mechanics, thermodynamics by statistical mechanics, classical by quantum mechanics. But did these theories become *wrong* after having been proved correct over such a long time? What actually happened was that they continued to provide the correct predictions, but only for a limited set of phenomena. For example, Newtonian mechanics became restricted to phenomena in which the velocities are small compared with the velocity of light. It became an *approximate theory*. Whether the approximate nature becomes apparent in a given instance depends entirely on the accuracy of the experiment.

Thus we learn that an established theory in a certain sense never becomes wrong; it only becomes restricted to a *domain of validity*. This domain of validity is characterized by *inequalities*, for example $(v/c)^2 \ll 1$. This means that if one carries out a measurement with a finite accuracy, δ, as one always does, the range of phenomena correctly described by the theory is clearly determined, e.g. all those which involve velocities such that $(v/c)^2 < \delta$. Improved accuracy, however, may give evidence of the approximate nature of the theory.

But there is another aspect to a theory. While the predictions of a theory will always remain correct when used in the validity domain appropriate to the given measurement, the foundations of the theory, its axioms and the underlying picture (model) may be radically modified by a more general theory: the notions of absolute space and time in Newtonian mechanics are abandoned in the special theory of relativity. But, since the predictions of Newtonian mechanics (within their validity limits) are also the predictions of relativistic mechanics, we can deduce the same results from *either* conceptual framework. The relativistic one, however, is more general and is therefore preferred. It supersedes the Newtonian. In this way, *the conceptual framework of every theory is eventually superseded.*

We can call a theory that agrees with another in all predictions within the latter's validity limits a *covering theory*. Special relativity is a covering theory of Newtonian mechanics. General relativity is a covering theory of special relativity as well as of Newtonian mechanics and Newtonian gravitation theory. Similarly, Maxwell's electrodynamics covers geometrical optics and physical optics; statistical mechanics covers thermodynamics and kinetic theory; and quantum mechanics is the covering theory for classical mechanics.

It is now clear why the validity limits of a theory are in general not known (apart from experimental indications in certain instances) until the covering theory is understood. The long fight between corpuscular optics and wave optics was, to a large extent, due to the lack of knowledge of the appropriate validity limits.

We also observe that there exist several *levels* of theory, one being the covering theory of the next. The highest-level theory is the most general one. The development of physical theory thus builds a *hierarchy* of theories: Newtonian mechanics—special relativity—general relativity, nonrelativistic classical mechanics—nonrelativistic quantum mechanics—relativistic quantum mechanics—relativistic quantum field theory.

———————————

This hierarchy attains special significance in view of our human limitations in studying the world around us. We are restricted to a very limited domain of the macroscopic and nonrelativistic classical world in which we function. All observations of nature must be reduced to this "approximation" of nature in order to be perceptible to us: the most complicated studies of nuclear structure, say, are eventually reduced to the macroscopic motion of a hand on a measuring instrument, to a humanly audible click, or to a light flash that is in the very narrow spectral region which we can see. This "translation" from the microscopic, relativistic, or quantum mechanical world to our human one *presupposes the interrelationship of the above hierarchies.*

As a particularly important example, consider quantum mechanics. This theory is meaningless unless it is accessible to observation and measurement. But any measurement involves the interaction of a quantum mechanical system with a classical one (our measuring instrument). In other words, we are at the classical, nonrelativistic end of a system which *must be* described *only partially* by quantum mechanics and *partially* by classical mechanics. This situation makes quantum mechanics incomplete without its classical approximation: the lower-level theory must be contained in the higher-level one as a suitable approximation in order that we can carry out the necessary measurements of the higher-level theory.

It is exactly this situation which permits us, as human beings, to study nature despite the fact that we are part of it.

There is a logical order in theoretical physics. Indeed, such an order is necessary if the endeavor of theoretical physics is to be meaningful and is to succeed in the direction of more and more general axiomatic deductive systems. With this in mind, it is obvious that the study of the relationship between two successive levels of theory, a theory and its minimal covering theory, is of great importance. It provides for the *logical coherence* which exists in theoretical physics but is so often ignored.

Clearly, a higher-level theory can never be derived from a lower-level one. It requires an additional act of invention to produce it. Conversely, it is essential that the lower-level theory be derivable from the covering theory. This must be true not so much with respect to the axiomatic framework, which is in general not a special case of the framework of the covering theory, but with respect to certain basic equations and postulates which contain all the predictive power of the lower-level theory.

The principle of equivalence contains the relationship between general and special relativity.* The correspondence principle dominates the relationship between quantum and classical mechanics. The history of these two principles indicates the importance of the study of the relationship between theories. It shows how one can be guided by such considerations toward a guess of a yet unknown covering theory. The principle of equivalence offers a particularly lucid example of this.

The relationship between classical and quantum mechanics is especially interesting. Here the conceptual frameworks are so radically different that they even employ a different logic. The Aristotelian logic, which pervades all of classical (i.e. nonquantum) physics, is replaced by a quantum logic, which appears strange to our common sense but can be clearly defined mathematically. As a consequence of this switch in logic, and especially as a consequence of the

---

* It can be so defined, however, that it contains more than that (cf. Part C of Chapter 3).

widespread lack of recognition of this fact, the beginner in quantum mechanics has considerable difficulties. Paradoxes arise easily when one uses common sense (i.e. Aristotelian logic, trained on the classical physics of our surroundings) to attack quantum mechanics.

1-3 A PLAN FOR THE PRESENTATION OF THE CLASSICAL THEORY OF CHARGED PARTICLES

The presentation of a physical theory should also place the theory in relation to other existing theories, indicate its derivation from the covering theories (if any) in the sense of the foregoing section, and derive from it lower-level theories (if any).

The classical theory of charged particles involves the special theory of relativity and, correspondingly, will be presented in a covariant formulation. It is therefore necessary to start with the basic ideas that enter into the make-up of relativistic theories, emphasizing those aspects which have direct bearing on our theory. This is done in Chapter 3. This chapter also contains a presentation of the foundations of relativistic particle mechanics.

The theory of charged particles is thus a relativistic theory in the sense of special relativity. It is built on two pillars, relativistic particle mechanics and Maxwell-Lorentz electrodynamics. A full chapter is therefore devoted to the latter also (Chapter 4). It is partly an application of the various concepts of special relativity discussed in Chapter 3. For example, the Noether theorem applied to Lorentz invariance is used here for the Maxwell-Lorentz Lagrangian.

Radiation is of special importance; already in classical physics has it an autonomous nature, but quantum physics crowns it by the existence of photons. Therefore, this subject matter is separated from the treatment in Chapter 4, and the following chapter is devoted to it (Chapter 5).

After introducing the two building blocks, the theory of charged particles proper is studied in Chapter 6 for systems of a single charged particle. Since the mathematically consistent formulation of this theory is not generally known, this chapter proceeds in an inductive way. First the usual difficulties are exhibited and the customary formulations are given which lead to the divergences of self-energy and self-stress; as one progresses, more and more of the difficulties disappear until a finite formulation is presented in Section 6-9. The last part of Chapter 6 is devoted to those applications that are of fundamental importance for the theory or exhibit particular techniques for solving the equations of motion.

In Chapter 7 the one-particle theory is generalized to a theory of any finite number of particles. Another generalization is that of a point charge to a finite charge distribution.

Having thus presented the theory, we turn to its relation to other levels of theory (Chapter 8). The nonrelativistic approximation is obtained as a special case, and so is the limit to a neutral particle. The relation to the classical covering theory (general relativity) is also indicated. It is clear that the considerations of Chapter 3 are relevant here and play an important role in the relation of this theory to its covering theory.

The second half of Chapter 8 is devoted to the connection of the classical charged particle theory with quantum physics. Here the validity limits of the classical theory come to the fore, and we discuss the question to what extent a classical description of charged elementary particles is meaningful. The insistence on classical particles without structure (point charges) in Chapter 6 is here justified by the classical approximation (limit) of the quantum mechanical covering theory.

The last chapter serves partly as a review and partly as a means of emphasizing the structure of the theory, its deductive nature, and its physical contents. No attempt is made at an axiomatization.

The theory of charged particles had a long and varied history in the last three quarters of a century. I believe that a full appreciation of it cannot be gained without some knowledge of its development. The following historical chapter therefore precedes the presentation of the theory.

CHAPTER 2

A Short History of the Classical Theory of Charged Particles

Through most of its history, the classical theory of charged particles was primarily a theory of the electron. This had its origin in the idea that the electron is an "atom of electricity" and is entirely electromagnetic in structure. With the discovery of quantum mechanics, interest in the classical electron theory diminished, but later it was studied again, partly to provide a basis for a quantum field theory of the electron. Consequently, the literature on this subject is very extensive. The purpose of the present chapter is not to present a historical study of all the many attempts, successes and failures, which preceded the present theory. Rather, this chapter will indicate the development of the basic ideas which led to the theory described in the following chapters. Why did this theory remain in an incomplete state for so long, and why did so many attempts at completing it fail?

2-1 BEFORE LORENTZ

Newtonian dynamics dealt with *extended bodies* or *mass points* and forces which act on them. These forces are assumed to be given. Their origin is in general not part of the theory. A notable exception to this statement is the force law of gravitation, which is specified with respect to both the affected body and the force-producing body (source).

It is this latter exception that led to the concept of a force *field*, so completely opposed to Aristotelian action by contact. This new idea was fostered by the discovery of various magnetic and electric phenomena* and resulted in the *action-at-a-distance* theories. When the mathematical description of continuous media was finally understood, especially through the works of Euler and the Bernoullis, classical field theory came into its own.

As a result of these developments the classical theory of charged particles as first conceived by Lorentz emerged as a hybrid theory of particles and fields: charged particles are interacting via an electromagnetic field. We shall now trace the developments up to Lorentz, emphasizing those features that will be of special importance to charged particle theory.

---

* Some of these discoveries which were so difficult to reconcile with contact action date back to pre-Newtonian times; see, for example, Gilbert's *De Magnete* (1600).

André-Marie Ampère (1775–1836) discovered the law of force between two currents in 1820. Correspondingly, he gave the science concerned with these phenomena the name *electrodynamics*. The law of induction was discovered by Michael Faraday (1791–1867) in the year 1831. These discoveries, combined with many other less celebrated ones, set the stage for an intensive study of electricity and magnetism, which culminated in the 1860's in Maxwell's classical field theory.

Most of the ideas at that time revolved around electricity as some kind of fluid or at least continuous medium. In 1845, however, Gustav T. Fechner suggested that electric currents might be due to *particles* of opposite charge which move with equal speeds in opposite directions in a wire. From this idea Wilhelm Weber (1804–1890) developed the first *particle electrodynamics* (1846). It was based on a force law between two particles of charges e_1 and e_2 and a distance r apart,

$$F = \frac{e_1 e_2}{r^2} \left[1 + \frac{r}{c^2} \frac{d^2 r}{dt^2} - \frac{1}{2c^2} \left(\frac{dr}{dt} \right)^2 \right].$$

This force seemed to fit the experiments (Ampère's law, Biot-Savart's law) but ran into theoretical difficulties and eventually had to be discarded when, among other things, the basic assumption of equal speeds in opposite directions was found untenable.

James Clark Maxwell (1831–1879) deserves the credit for having combined into one theory* (1864) the large compilation of empirical laws in electro-dynamics accumulated by his ingenious predecessors. He succeeded in reducing the problem to the knowledge of the charge and current distributions and two types of material properties, viz., the dependence of H on B and the dependence of D on E. The last two properties are usually expressed by two material "constants," the magnetic permeability μ and the dielectric constant ϵ. From this knowledge Maxwell's famous equations permit the computation of the electromagnetic fields B, E, H, and D for any given charge and current distribution.

The field concept has thus been carried to fruition. To be sure, Maxwell was led to it only by abstraction from the very dynamical model of an elastic ether, using first mechanical and hydrodynamic analogies. And it was not until years later, after Einstein's ideas gradually became accepted, that the notion of an ether died its slow death.

Particle electrodynamics was revived in 1881 by Hermann von Helmholtz (1821–1894) and by G. Johnstone Stoney. (The latter also suggested the name *electron* for the negatively charged particles.) The acceptance of atoms as the building stones of matter made the acceptance of electrons as "atoms of electricity" quite plausible by that time.

---

* *Phil. Trans. Roy. Soc.* London, **155**, 459 (1865).

In the same volume in which Stoney's paper appeared, Sir Joseph J. Thomson (1856–1940) wrote on the electromagnetic mass of a charged particle.* He observed that if such a particle is in uniform motion with velocity v, its electromagnetic field will have a kinetic energy

$$T_{\text{elm}} = f \frac{e^2}{Rc^2} \frac{v^2}{2},\qquad (2\text{–}1)$$

where f is a numerical factor of order 1 which depends on the charge distribution inside the spherical charge† of radius R and total charge e. Consequently, one can identify

$$m_{\text{elm}} = f \frac{e^2}{Rc^2} \qquad (2\text{–}2)$$

as an "electromagnetic mass." He ascribed this electromagnetic mass to the fact that the particle has a charge. If it were neutral, its mass m_0 would yield a kinetic energy

$$T_{\text{neutral}} = m_0 \frac{v^2}{2}.$$

As a charged particle its total kinetic energy is

$$T = (m_0 + m_{\text{elm}}) \frac{v^2}{2}. \qquad (2\text{–}3)$$

The observed mass of the charged particle is therefore

$$m \equiv m_{\text{exp}} = m_0 + m_{\text{elm}}. \qquad (2\text{–}4)$$

The result (2–2) appeared very encouraging because it paved the way for a purely electromagnetic theory of the electron: if the electron is an atom of electricity, is it not likely to have a purely electromagnetic mass? If we set $m_0 = 0$, Eqs. (2–4) and (2–2) combine to yield a prediction for the radius of the electron,

$$R = f \frac{e^2}{mc^2}. \qquad (2\text{–}5)$$

Until the development of quantum mechanics this relation was considered correct at least in order of magnitude, since there was no good experimental or theoretical reason to question it.

The electron was discovered by J. J. Thomson in 1897 when he showed‡ that cathode rays were deflected by an electrostatic field. This study led to the first

---

* *Phil. Mag.* (5) **11**, 229 (1881).

† In particular, for a surface charge $f = \frac{2}{3}$, for a uniform volume charge $f = \frac{4}{5}$, provided one uses CGS and Gaussian units, as we shall do throughout this book.

‡ *Phil. Mag.* **44**, 298 (1897); first announcement in *Nature* **55**, 453 (1897).

measurement of the ratio e/m for the electron* 10^7 in Gaussian units, which is not far from the correct value of 1.76×10^7. Since the magnitude of the charge of the electron must be the same as that of a hydrogen ion, it was soon realized that this result required the electron to be much lighter than even the hydrogen atom, a thought that, understandably, did not find ready acceptance.

2-2 THE WORK OF LORENTZ AND ABRAHAM

Meanwhile, Hendrik Antoon Lorentz (1853–1928) embarked on a most ambitious program.† Like others before him, he constructed a particle electrodynamics. But his theory went far beyond that of his predecessors, because he meant to account for all the macroscopic phenomena of electrodynamics and optics in terms of the microscopic behavior of electrons and ions.

His fundamental interaction differed essentially from the previous ones; it was not a direct interaction between the charges but an interaction via the electromagnetic field. Maxwell's equations determined the field produced by an electron or any other charged particle when the current density of that moving charge is taken as

$$\mathbf{j} = \rho\mathbf{v}, \tag{2-6}$$

where ρ is the charge density of that particle,

$$e = \int \rho \, d^3\mathbf{r}. \tag{2-7}$$

Lorentz therefore had to find the force which electromagnetic fields $\mathbf{E}$ and $\mathbf{B}$ exert on a charged particle. If that force is known, the field produced by one particle can act on another particle, producing a particle-particle interaction via the field.

Thus the basis of Lorentz's particle electrodynamics was the famous expression for the force density

$$f = \rho \left(\mathbf{E} + \frac{\mathbf{v}}{c} \times \mathbf{B} \right). \tag{2-8}$$

This law of force, together with Maxwell's equations as applied to electrons in a vacuum (*Maxwell-Lorentz equations*), then determined the interaction of the charged building stones of matter, *viz.* of the electrons and the ions.

As the simplest model of matter Lorentz pictured atoms or ions to which the electrons are bound elastically. The fundamental physical system is thus a charged harmonic oscillator. Such an oscillator emits radiation at the rate of

$$\mathcal{R} = \frac{2}{3} \frac{e^2}{c^3} \mathbf{a}^2 \tag{2-9}$$

---

* W. Kaufmann, *Ann. Physik* **61**, 544 (1897), also and independently observed the magnetic deflection and obtained data for the determination of this ratio.

† *Arch. Ne'erland, Sci. Exact. Nat.* **25**, 363 (1892). An expanded and more complete presentation is Lorentz's article in *Enzykl. Math. Wiss.* **V**, 1, 188 (1903).

ergs/sec if $\mathbf{a}$ is its acceleration, as was shown by Sir J. Larmor.* Such an energy loss is equivalent to a damping force $\mathbf{F}$; the work done by it

$$-\int \mathbf{F} \cdot d\mathbf{r} = W = \int \Re \, dt = \frac{2}{3}\frac{e^2}{c^3}\int \mathbf{a} \cdot \frac{d\mathbf{v}}{dt}\, dt = \frac{2}{3}\frac{e^2}{c^3}\int \left(\frac{d}{dt}(\mathbf{a}\cdot\mathbf{v}) - \frac{d\mathbf{a}}{dt}\cdot\mathbf{v}\right) dt.$$

Since $\mathbf{a}$ and $\mathbf{v}$ are periodic functions for a harmonic oscillator, they will be almost periodic when the damping force is very small. In that case, the first term in the integrand gives no contribution for integration over an integral number of periods. Since $d\mathbf{r} = \mathbf{v}\,dt$, comparison of the integrands gives an approximate average damping force

$$\mathbf{F} = \frac{2}{3}\frac{e^2}{c^3}\frac{d\mathbf{a}}{dt}. \tag{2-10}$$

This result also was first obtained by Lorentz.

The equation for the elastically bound electron including the damping force due to radiation emission is now

$$m\ddot{\mathbf{r}} = k\mathbf{r} + \frac{2}{3}\frac{e^2}{c^3}\dddot{\mathbf{r}}, \tag{2-11}$$

where a dot was used to indicate the time derivative. This equation played a very important role in the particle electrodynamics and optics initiated by Lorentz.

Lorentz's theory was a replacement of the *macroscopic* theory of Maxwell by a *microscopic* theory. It accounted for such things as the Zeeman effect, the dispersion of light, and the propagation of light in moving bodies, thereby including Fresnel's wave theory. But most important for the structure of physical theories, Lorentz showed that Maxwell's macroscopic theory can be *derived* from this simple model by a suitable averaging process, thereby ensuring an entirely electromagnetic origin of ϵ and μ. He also laid the foundations for Sommerfeld's electron theory of metals, accounting for both electric and thermal conductivity. While the details of this program could not be carried through until the days of quantum mechanics, the richness of electromagnetic interactions was by then fully apparent and conceptually established as the only fundamental interaction down to the atomic level.

Lorentz called his microscopic approach† *the theory of electrons* and he included in it also a study of the electron itself. This attempt at accounting for the structure of the electron, in which Max Abraham among others contributed so much, was not very successful. Let us follow this work in some detail.

---

* *Phil. Mag.* **44**, 503 (1897).

† H. A. Lorentz, *The Theory of Electrons and Its Applications to the Phenomena of Light and Radiant Heat*, Second Edition, Dover Publications, Inc., New York, 1952. The first edition of these 1906 lectures by Lorentz appeared in 1909.

The first model of the electron was that of a rigid sphere with spherically symmetric charge distribution (usually assumed to be distributed uniformly over either the surface or the volume of the electron). This model was studied in detail by Abraham.* The energy was found to be the same as that obtained by J. J. Thomson, Eqs. (2–1) and (2–3), so that a purely electromagnetic electron would have a radius given by (2–5). Abraham's basic assumption for the momentum of the electron was that it is given by Poynting's vector

$$\mathbf{S} = \frac{c}{4\pi}\,\mathbf{E} \times \mathbf{B}, \tag{2-12}$$

which accounts for the energy flux density of radiation. The relation between the momentum of the Coulomb field of the moving electron and $\mathbf{S}$ was assumed to be

$$\mathbf{p} = \frac{1}{c^2}\int \mathbf{S}\,d^3\mathbf{r}. \tag{2-13}$$

We shall see in Section 6–3 that this assumption was incorrect.

The above assumption permitted the computation of the momentum due to the Coulomb field of an electron moving with velocity $\mathbf{v}$. The result was

$$\mathbf{p}_{elm} = \tfrac{4}{3}m_{elm}\mathbf{v}. \tag{2-14}$$

Since the momentum of an uncharged particle is clearly

$$\mathbf{p}_{neutral} = m_0\mathbf{v} \tag{2-15}$$

with m_0 the mass of this particle, the total momentum of the electron is

$$\mathbf{p} = (m_0 + \tfrac{4}{3}m_{elm})\mathbf{v} = m\mathbf{v}, \tag{2-16}$$

in analogy to Eqs. (2–3) and (2–4).

A moving charge thus has not only an energy but also a momentum due to its surrounding field. However, the usual relationship between the momentum and the velocity of a particle, (Eq. 2–15), is apparently not valid for the electromagnetic case (Eq. 2–14). An additional factor $\tfrac{4}{3}$ is present.† This factor was to plague the theory for a long time, especially when Lorentz invariant formulations were being attempted: it is inconsistent with this invariance property.

One important feature of the theory of the electron structure was the exact evaluation of the force that an electron exerts on itself. Apparently, this force

---

* M. Abraham, *Ann. Physik* **10**, 105 (1903).

† The careful reader will note that we could have defined the electromagnetic mass on the basis of (2–14), $m'_{elm} = \tfrac{4}{3}m_{elm}$. Then the kinetic energy (2–1) would have an incorrect relation to the velocity. However, (2–1) is correct, while (2–14) will be found incorrect (Section 6–3).

must be present, since by Coulomb's law each part of the charged sphere repels all other parts. According to Lorentz's fundamental force law, this self-force is

$$\mathbf{F}_{self} = \int \rho \left(\mathbf{E} + \frac{\mathbf{v}}{c} \times \mathbf{B} \right) d^3r, \tag{2-17}$$

$\mathbf{E}$ and $\mathbf{B}$ being the fields produced by the charge density ρ itself according to the Maxwell-Lorentz equations. A lengthy calculation yields the result* first obtained by Lorentz†

$$\mathbf{F}_{self} = -\frac{2}{3c^2} \sum_{n=0}^{\infty} \frac{(-1)^n}{n!} \frac{d^n \mathbf{a}}{c^n \, dt^n} \iint \frac{\rho(\mathbf{r})\rho(\mathbf{r}')}{|\mathbf{r} - \mathbf{r}'|} |\mathbf{r} - \mathbf{r}'|^n \, d^3r \, d^3r' \tag{2-18}$$

$$= -\frac{4}{3c^2} W_{self}\, \mathbf{a} + \frac{2}{3} \frac{e^2}{c^3} \dot{\mathbf{a}} - \frac{2e^2}{3c^2} \sum_{n=2}^{\infty} \frac{(-1)^n}{n!} \frac{d^n \mathbf{a}}{c^n \, dt^n} O(R^{n-1}). \tag{2-19}$$

In this expression we used (2-7) and

$$W_{self} \equiv \frac{1}{2} \iint \frac{\rho(\mathbf{r})\rho(\mathbf{r}')}{|\mathbf{r} - \mathbf{r}'|} \, d^3r \, d^3r'. \tag{2-20}$$

The symbol $O(R^{n-1})$ indicates an expression of the order of magnitude of R^{n-1}, R being the electron radius.

The self-force can therefore be written as an infinite series. The first term of this series is proportional to the acceleration of the particle; the second term is proportional to the derivative of the acceleration and is independent of the electron radius or charge distribution. The higher terms all depend on the electron structure, i.e. its radius and charge distribution. In the special case in which the electron oscillates with circular frequency $\omega = 2\pi\nu$, these higher terms can be written

$$\frac{2e^2}{3c^3} \dot{\mathbf{a}} \sum_{n=1}^{\infty} \frac{(-1)^n}{(n+1)!} \alpha_n \left(\frac{R}{\lambda} \right)^n, \tag{2-21}$$

where $\lambda \equiv c/\omega$ and the α_n are numbers of order one. These terms are therefore small compared with the second term in the series (2-19) whenever $R \ll \lambda$. They vanish in the limit $R \to 0$.

If the electron is to be of purely electromagnetic structure and is subject to some external force $\mathbf{F}_{ext}$, its motion must be such that the total force vanishes,

$$\mathbf{F}_{self} + \mathbf{F}_{ext} = 0. \tag{2-22}$$

---

* A detailed derivation can be found in J. D. Jackson, *Classical Electrodynamics*, John Wiley and Sons, Inc., New York, 3rd edition, 1999, Section 16.3.
† Loc. cit. (p. 12), Note 18 on p. 252.

This equation has indeed a similarity to Newton's equation of motion because the leading term in (2-19) is proportional to **a**. In fact, if we put

$$m_{\text{self}} = \frac{W_{\text{self}}}{c^2}, \tag{2-23}$$

this equation becomes

$$\frac{d}{dt}\left(\frac{4}{3}\, m_{\text{self}}\mathbf{v}\right) - \frac{2}{3}\frac{e^2}{c^3}\,\dot{\mathbf{a}}\,(1 + O(R)) = \mathbf{F}_{\text{ext}}. \tag{2-24}$$

A comparison of (2-20) with J. J. Thomson's calculation (2-2) shows that

$$m_{\text{self}} = m_{\text{elm}}, \tag{2-25}$$

so that the first term on the left of (2-24) is just the time derivative of the electromagnetic momentum (2-14), including the factor $\frac{4}{3}$. The second term is identical with Lorentz's damping force (2-10) when the structure-dependent terms are ignored. In this approximation (2-11) becomes a special case of (2-24) with $m = \frac{4}{3}m_{\text{elm}}$. The ad hoc identification of a mass with $1/c^2$ times an energy should also be noted. It arises here in a nonrelativistic calculation and hardly suggests the much deeper meaning of this kind of relationship that the special theory of relativity demanded later on.

The electron theory based on these equations faced some obvious difficulties:

(1) Equation (2-24) is not an equation of motion in the spirit of Newtonian mechanics. It contains derivatives of the acceleration to all orders [cf. Eq. (2-18)]; and even when the structure-dependent terms are dropped, it is a third-order differential equation for the position **r** instead of a second-order equation. Consequently, initial position and velocity are not sufficient to determine the motion.

(2) If one tries to eliminate the structure terms rigorously by assuming a point electron ($R \rightarrow 0$), the self-energy W_{self} diverges.

Nevertheless, there was an apparent success of the purely electromagnetic electron, as exhibited by the derivation of the Eq. (2-24) from (2-22). But it was of short duration. In 1905 Albert Einstein (1879–1955) provided a rational basis for the Lorentz transformations which were found independently by Lorentz* and by Einstein.† He established the foundations of the special theory of relativity and thereby provided the validity limits of Newtonian mechanics to $(v/c)^2 \ll 1$. It became necessary to revise the nonrelativistic Abraham-Lorentz theory of the electron and make it relativistically valid.

---

* H. A. Lorentz, *Proc. Amsterdam Acad. Sci.* **6**, 809 (1904).
† A. Einstein, *Ann. Physik* **17**, 891 (1905).

2-3   THE IMPACT OF RELATIVITY

In the same year in which the Lorentz transformations were first published, 1904, Abraham realized the inconsistency of his rigid purely electromagnetic electron with these transformations.* First, the Lorentz-Fitzgerald contraction of lengths in the direction of motion is a consequence of the Lorentz transformation and would thus require the rigid electron to contract into a spheroid. This was not regarded to be serious, because at that time the contraction was thought to be due to the earth's motion through the ether, so that one could still uphold a rigid model so long as there was no motion relative to the ether.

What seemed to make things much worse, however, was the appearance of the factor $\frac{4}{3}$ in (2–14) and (2–24). Relativity requires the relations

$$\mathbf{p} = \frac{m\mathbf{v}}{\sqrt{1 - v^2/c^2}}, \qquad E = \frac{mc^2}{\sqrt{1 - v^2/c^2}},$$

for every free particle. Thus,

$$\mathbf{p} = \frac{E}{c^2}\,\mathbf{v} \tag{2–26}$$

is required by the Lorentz transformations, while

$$\mathbf{p} = \frac{4}{3}\,\frac{W_{\text{self}}}{c^2}\,\mathbf{v} \tag{2–27}$$

is the result obtained for the purely electromagnetic electron. Equation (2–26) is valid irrespective of the magnitude of the velocity, so long as E is the total energy of the particle (including the rest energy).

The first conclusion was, therefore, that relativity made things worse rather than better, at least as far as electron theory was concerned: the inconsistency (2–26), (2–27) requires giving up the purely electromagnetic electron: part of the electron's self-energy must be of nonelectromagnetic nature. At this point, however, it was realized that the nonrelativistic theory was not so satisfactory either, because the various parts of the charged sphere must repel one another according to Coulomb's law, giving rise to an *unstable* electron.

An ingenious solution was suggested by Poincaré.† The Coulomb repulsion, which is responsible for the instability of the electron, can be compensated by nonelectromagnetic cohesive forces, a kind of negative pressure. These forces must be postulated ad hoc, to be sure, but the amount of pressure necessary to provide stability also exactly compensates $\frac{1}{3}W_{\text{self}}$ in (2–27) and gives

$$\mathbf{p} = \left(\frac{4}{3}\,\frac{W_{\text{self}}}{c^2} - \frac{W_{\text{coh}}}{c^2}\right)\mathbf{v} = \frac{W_{\text{self}}}{c^2}\,\mathbf{v}. \tag{2–28}$$

---

* M. Abraham, *Phys. Zeits.* **5**, 576 (1904).

† H. Poincaré, *Comptes Rend.* **140**, 1504 (1905) and *Rend. Circ. Mat. Palermo* **21**, 129 (1906).

because $W_{coh} = \frac{1}{3}W_{self}$. This is consistent with the Lorentz transformations. The Poincaré cohesive forces thus have a double purpose: they make the electron stable and at the same time bring the theory in accord with the special theory of relativity.

It seems appropriate here to leave the historical sequence of events temporarily and to follow the interesting fate of the factor $\frac{4}{3}$ to date.

In 1922 Enrico Fermi (1901–1954) published several papers* which apparently went largely unnoticed. He showed that a careful examination of the field carried along by a relativistic charged particle has an energy which in first approximation agrees with (2–20), but which has a momentum

$$\mathbf{p}_{elm} = \frac{W_{self}}{c^2}\mathbf{v} \qquad\qquad (2\text{–}29)$$

without the factor $\frac{4}{3}$ of (2–14) or (2–24). The implication of this work was that in the past the Lorentz transformations had not been applied consistently or that the definition of momentum (2–13) had not been correct.

Since this result went unnoticed, it was bound to be rediscovered. This happened in 1936 when the same result was obtained by a different line of reasoning.† This paper was also forgotten and did not find its way into the standard references or text books. In 1949 Kwal‡ showed in a short note that a consistent application of the transformation of the electromagnetic energy-momentum tensor is bound to lead to (2–29). Finally, the result was discovered for a fourth time by the present author in 1960,§ again without the knowledge of any of the previous papers. The conclusion was that it was indeed the definition (2–13) of the momentum of the electromagnetic field which was to blame for the resulting incorrect transformation properties. That definition is valid for free fields (radiation fields) but not for bound fields such as Coulomb fields. We shall return to this point in a more quantitative way in Section 6–3.

The result (2–29) seems to leave no room for the Poincaré cohesive forces: There is no relation between the need for cohesive forces and the factor $\frac{4}{3}$. We seem to have a relativistic but unstable electron. A careful study of non-electromagnetic cohesive forces shows, however, that one can construct such forces in various ways so that the electron is stable in a relativistically invariant fashion.‖ But let us now return to the effect of the newly discovered theory of special relativity on the electron theory.

When the meaning of the Lorentz transformations became better understood primarily because of the work of Einstein, the rigid Abraham electron was

---

* E. Fermi, *Physik. Z.* **23**, 340 (1922) and *Atti Accad. Nazl. Lincei* **31**, 184 and 306 (1922).

†W. Wilson, *Proc. Phys. Soc.* (London) **48**, 736 (1936).

‡ B. Kwal, *J. Phys. Radium* **10**, 103 (1949).

§ F. Rohrlich, *Am. J. Phys.* **28**, 639 (1960).

‖ H. M. L. Pryce, *Proc. Roy. Soc.* (London) A**168**, 389 (1938).

abandoned in favor of a "compressible" electron whose shape became that of an oblate spheroid under Lorentz transformations.   Its rather complicated charge distribution and its fields were computed in detail by Lorentz. The corresponding momentum, starting with (2–13), led to the expression (2–14) times a complicated function of $(v/c)^2$ which depended on the charge distribution.   This result could be understood as an electromagnetic mass m_{elm} of complicated velocity dependence which has to be inserted in (2–24).

Meanwhile, Abraham succeeded in giving a relativistic generalization of the second term in (2–24) which is due to the energy loss by radiation and which became known as the "radiation reaction" term. His generalization of this term was* the four-vector

$$\Gamma^\mu = \frac{2}{3} \frac{e^2}{c^3} \left(\frac{d^2 v^\mu}{d\tau^2} - \frac{dv^\alpha}{d\tau} \frac{dv_\alpha}{d\tau} v^\mu \right), \qquad (2\text{--}30)$$

which became known as the Abraham four-vector of radiation reaction.† This result was also found by v. Laue.‡ It played an important role in the theory ever since.

Equation (2–24) could thus be written in a relativistically invariant way although, as mentioned above, the first term was not computed correctly.

The theory of the electron had now come to an impass. While it appeared possible to formulate it in agreement with special relativity, at least when cohesive forces were added, the idea of a purely electromagnetic electron had failed. Furthermore, cohesive forces were necessary to ensure stability of any finite-size electron. Point electrons did not yield a meaningful theory because their self-energy diverged. As a consequence, the structure-dependent terms in (2–24) apparently had to be taken seriously.

2–4   ATTEMPTS AT MODIFICATION AND THE EVENTUAL FORMULATION OF THE THEORY FOR POINT CHARGES

The theory of Lorentz, Abraham, and Poincaré had ground to a halt before 1910 and was to remain in this incomplete state until Dirac made important progress in 1938.

In the meantime numerous attempts were made at modifications of the apparently unsuccessful theory. Only some of these attempts shall be mentioned here.

(a) Early modifications.   The theory consists of the Maxwell-Lorentz field equations, the Lorentz force law, and an assumption concerning the structure of the electron. From these one tries to obtain an equation of motion for the

---

* M. Abraham, *Theorie der Elektrizität*, Vol. II, Springer, Leipzig, 1905.

† For simplicity, we use here covariant notation which is not the notation used by Abraham. The reader not familiar with covariant notation should turn to Appendix 1.

‡ M. v. Laue, *Ann. Physik* **28**, 436 (1909).

electron which fully takes account of the loss of energy and momentum due to the emission of radiation (radiation reaction). Hence there are three obvious places where modifications can be made: the electron structure, the field equations, and the Lorentz force law. The whole theory, however, must be Lorentz invariant. This puts severe restrictions on the possible modifications but still leaves a great deal of freedom.

Perhaps the first of these attempts was due to Mie.* He suggested a modification of the field equations that would assure the stability of the electron by means of electromagnetic fields alone. To this end he admitted the occurrence of D and H in addition to E and B, in distinction to Lorentz's idea. Modified field equations are then obtained by modifying the action principle from which these equations can be derived. The hope was that the solutions of the new field equations would yield fields which differ from those of the Lorentz theory mainly inside the electron. No modification could be found which gave satisfactory results. One difficulty was that any modification of the desired kind led to nonlinear equations, causing mathematical problems and making it hard to study their implications.

Much later, an interesting example of Mie's theory was given by Born and Infeld.† Exact solutions could be obtained for certain simple problems; furthermore, it contained a parameter which, when chosen small enough, reproduced the Maxwell-Lorentz equations to arbitrary accuracy. The corresponding Lagrangian is

$$\mathcal{L} = b^2 \left(\sqrt{1 - \frac{1}{8\pi b^2} F^{\alpha\beta} F_{\alpha\beta}} - 1 \right).$$

For small b^{-1}, the Lagrangian for the usual field equations is obtained [cf. Eq. (4-111)]. The nonlinear nature of this theory is evident. For field strengths of order b or larger, essential modifications set in, but these are too high to be observed. Of course, modifications of this type are legion and such attempts therefore remain unsatisfactory as long as there is no experimental evidence in favor of any one of them.

It is interesting that certain predictions of quantum electrodynamics can be expressed by means of a modification of the classical Maxwell-Lorentz equations, which corresponds to terms in the Lagrangian that are of fourth order in the field strengths (interaction of "light with light"). But this does not provide for a consistent classical theory which exceeds the original domain of validity of classical electrodynamics. Rather, it should be regarded as an ad hoc phenomenological extension of certain equations of the classical theory which takes into account in an effective and approximate way certain predictions of a higher-level theory. These predictions could not be obtained within the classical theory.

---

* G. Mie, *Ann. Physik* **37**, 511 (1912); **39**, 1 (1912); **40**, 1 (1913).

† M. Born, *Proc. Roy. Soc.* (London) **A143**, 410 (1934); M. Born and L. Infeld, *ibid.* **144**, 425 (1934); **147**, 522 (1934); **150**, 141 (1935).

The discovery of quantum mechanics in the mid-twenties at first channeled interests and energies into a different direction. Classical electron theories were forgotten by almost all except general relativists who were searching for a theory that combined gravitational and electromagnetic interactions. After the very ingenious but unsuccessful attempt by Weyl* in 1918, these endeavors became strongly directed toward unified field theory. We shall not pursue them here.

The establishment of quantum mechanics and of the corresponding validity limits of classical physics below certain small dimensions later also resulted in frequent apologies in papers on the classical theory of charged particles. It was usually emphasized that these investigations were made with the intention of future quantization leading to a quantum field theory. Seldom was the purpose the completion of the classical theory *per se*.

There is, however, a group of physicists who did not accept quantum mechanics and its probabilistic interpretation† and who felt that a classical theory lies behind quantum mechanics. For those the classical electron theory and the classical electron structure has a very different meaning. Additional degrees of freedom, not appearing in quantum mechanics, have to be discovered.

In this connection the work of Wessel‡ should be mentioned. He tried to understand the meaning of the occurrence of a third time derivative of position in the basic equation (2–24). What additional degrees of freedom are indicated by this? Naturally, the discovery of the electron spin was suggestive and many attempts were made to construct a classical theory for it. However, we shall not pursue this line of investigation but refer the reader to Sections 7–4 and 8–5, where the electron spin is discussed.

(b) The point charge. We must now interrupt the history of the modified theories and turn to the development of the point charge theory. This theory is not a modification of the Lorentz theory but, rather, a special case of it, viz. the limit $R \to 0$ of the spherical electron. In the first decennium of this century not much attention was paid to this limit because of the concern with the diverging electromagnetic self-energy. With the development of quantum mechanics, physicists began to feel that too much emphasis may have been put on the self-energy problem and that progress could be made in a point electron theory in spite of it. The advantage of having all structure-dependent terms in (2–24) vanish was certainly attractive.

An interesting idea was suggested by Fokker.§ He noticed that since the electromagnetic field plays the intermediary for the interaction between charged particles, one might try to eliminate it completely. When this is done in a system of point charges, the following emerges. The problem of infinite fields

---

* H. Weyl, *Math. Z.* **2**, 384 (1918), *Ann. Physik* **59**, 101 (1919), and other papers.
† Einstein's famous quote, "God does not play dice," is characteristic.
‡ W. Wessel, *Z. f. Physik* **92**, 407 (1934) and papers thereafter.
§ A. D. Fokker, *Z. f. physik* **58**, 386 (1929).

at the position of a point charge is absent because all fields have been eliminated. Furthermore, it is very easy to formulate this theory in such a way that self-interactions no longer occur at all. The self-energy problem is thus completely eliminated. The physical picture is now that of the old *direct interaction* without intermediary; there is, of course, no objection to this in principle other than personal taste. The main difficulty with this otherwise brilliant solution of the problem is the absence of radiation. A solution to this difficulty was suggested much later by Wheeler and Feynman (1945) by a suitable supplementary assumption of absorption. We shall present this theory in some detail in Section 7-3, where it will be compared with the theory studied in the remainder of this book. Suffice it to say here that the absorption assumption seems somewhat contrived and the established usefulness of the concept of photons as physical particles seems to favor a more lenient approach to the existence of electromagnetic fields.

In the early thirties, the Fokker approach did not appear to be very satisfactory. The return to the Maxwell-Lorentz equations together with the point particle was indicated. In 1933 Wentzel discovered that in this framework the fields are not necessarily divergent at the position of a point charge.* A convergent result can be obtained by a careful limiting process. The same method also gave a convergent momentum and energy of the field.

This ray of hope forecast the great contribution in 1938† by Dirac. Using only the conservation laws of momentum and energy and the Maxwell-Lorentz equations, he showed that a point charge must satisfy the differential equation

$$\dot{B}^\mu + m_{\text{elm}}\dot{v}^\mu = F^\mu_{\text{ext}} + \Gamma^\mu.$$

The external force F^μ_{ext} is the Lorentz force due to an external electromagnetic field. Γ^μ is the Abraham four-vector of radiation reaction (2–30), and B^μ is an undetermined four-vector whose time derivative‡ $\dot{B}^\mu$ must be orthogonal to the velocity four-vector v^μ. Of the infinity of possible choices for B^μ Dirac chose $B^\mu = m_0 v^\mu$, where m_0 is a neutral mass in the sense of Eq. (2–4). He justified this choice in the following way: "There are other possible expressions . . . but they are all so much more complicated . . . that one would hardly expect them to apply to a simple thing like an electron."

The result of this choice was that the equation became

$$m\dot{v}^\mu = F^\mu_{\text{ext}} + \Gamma^\mu, \tag{2–31}$$

where (2–4) was used. This equation is exactly the relativistic generalization of Lorentz's equation (2–24) when applied to a point electron. To be sure, the electromagnetic mass m_{elm} is infinite in the limit $R \to 0$, but it is lumped

---

* G. Wentzel, *Z. f. Physik* **86**, 479 and 635 (1933); **87**, 726 (1934).
† P. A. M. Dirac, *Proc. Roy. Soc.* (London) **A167**, 148 (1938).
‡ The dot here denotes differentiation with respect to the *proper* time.

into m_0 to yield the observed mass m, ignoring its divergent nature. It should also be noted that m_{elm} is a relativistic invariant and that the disturbing factor $\frac{4}{3}$ did not arise in the completely covariant calculation of Dirac: the momentum and the energy of the Coulomb field form a four-vector, $m_{elm}v^\mu$.

Equation (2–31) is called the Lorentz-Dirac equation. It is meant to be a rigorously valid equation for a classical point charge. Dirac thus provided an equation which became one of the cornerstones of the classical theory of charged particles. But this equation is not without difficulties. The most important one was discussed in detail by Dirac: the appearance of a third derivative of position [see Eq. (2–30)] does not permit a determination of the solutions of this equation in terms of the Newtonian initial conditions of position and velocity. A whole family of solutions results, and, most embarrassingly, all but one of these solutions are physically meaningless. They give the electron a velocity which increases asymptotically (in the distant future) to the velocity of light irrespective of the applied forces; it happens even when no force at all is acting. These *runaway solutions* became of great concern to everyone working with the Lorentz-Dirac equation.

Solutions to this problem were suggested by various authors. However, these solutions seemed to be based on ad hoc assumptions. Dirac himself noticed that the requirement of an asymptotically vanishing acceleration would suffice. Eliezer and Mailvaganam* suggested that Γ^μ should vanish asymptotically; Bhabha† required that the solutions be regular near $e = 0$, so that they can be expanded in power series in the charge. This last requirement indeed does away with the undesirable solutions, since they all seem to have essential singularities at $e = 0$; but it remains to be proven that all physical solutions can be expanded in a power series in e.

Another difficulty of the Lorentz-Dirac equation is related to causality. It appears that the *physical* solution leads to a violation of causality over very short time intervals; e.g. an acceleration occurs prior to the applied force. It seems difficult to accept a theory which leads to such results. This problem will be studied in detail in Section 6–7 and in Part C of Chapter 6.

Finally, Dirac's work also showed that it is of great advantage not to formulate the theory in terms of the customary retarded fields (which are the ones measured in a typical experiment). Very convenient fields are obtained by combining the retarded and the advanced solutions of the field equations $F_\pm^{\mu\nu} = \frac{1}{2}(F_{ret}^{\mu\nu} \pm F_{adv}^{\mu\nu})$, where $F^{\mu\nu}$ is the field strength tensor. This fact will play an important role in the later development of the theory.

(c) The extended charge. The divergent electromagnetic mass, the runaway solutions, and the causality violation were responsible for a considerable lack of enthusiasm for Dirac's formulation of the electron theory. Attempts were

---

* C. J. Eliezer and A. W. Mailvaganam, *Proc. Cambridge Phil. Soc.* **41**, 184 (1945).
† H. J. Bhabha, *Phys. Rev.* **70**, 759 (1946).

made to return to the finite, extended charge distribution. But Dirac's work was of sufficient influence to make the point charge the starting point for these studies. They were begun* by Bopp in 1940 and were actively pursued over many years by him and his coworkers.†

The charge density $\rho(\mathbf{r})$ of a point charge can be expressed by the three-dimensional δ-function, $\rho(\mathbf{r}) = e\delta_3(\mathbf{r} - \mathbf{z})$, where $\mathbf{z}$ is the position of the charge. A covariant relativistic point charge theory will characteristically contain a $\delta_4(x - z)$ in the current density. An extended charge can thus be obtained easily by the replacement

$$\delta_4(x - z) \to f(x - z), \tag{2–32}$$

where the function f must have the same transformation and normalization properties as the δ-function. Physically, $f(x)$ characterizes the shape of the charge. For this reason, theories of this kind are often called covariant form factor theories. They assure a finite self-energy; they can also yield a stable electron, provided suitable terms are added to compensate for the Coulomb repulsion (Poincaré cohesive forces). But they are arbitrary because the form factor f is not known from experiment.

The situation is actually even worse than that: the form factor cannot be determined by *classical* experiments even in principle. The size of the electron is in the quantum domain and outside the applicability limits of the classical theory. For this reason, many extended electron theories were constructed with the intention of eventually taking the point particle limit.‡ We shall see in Section 7–4 that the modification (2–32) of the point particle current is not consistent unless this limit is taken or unless the acceleration is restricted in a certain way. This has to do with the Thomas precession of special relativity [cf. Eqs. (7–41) and (7–38′)].

Very recently, a novel idea was proposed by Prigogine and Henin,§ which in some sense falls into this type of theories. They introduced a form factor (cut-off) in the frequency spectrum of the fields produced by the charged particles. Requiring the validity of the Maxwell-Lorentz equations, this restriction on the frequency spectrum implies a certain extended charge and current distribution. The result assures a finite self-energy. The charge structure of the particle is now dependent on its motion and has no absolute meaning, even in the comoving system.

(d) The recent emergence of a satisfactory point charge theory. In the late 1940's quantum field theory and especially quantum electrodynamics had made

* F. Bopp, *Ann. Physik*, **38**, 345 (1940); **42**, 572 (1943) and later papers.

† See also H. McManus, *Proc. Roy. Soc.* (London), **A195**, 323 (1948); Bohm, Weinstein, and Kauts, *Phys. Rev.* **76**, 867 (1949).

‡ R. P. Feynman, *Phys. Rev.* **74**, 939 (1948); H. Lehmann, *Ann. Physik* **8**, 109 (1951).

§ I. Prigogine and F. Henin, *Physica* **28**, 667; **29**, 286 (1962).

great strides, stimulated largely by new experimental results. The new computational techniques and the increased sophistication permitted a new look at the classical theory of charged particles that indicated promise of success. In quantum mechanics and quantum electrodynamics the electron has no charge distribution; it is a point particle. Is it not necessary that the corresponding classical theory be also a point particle theory? This reasoning and the exploitation of analogies between the classical and the quantum field-theoretical descriptions of the electromagnetic field initiated valuable studies in classical point charge theory.

In 1951 Gupta* observed that the divergent self-energy effects (self-mass and instability) can be removed by a subtraction procedure involving the fields $F_+^{\mu\nu}$ that were used so successfully by Dirac. This was followed in 1955 by an important paper on the role of the asymptotic condition. No doubt stimulated by the analogous problem in quantum electrodynamics, Haag† emphasized the fundamental importance of such a condition in the axioms of any field theory. This point assures the absence of runaway solutions of the Lorentz-Dirac equation: they need no longer be removed by ad hoc assumptions. He thus put new emphasis on Dirac's work and removed one of the main objections to it.

The Lorentz-Dirac differential equation can be combined with the asymptotic postulate to yield an integrodifferential equation. This was first shown for the nonrelativistic case by Haag (*loc. cit.*) and by Iwanenko and Sokolow‡ and was later extended to the relativistic equations.§ The resulting equation of motion poses a much more difficult mathematical problem than did the corresponding Newtonian equations. The establishment of an existence proof for the solutions of this equation under a large class of external forces was therefore a very important contribution to the theory.‖

The difficulties remaining in the Lorentz-Dirac equation, were considerable: the solutions of that equation include physically meaningless results (Section 6 C). Only early in the present century were these difficulties finally resolved. The Supplement to the present edition is devoted to that resolution. Sections 9-4 and 9-5 summarize the present satisfactory state of the theory.

---

*S. N. Gupta, *Proc. Phys. Soc.* (London), **A64**, 50 (1951). See also J. W. Gardner, ibid., p. 426.

† R. Haag, *Z. f. Naturf.* **10a**, 752 (1955).

‡ D. Iwanenko and A. Sokolow, *Klassische Feldtheorie*, Akademie-Verlag, Berlin, 1953; Russian edition, (1949).

§ F. Rohrlich, *Ann. Phys.* (N.Y.), **13**, 93 (1961).

‖ J. K. Hale and A. P. Stokes, *J. Math. Phys.* **3**, 70 (1962).

The results of these sections were first published by me in *Phys. Rev. Lett.* **12**, 375 (1964). After completion of this work, a paper by M. Schönberg was brought to my attention [*Phys. Rev.* **69**, 211 (1945)], containing an action principle which is similar to the one presented in Section 7–1 but which does not permit the derivation of the inhomogeneous Maxwell-Lorentz equations. That paper perhaps received less attention than it deserved because of its considerations of such unphysical topics as negative kinetic energies and radiationless modes of point charges.

CHAPTER 3

Foundations of Classical Mechanics

The laws of motion of classical mechanics have both general and specific aspects. The general aspects are common to all of classical dynamics, irrespective of the particular type of interaction responsible for the motion of a system. The specific aspects vary with the types of forces under consideration: electrostatic, magnetostatic, gravitational, etc. This chapter is concerned with the *general* features of classical motion and with various assumptions and definitions which enter all considerations of a dynamical nature. The gravitational interactions, however, enter these considerations in a special way; some of their specific features *will* have to be considered in the context of the general foundations of classical mechanics.

A. The Equivalence of Static Coordinate Systems

3–1 REFERENCE FRAMES

A rigid body is said to move if it changes its position or orientation; a nonrigid body may also change its shape. This statement can be made precise by choosing a *coordinate system* in our three-dimensional space so that each point of the body is associated with three numbers. If the body moves, some or all of the points will, in the course of time, be associated with different numbers; in general, the coordinates of a point will be functions of time. Obviously, these functions will be continuous. Motion can then be characterized generally by the observation that not all coordinates of all points of the body remain constant in time.

From these simple considerations follows a very basic conclusion, viz. that it is necessary to specify a coordinate system. This necessity also follows from the question, "relative to what?," which must be answered in order that the statement "this body is moving," have meaning. For this reason one often uses the notion *reference frame* synonymously with *coordinate system*. We shall adopt this habit.

The need for a reference frame immediately raises a difficulty. If two *observers** use different coordinate systems, they will in general not agree with each other on whether or not a body is in motion. Indeed, this will happen in exactly those cases in which the transformation that leads from one coordinate system to the other depends on time. We then say that the coordinate systems are in motion relative to each other. Since confusion can arise when the same observer uses two different coordinate systems, we shall not allow this to happen. We postulate that there is a different observer associated with every coordinate system, and vice versa, each observer can construct only one coordinate system. This one-to-one correspondence will permit us to use observer and coordinate system interchangeably in certain contexts.

3-2 THE EUCLIDEAN GROUP

Two coordinate systems will be related in an interesting fashion only when they are in motion with respect to each other; only then is there a question of whether the two observers will describe a given phenomenon in the same way. When two coordinate systems are not in relative motion, i.e. when the transformation connecting them is time independent, the above question has a trivial answer. Thus, it is obvious that a physical phenomenon does not depend on the choice of Cartesian coordinates (x, y, z) rather than spherical coordinates (r, θ, φ) to specify a point in a three-dimensional Euclidean space. This trivial difference of coordinate systems can be eliminated conveniently by a suitable mathematical notation: one specifies a point not by three coordinates $(x, y, z; r, \theta, \varphi;$ etc.) but by a three-vector $\mathbf{r}$. This vector can then be referred to any one of a set of trivially equivalent systems. These are the physical implications of the well-known fact that the laws of mechanics can be written in vector notation. This still leaves two types of equivalent sets of coordinate systems to be specified explicitly, viz. those that differ from each other only by a relocation of the origin (translation by a fixed distance),

$$\mathbf{r'} = \mathbf{r} + \boldsymbol{\alpha} \qquad (\boldsymbol{\alpha} = \text{const}), \qquad (3\text{--}1)$$

and those that differ from each other only by a relocation of a reference direction (rotation by a fixed angle),

$$\mathbf{r'} = \mathsf{R} \cdot \mathbf{r}. \qquad (3\text{--}2)$$

Here R is a dyadic whose coefficients are constants and form an orthogonal matrix with determinant $+1$,

$$\widetilde{R} = R^{-1}, \qquad \det R = 1, \qquad (3\text{--}3)$$

so that the transformation inverse to (3–2) is

$$\mathbf{r} = \widetilde{\mathsf{R}} \cdot \mathbf{r'}. \qquad (3\text{--}2')$$

---

* A concise definition of the often used concept *observer* is difficult to give. The following sentences can be regarded as defining this notion in context.

PROBLEM 3-1

Write the dyadics R and R̃ in component form and specialize them to the rotation by an angle α in the positive sense about an axis which in polar coordinates is given by $\theta = 30°$, $\varphi = 0°$.

The transformations (3-1) constitute a representation of the three-dimensional *translation* group; the transformations (3-2) constitute a representation of the three-dimensional *rotation* group. These two groups together, i.e. their group product, constitute the three-dimensional *Euclidean group*.

The above physically trivial equivalence of coordinate systems can now be expressed in the much less trivial precise form: *The laws of physics should be invariant under the three-dimensional Euclidean group.*

The invariance of the laws of physics under translation of the coordinate system (3-1) is sometimes expressed in the more intuitive but physically less meaningful way: space is homogeneous. Similarly, the rotational invariance under the transformations (3-2) is said to be an expression of the isotropy of space. Thus, the invariance under the Euclidean group can be regarded as the mathematical statement of the homogeneity and isotropy of space.

B. The Relativity of Velocity

3-3 FORCE AND THE LAW OF INERTIA

Is the motion of a body the result of the influence of other bodies? Not necessarily. A precise criterion is offered by the *law of inertia* (Newton's first law of motion), which was discovered by Galileo Galilei (1564-1642). It can be stated as follows: A body remains at rest or in motion with constant velocity if and only if it is not subjected to the influence of other bodies. This law assures us that the *change* of velocity, i.e. the acceleration, can be used as a criterion for the presence or absence of the interaction of bodies. We are thus led to the study of the time derivatives of the position coordinates.

In the mechanics of Sir Isaac Newton (1642-1727) it is convenient to express the interaction of bodies by *forces*. The force acting on a mass point is a vector function of position, velocity, and certain characteristic properties of the mass point which describe completely the influence of all other bodies on that mass point. It is completely determined by these other bodies. Unless a theory is available which permits the computation of this function **F** from the various force-producing bodies, **F** must be assumed to be given, Newtonian mechanics being restricted to the study of the effect of **F** on a body (collection of mass points). Newton's law of gravitation and Lorentz's law of electromagnetic forces are examples of forces which can be computed from the force-producing bodies.

It should be noted that in its original formulation Newtonian mechanics admitted only forces which are independent of velocity and which act in a line joining two interacting mass points. With the discovery of magnetic forces which satisfy neither of these two requirements, this framework had to be extended.

PROBLEM 3-2

Discuss how the action of a magnetic field on an electrically charged mass point violates Newton's original idea of force.

In order to put these considerations into mathematical language we consider a mass point which is not necessarily at rest so that its position vector $\mathbf{r}$ may depend on the time t. Its velocity is defined by $\mathbf{v}(t) = d\mathbf{r}(t)/dt$, and its acceleration by $\mathbf{a}(t) = d^2\mathbf{r}(t)/dt^2$. The law of inertia provides the criterion

$$\mathbf{a}(t) = 0 \quad \text{if and only if} \quad \mathbf{F} = 0. \tag{3-4}$$

Consider now the *equation of motion* of a mass point. By definition, this is a *second-order* differential or integrodifferential equation of the position vector:

$$V(t, \mathbf{r}, \mathbf{v}, \mathbf{a}; \lambda) = 0. \tag{3-5a}$$

The properties of the mass point (mass, charge, etc.) are summarily denoted by λ. If this equation can be solved for $\mathbf{a}$, it will have the form

$$\mathbf{a}(t) = \mathbf{f}(t, \mathbf{r}, \mathbf{v}; \lambda). \tag{3-5b}$$

The equation must be of second order, since motion at time $t > t_0$ (or $t < t_0$) must be uniquely determined by the specification of the position and the velocity at $t = t_0$, $\mathbf{r}_0 = \mathbf{r}(t_0)$ and $\mathbf{v}_0 = \mathbf{v}(t_0)$. This is a physical requirement. (See Section 3-15.)

If the equation of motion is of the form (3-5b), the law of inertia requires that the right-hand side vanish when $\mathbf{F} = 0$. In any case, it must be a known function. In Newtonian dynamics $\mathbf{f}$ is identified with the *specific* force or the force per unit mass. Thus, Newton's second law of motion is written

$$m\mathbf{a}(t) = \mathbf{F}(t, \mathbf{r}, \mathbf{v}; \lambda), \tag{3-6}$$

which is obviously a special case of (3-5) and implies the law of inertia.

According to Newton's original ideas $\mathbf{F}$ should not depend on $\mathbf{v}$, which restriction had to be abandoned (cf. Problem 3-2). The independence of $\mathbf{F}$ from $\mathbf{a}$, however, is a much more stringent requirement. Obviously, the concept of force loses its meaning if it depends on the acceleration. When such "forces" have to be introduced in noninertial systems (see Section 3-9), the name *pseudoforces* is used. Nevertheless, it has become customary to speak of the *force of radiation reaction* in electrodynamics, although that quantity also depends on the acceleration (see beginning of Section 6–7).

3–4 GALILEAN INVARIANCE

We now recall the conclusions reached in Section 3–2. The relative nature of motion seems to make the law of inertia a *relative* criterion: Whether a body is subjected to forces, i.e. is accelerated, depends on the observer. For example, a comoving observer will see no acceleration. Forces appear to have no absolute significance. However, our physical intuition does not seem to agree with such a conclusion.

This situation was known to Newton, but he had no difficulties resolving the apparent dilemma. He postulated the absolute nature of space and time and thereby established a stage on which the events of nature are supposed to take place. Thus, according to the law of inertia, forces act if and only if a body is accelerated *relative to absolute space*. Forces are absolute in Newton's dynamics.

The Newtonian postulate of absolute space appears absurd today, but it was very much in keeping with the absolutist philosophy that dominated the thinking of western culture for such a long time before Newton. In fact, people like Galileo, Copernicus, and Newton were the first to take a step out of that framework. Today, we are much more willing to learn from experience. For example: two observers in uniform motion relative to each other (one of them inside a vehicle, say) see exactly the same phenomena take place under the same conditions. They cannot decide by any experiment which one of them is in motion relative to absolute space. This type of experience makes absolute space appear meaningless to us intuitively.

Concerning absolute time it appears that we have at least the freedom of choosing the time instant which we call $t = 0$: the origin of the time coordinate is arbitrary. This means that the translation invariance under space translations (3–1) can be extended to the four-dimensional group including time:

$$t' = t + \alpha^0 \qquad (\alpha^0 = \text{const}). \tag{3–7}$$

Obviously, if we accept the law of inertia, the postulate of absolute space and time is unnecessarily strong for ensuring this law to be a criterion for absolute forces. All we need is absolute acceleration. We do not need absolute velocity.

Indeed, Newton's laws of motion relative to absolute space are identical to those relative to a frame of reference moving with constant velocity relative to absolute space. If S refers to absolute space [position vector $\mathbf{r}(t)$] and S' refers to a coordinate system in uniform motion relative to S [position vector $\mathbf{r}'(t)$], then

$$\mathbf{r}' = \mathbf{r} - \mathbf{v}t. \tag{3–8}$$

It is easy to prove that Newton's dynamics is invariant under the transformations (3–8), provided the forces are not velocity dependent.

PROBLEM 3-3

Show that Newton's three laws of motion are invariant under (3-8), i.e. have the same form in the primed and unprimed reference system for velocity-independent forces.

The transformations (3-8) form a group which is known as the *restricted Galilean group*. The invariance of Newton's dynamics under this group assures a *relativity of velocity*. This means that the effect of forces on the motion of bodies does not depend on the velocity of the observer relative to absolute space; it can depend on his acceleration, however.

This conclusion establishes the equivalence of a threefold infinity (three components of **v**) of reference systems. The latter, together with the sixfold infinity (three components of **α** and three Euler angles) of reference systems, which are equivalent by invariance under the three-dimensional Euclidean group [Eqs. (3-1) and (3-2)], and the invariance under time translation (3-7), forms a tenfold infinity of equivalent systems known as the *Galilean inertial systems* because Galileo's law of inertia is valid relative to them. The corresponding 10-parameter group is the *full (proper) Galilean group*. It contains the Euclidean group as a subgroup. The invariance under this Galilean group of Newtonian dynamics is the precise statement of the *Galilean principle of relativity*. It is a relativity of velocity. Note that the often used term *nonrelativistic mechanics* is a misnomer.

3-5 INERTIAL FRAMES OF REFERENCE

So far, however, Galileo's law of inertia is still an empty statement. Obviously, we must find a way of testing the validity of this law. According to (3-4) the crucial point here is the construction of a force-free situation. A particle in force-free surroundings should have no acceleration. This can be tested experimentally. Having established the validity of the law of inertia in this way, we can use the above definition of *inertial system* and identify the coordinate system, in which the force-free situation was achieved, with an inertial system.

Thus we study the interaction of bodies with one another and we learn how to arrange bodies in such a way that they do not influence one another. We consider one particular body B which is nonrotating and we arrange all other bodies in such a way that the forces they produce do not act on B. A coordinate system anchored in B is then called inertial. All other inertial systems are obtained from this one by one of the transformations of the Galilean group.

A two-dimensional realization of such a system is provided by an orthogonal rectilinear coordinate system marked on a horizontal sheet of ice. The horizontal position ensures that the force of gravity is made ineffective; the ice is assumed to be sufficiently smooth to suppress the frictional force of bodies sliding on it (within the accuracy of the experiment). Of course, one must neglect the rotation of the earth. The centrifugal effects can be eliminated by locating the sheet on the equator. The Coriolis effects, however, would still be there unless one restricted the motion to a meridian. In addition, one must neglect the effects, due to the revolution of the earth around the sun, of the motion of the solar system inside our galaxy, etc.

A three-dimensional realization to good approximation (neglecting, for example, the proper motion of fixed stars) is a coordinate system anchored on the fixed stars. If all celestial bodies are sufficiently far removed, no gravitational forces will act on a given body B. If electromagnetic forces are also made ineffective (by assuring that B has sufficiently small electric charge and magnetic moment densities), there will be no classical forces acting on B. It is then a matter of experimental tests to see whether B indeed remains at rest or in uniform motion relative to the fixed stars.

It is important to realize the difference between an inertial system and absolute space: If the fixed stars were identified with absolute space, the relativity of velocity would be lost. As it is, however, the relativity of velocity requires that there be no physical distinction between uniform motion of a body relative to the fixed stars and uniform motion of the fixed stars relative to a body. This physical indistinguishability is not trivial. When the same criterion will be used for the relativity of acceleration, it will have far-reaching consequences (Section 3–14).

3–6   LORENTZ INVARIANCE

Electromagnetic fields afford a peculiar difficulty; viz. they propagate in a vacuum with a velocity, c, which is the same for two observers moving with constant velocity relative to each other.

PROBLEM 3–4

Review the experiments leading to the conclusion that the velocity of light in a vacuum is independent of the velocity of the source and of the observer.

This result is inconsistent with the nonrelativistic addition of velocities (three-vector addition). On the other hand, it is not inconsistent with the relativity of velocity. It only limits the velocity attainable by an object. Einstein concluded from this that Galilean invariance of dynamics must be abandoned and must be replaced by a different expression for the relativity of velocity.

The corresponding new symmetry property must be consistent with the experimentally established constancy of the velocity of light, with Galileo's law of inertia, and with the fact that Newtonian mechanics with its Galilean invariance is, after all, a pretty good theory, having (up to 1905) never given rise to a serious discrepancy with experiment* except possibly for electro-

---

* After Laplace only some very small effects remained unexplained, like the discrepancy in the perihel motion of mercury.

magnetic phenomena. One can find the required symmetry property by asking for the most general linear Cartesian coordinate transformation which leaves Maxwell's equations invariant. This transformation involves also a transformation of time. It is the inhomogeneous Lorentz group,* i.e. the group of transformations which contains the Euclidean group and the homogeneous Lorentz transformation

$$\mathbf{r}' = \gamma(\mathbf{r} - \mathbf{v}t) - (\gamma - 1)\left(\mathbf{r} - \frac{\mathbf{v}\,\mathbf{r}\cdot\mathbf{v}}{v^2}\right),$$

$$t' = \gamma\left(t - \frac{\mathbf{r}\cdot\mathbf{v}}{c^2}\right), \qquad \gamma = \left(1 - \frac{v^2}{c^2}\right)^{-1/2}.$$

(3-9)

Indeed, this group satisfies all the requirements. As the invariance group of Maxwell's equations, it ensures that if these equations are valid relative to a reference frame S, they will also be valid relative to every system S' which is related to S by a transformation belonging to the Lorentz group. In particular, therefore, the velocity of light will have the same value in S' as it has in S. Galileo's law of inertia is still valid, the transformation group ensuring again a *relativity of velocity* but leaving acceleration as an absolute quantity. Finally, for small velocities characterized by $v \ll c$ the Lorentz transformations approach the Galilean transformations. This gives a precise specification of the validity limits of Newtonian mechanics.

Based on Lorentz invariance, the special theory of relativity provides a new kinematics and dynamics which reduce to those of Newton in the low velocity limit. The general structure (3-5) of the equation of motion is still correct. But this structure is conditioned to a considerable extent by the particular notation used. The one used in (3-5) and (3-6) is that of three-vectors. This notation is no longer particularly convenient because, in the kinematics of special relativity, velocities, accelerations, etc. no longer add like three-vectors.

The notation most appropriate for a Lorentz invariant theory is that of four-vectors (and corresponding higher-order tensors) rather than three-vectors, dyadics, etc. This is a consequence of the involvement of time as just another coordinate in the Lorentz transformation. Newton's absolute time is thereby abandoned and physical meaning can be ascribed only to the four-dimensional space-time continuum. To ensure this, one considers a four-dimensional Cartesian coordinate system ct, x, y, z, or simply x^μ ($\mu = 0, 1, 2, 3$). The corresponding four-dimensional space is not Euclidean space because the scalar product of two vectors is not positive definite. It is called a *pseudo-Euclidean space*.†

---

* The extension of this group to include inversions of space and time will be ignored here since it is not essential for the present purpose.

† The reader not familiar with four-vector notation should study Appendix 1 before proceeding with the main text.

The quantity which plays the same role as t in Newtonian mechanics is now the proper time τ, which is an invariant defined by

$$c^2 \, d\tau^2 = c^2 \, dt^2 - dx^2 - dy^2 - dz^2. \tag{3-10}$$

Thus, the position four-vector x with components $x^\mu(\tau)$ leads to the four-velocity, $v^\mu(\tau) = dx^\mu/d\tau$ and to the four-acceleration, $a^\mu(\tau) = d^2x^\mu/d\tau^2$. The equation of motion (3-5) can then be written in four-vector notation,

$$V^\mu(\tau; x, v, a; \lambda) = 0, \tag{3-11a}$$

or, when solvable for a,

$$a^\mu(\tau) = f^\mu(\tau; x, v; \lambda). \tag{3-11b}$$

The Newtonian concept of force can now be generalized suitably to refer to a four-vector F^μ. If this is done, (3-11) can be written

$$ma^\mu(\tau) = F^\mu, \tag{3-12}$$

where m is the rest-mass. What was said in connection with (3-5) concerning the necessity of a *second*-order equation is still valid, and so are the remarks concerning the dependence of F on the acceleration.

The inclusion of time as a fourth coordinate is not entirely new to us. We have seen earlier that the arbitrariness of the choice of origin of space as well as time coordinates leads to the invariance of the laws of physics under a *four-dimensional* translation group, viz. the transformations (3-1) and (3-7). In four-vector notation these transformations combine to read

$$x'^\mu = x^\mu + \alpha^\mu \qquad (\alpha^\mu = \text{const}). \tag{3-13}$$

The corresponding generalization of the three-dimensional rotation group must be done with proper attention to the indefinite character of the scalar product. The transformation (note the summation convention)

$$x'^\mu = \alpha^\mu_{\ \nu} x^\nu \tag{3-14}$$

must then be restricted by

$$\det \alpha^\mu_{\ \nu} > 0 \qquad \det \alpha^0_{\ 0} > 0 \tag{3-15}$$

and is related to the inverse transformation

$$x^\mu = \alpha^\mu_{\ \nu} x'^\nu \tag{3-16}$$

by

$$\alpha^\mu_{\ \lambda} \alpha_\nu^{\ \lambda} = \delta_\nu^{\ \mu}. \tag{3-17}$$

Equations (3-15) and (3-17) are the appropriate generalizations of (3-3). In fact, if all components are restricted to space coordinates only, the three-dimensional rotation can easily be seen to be a subgroup of the rotations (3-14).

The most important and interesting observation concerning (3-14), however, refers to those transformations which mix space and time coordinates. They are seen to be the Lorentz transformations (3-9) or a mixture of these with three-dimensional rotations. Thus, the whole group of homogeneous Lorentz transformations, together with three-dimensional rotations is expressed by (3-14). Since $\alpha^{\mu}{}_{\nu}$ has 6 linearly independent components, the inhomogeneous Lorentz group (3-13) and (3-14) is a 10-parameter group. This enumeration agrees with the more physical listing of four translation parameters (α^{μ}), three rotation parameters (e.g. three Eulerian angles), and three velocity components ($\mathbf{v}$ in Eq. 3-9).

The invariance of the laws of physics under the 10-parameter Lorentz group is the precise statement of the *Lorentz relativity of velocity* or *principle of relativity of the special theory of relativity*. These laws encompass those of mechanics as well as of electrodynamics, while the Galilean relativity principle is valid in mechanics (within the Newtonian validity limits $\nu^2 \ll c^2$) and in Newton's theory of gravitation.

The definition of an inertial system is not affected when one passes from Galilean invariance to Lorentz invariance. It is still anchored in the fixed stars. However, the equivalence class is now determined by the Lorentz rather than the Galilean transformations.

C. The Principle of Equivalence

We have seen that in defining an inertial coordinate system one has to look for a force-free situation. This is relatively easy to do for electromagnetic forces where it is a matter of electric neutrality and absence of higher electromagnetic moments. However, the other fundamental classical force, the gravitational force is much more difficult to dodge, as the examples of the previous sections show. The reason is that this force depends on the mass of a body in such a way that the corresponding gravitational acceleration is *independent* of its mass. It can therefore not be eliminated by letting the mass go to zero. This is in sharp contrast to the electromagnetic forces which can be eliminated by letting the charge go to zero.*

The principle of equivalence tells us how and to what extent one can dodge the gravitational force. But before discussing it we must clarify the difference between three notions all loosely called "mass."

---

* The existence of higher inertial moments analogous to electromagnetic multipole moments is not of interest at this level of investigation.

3-7 THREE KINDS OF MASS

Given a ponderomotive force, e.g. the tension of an extended spring, it will accelerate a body by an amount *inversely* proportional to its mass. This mass is called *inertial mass*:

$$\frac{m_1^{\text{inertial}}}{m_2^{\text{inertial}}} = \frac{a_2}{a_1}. \tag{3-18}$$

A gravitational pull can be balanced by a nongravitational force. However, the force needed depends on the mass of the body to be balanced. For two bodies, B_1 and B_2, the forces needed are *directly* proportional to the masses; these masses are called *passive gravitational masses*:

$$\frac{m_1^{\text{pass grav}}}{m_2^{\text{pass grav}}} = \frac{F_1}{F_2}. \tag{3-19}$$

For a given gravitational field strength, the passive gravitational mass determines the corresponding gravitational force acting on that mass. If the field strength **f** is derived from a potential ϕ,

$$\mathbf{F} = m_{\text{grav}}^{\text{pass}} \mathbf{f} = -m_{\text{grav}}^{\text{pass}} \nabla\phi. \tag{3-20}$$

The third type of mass is the *active gravitational mass*. It is the mass that determines how large a gravitational potential and field strength are produced by a body. According to Newton's law of gravitation,

$$\phi = + G \frac{m_{\text{grav}}^{\text{active}}}{r}, \tag{3-21}$$

and

$$\mathbf{f} = -\nabla\phi = G \frac{m_{\text{grav}}^{\text{active}}}{r^2}. \tag{3-22}$$

In these equations, G is Newton's gravitational constant. If first body B_1 and then B_2 are placed at a given distance from a body B, then the gravitational forces acting on B due to B_1 and B_2, respectively, are in the ratio

$$\frac{F_1}{F_2} = \frac{m_1^{\text{active grav}}}{m_2^{\text{active grav}}}. \tag{3-23}$$

The three equations (3-18), (3-19), and (3-23) permit one to establish a mass scale for each of the three kinds of masses by well-defined basic experiments. After choosing a unit for each of the three scales, one has a complete operational definition for m_{inertial}, $m_{\text{grav}}^{\text{passive}}$, and $m_{\text{grav}}^{\text{active}}$.

When Newton's law of action and reaction (his third law of motion) is applied to the gravitational force between two bodies, we have

$$G \frac{m_1^{\text{pass grav}} m_2^{\text{act grav}}}{r_{12}^2} = G \frac{m_2^{\text{pass grav}} m_1^{\text{act grav}}}{r_{12}^2}. \tag{3-24}$$

If we choose $m_1^{\text{pass grav}}$ to be the unit of passive gravitational mass and $m_1^{\text{active grav}}$ to be the unit of active gravitational mass, we obtain $m_2^{\text{active grav}} = m_2^{\text{pass grav}}$. Comparing all masses with m_1 we find

$$m_{\text{grav}}^{\text{active}} = m_{\text{grav}}^{\text{passive}}. \qquad (3\text{--}25)$$

This is a direct consequence of the law of action and reaction. If (3-25) is valid, we can simply write m_{grav} and need no longer distinguish between active and passive mass.

As long as we stand in the framework of Newton's gravitation theory, (3-25) is obviously valid. If we go to a more general gravitation theory, Eq. (3-25) will be restricted to the domain of validity outlined by this theory for its Newtonian approximation (see Section 1-2). Thus, in the general theory of relativity (3-25) will be valid *locally*. It will always be valid for a point particle. The generalization of (3-25) to extended masses in general relativity is not trivial and is connected with the problem of defining energy over a finite space-time region in a covariant way. We shall not be concerned with this problem since the dimensions of the particles we consider will be sufficiently small. Equation (3-25) will therefore be valid for us.

3-8 FREE FALL

We must now study the relationship between two types of mass, m_{inertial} and m_{grav}. Again within the framework of Newtonian physics, the *principle of equivalence* states that the units of m_{inertial} and of m_{grav} can be so chosen that

$$m_{\text{inertial}} = m_{\text{grav}}. \qquad (3\text{--}26)$$

This equality can be applied to the equation of motion of a body moving under the influence of gravitational forces only. Such a motion is known as *free fall*. According to Newton's second law of motion, the equation of free fall is

$$m_{\text{inertial}}\, \mathbf{a} = m_{\text{grav}}\, \mathbf{f}, \qquad (3\text{--}27)$$

where $\mathbf{f}$ is the resultant gravitational field strength acting on the body. The principle of equivalence in the form (3-26) tells us that

$$\mathbf{a} = \mathbf{f}, \qquad (3\text{--}28)$$

which means that the acceleration of a body in free fall is *independent of the inertial mass* of that body. This law was also first discovered by Galileo.

As an example, consider a satellite circling the earth at a certain distance r from its center. The satellite's acceleration has the magnitude $a = \omega^2 r$, where ω is its angular velocity of revolution. Its period is therefore $\tau = 2\pi/\omega$. From Galileo's law (3-28) it follows now that this period is the same for all satellites circling at the distance r, independent of their inertial mass.

Obviously, in Newtonian physics the statement (3–26) implies (3–28) and vice versa. Therefore, the principle of equivalence in nonrelativistic physics can also be expressed as follows: *the equations of motion of a body in free fall are independent of its mass.*

The special theory of relativity adds nothing to Newton's theory of gravitation. However, when combined with it, the principle of equivalence receives a richer meaning. Special relativity teaches the equivalence of inertial mass and energy. It follows that the principle of equivalence (3–26) can be extended to

$$m_{\mathrm{grav}} = E/c^2, \tag{3–29}$$

where E is any form of energy whatsoever. A body whose rest energy $m_{\mathrm{inertial}}c^2$ consists of kinetic and potential energy [the latter making a negative (!) contribution] has a gravitational mass corresponding to the sum of these energies divided by c^2.

In general relativity an essential generalization of Newton's theory of gravitation is achieved. It can be characterized as a geometrization of the gravitational interaction. This means that the latter is no longer described as a certain force produced by one body (B_1) on another (B_2), but as an intrinsic property of space, viz. *space curvature*, produced by the presence of B_1; the second body, B_2, is accelerated because it finds itself in a curved space. Thus, the "force" acting on B_2, being due to a property of space, is completely independent of the nature of B_2. In particular, it does not depend on its (inertial) mass, but it also does not depend on its shape or chemical composition. It must be added, however, that B_2 must be a nonrotating test body; gravitational effects due to the rotation of B_2 and due to its $m_{\mathrm{grav}}^{\mathrm{active}}$ must be negligible.

This geometrization of gravitation suggests itself if one generalizes the principle of equivalence of Newtonian gravitation theory; historically, this principle was in fact the seed from which general relativity evolved. The *generalization* of the principle follows easily from (3–28) and (3–29). One postulates as a principle that

> *The equations of motion of a nonrotating test body in free fall in a gravitational field* is *independent of the energy content of that body.* (E)

While we know this principle to be true in Newtonian approximation, it must be confirmed by experiment to be acceptable outside the validity limits of that approximation. We shall return to these experiments below (Section 3–11).

The mathematical statement of the (generalized) equivalence principle is the requirement that the equation of motion (3–11) for a test body in a gravitational field be independent of its energy content, irrespective of the nature of this energy (gravitational, electromagnetic, or other). Since all the various contributions to the internal energy of a body are included in a measurement of m_{inertial} and, in fact, cannot be differentiated experimentally, the principle means simply independence of (3–11) from m_{inertial}. To the extent that the charge of a body contributes to its energy content, this means also independence from the body's charge. However, the charge also is responsible for the emission of electromagnetic radiation which *can* affect the equations of motion. Thus, the question concerning the extent to which the principle of equivalence implies that the equations of motion are independent of the charge is left open at this point. It will be discussed in detail in Section 8-3. The emission of gravitational radiation will also affect the equation of motion but this radiation is neglected by assumption (definition of a test body).

3-9 FICTITIOUS FORCES AND APPARENT GRAVITATIONAL FIELDS

Newton's second law of motion is not valid in a noninertial coordinate system. One special noninertial system, however, is so important that we are well acquainted with the modifications necessary to obtain from Newton's second law of motion an equation of motion valid in that special system: in a rotating coordinate system a centrifugal and a Coriolis force must be added to the forces otherwise present. These two forces are *fictitious* in the sense that they are absent in an inertial system; also, they depend on the velocities, which is contrary to the requirements of Newtonian mechanics. This theory offers no explanation for these "forces" other than a demonstration by means of a transformation that they must be there. It is proven that the acceleration $\mathbf{a}'$ of a mass point in a system rotating with constant angular velocity ω differs from the acceleration $\mathbf{a}$ in an inertial system by two additive terms which, of course, have the dimensions of acceleration,

$$\mathbf{a} = \mathbf{a}' + 2\omega \times \mathbf{v}' + \omega \times (\omega \times \mathbf{r}'). \tag{3-30}$$

When the left-hand side of Newton's equation of motion (3-7) is transformed to the rotating coordinate system, these two accelerations occur multiplied by m_{inertial}, thus yielding the dimensions of a force.

In S: $m_{\text{inertial}} \mathbf{a} = \mathbf{F}.$

In S': $m_{\text{inertial}} \mathbf{a}' = \mathbf{F} + \mathbf{F}_{\text{fictitious}},$ (3-31)

$\mathbf{F}_{\text{fictitious}} = -2m_{\text{inertial}} \omega \times \mathbf{v} - m_{\text{inertial}} \omega \times (\omega \times \mathbf{r}').$

The equivalence principle (3-26) permits us to give a different interpretation to these fictitious forces: they have the structure of gravitational forces because they are proportional to the gravitational mass of the body on which they act. Since their origin is not known and since this proportionality is the only criterion for the identification of forces as gravitational, the presence of centrifugal and Coriolis forces *can never be distinguished* from the presence of certain appropriate gravitational forces. These gravitational forces, however, have the property that they disappear when the observer moves in a certain way (viz. when he does not rotate).

Next, consider a static homogeneous gravitational field, as would be the earth's field when observed only over a spatial region of dimensions small compared with the earth's radius R. On the surface of the earth a mass point will experience a force

$$\mathbf{F} = -G \frac{M m_{\text{grav}}}{R^2} \hat{\mathbf{k}} = -m_{\text{grav}} g \hat{\mathbf{k}}, \qquad (3\text{-}32)$$

where M is the active gravitational mass of the earth, and $\hat{\mathbf{k}}$ is a unit vector pointing radially away from the earth's center. An observer at rest on the surface of the earth can measure the acceleration of gravity by measuring the acceleration of that mass point in free fall. The result will be g. If, however, this observer is accelerated in the direction $\hat{\mathbf{k}}$ with acceleration a, the same measurement will produce $g' = g + a$. A moment's thought convinces one that there is no measurement that the observer can perform on that mass point which would tell him how much of g' is due to the *true* gravitational acceleration and how much is due to some force $\mathbf{F}'$ pulling the observer and his laboratory away from the earth. The *apparent* gravitational field produced by the force $\mathbf{F}'$ cannot be distinguished from a true gravitational field. However, another observer, who is freely falling in the earth's field and who is not exposed to $\mathbf{F}'$, will see no gravitational acceleration at all: an object dropped by him will remain at rest with respect to him. Thus, the force of a static homogeneous gravitational field cannot be distinguished from a constant force pulling the observer and his laboratory in the opposite direction but can be made to vanish by falling freely in it.

In Newtonian physics *all* gravitational forces are absolute. Neglecting the motion of the earth, the observer on the surface of the earth is an inertial observer (at rest relative to the fixed stars) who decides that the earth's gravitational acceleration is "really" g and not g'. The measurement of g' is made in a noninertial system in which a *fictitious force* $ma\hat{\mathbf{k}}$ had to be added in order to make Newton's second law valid.

Consider an observer whose laboratory is so small that within the accuracy of his measurements the earth's field appears completely homogeneous. Is he justified in this distinction between the true g and the apparent g'? A negative answer to this question will lead us to abandon the definition of inertial systems adopted for Galilean invariant and Lorentz invariant physics. A new definition of inertial systems will emerge which is consistent with the characteristic features of gravitational forces. These forces are always present and therefore play an essential role in this definition (Section 3-12).

The above two examples of noninertial systems, one rotating with constant angular velocity, the other accelerated with constant acceleration, suggest the following generalizations. Newton's equations of motion can be modified so as to be valid in noninertial coordinate systems. This modification consists of the addition of suitable fictitious forces. On the basis of the principle of equivalence, (3-26), these forces cannot be distinguished from certain corresponding gravitational forces by any experiment on the mass points under consideration.

3-10 EINSTEIN'S PRINCIPLE OF EQUIVALENCE

This equivalence between fictitious forces and gravitational forces was first fully recognized by Einstein. He fully appreciated the importance of this equivalence for the construction of a theory of gravitation which is consistent with the special theory of relativity; (Newtonian gravitation theory is Galilean invariant and not Lorentz invariant). For this reason he called this equivalence between accelerated observers and observers in gravitational fields *the principle of equivalence.* Since this formulation of the principle of equivalence is based on considerations in Newtonian physics, it will not come as a surprise that it has only limited validity. Now that we have the desired generalization of gravitation theory (viz. Einstein's general theory of relativity), its limitations are evident, as will be shown below. For this reason we use the statement (E) of Section 3-8 above as the general statement of the principle of equivalence. In Newtonian physics, the equality of inertial and gravitational masses is then a consequence, and Einstein's formulation of the equivalence principle follows.

In the general theory of relativity the mathematical framework is Riemannian geometry in a four-dimensional Riemann space* with indefinite metric $g_{\mu\nu}$. A gravitational field is present at a given point *if and only if space is curved* at that point. Thus, the space curvature (more precisely the curvature tensor) is a unique criterion for the presence of a gravitational field. If no gravitational field is present at a point or in a region, the Riemann space is flat there.

Just as every sufficiently continuous curved surface has a tangent plane at every point, so does the Riemann space have a flat tangent space at every point. This space is the four-dimensional pseudo-Euclidean space of special relativity. Therefore, at every point P of a Riemann space a coordinate system can be chosen whose metric at this point coincides with that of special relativity, $g_{\mu\nu}(P) = \eta_{\mu\nu}$ (local geodesic coordinate system). This coincidence will be restricted to P if and only if there is a nonvanishing curvature at that point. If the space is flat over a finite† domain D, then there always exists a coordinate system for which $g_{\mu\nu} = \eta_{\mu\nu}$ everywhere in D.

A coordinate system (observer) is specified by the tensor $g_{\mu\nu}(x)$ as a function of the four-vector $x = \{x^\alpha\}$. A transformation from one coordinate system to another will in general involve the time; the two corresponding observers are in relative motion. An example is the transformation from a rotating (S') to a nonrotating (S) coordinate system. If the fictitious forces in S' are considered to be gravitational forces in Newtonian physics, they are "transformed away" in S, corresponding to flat space in general relativity. But flat space remains flat no matter what coordinate transformation is carried out. Consequently, according to general relativity, the fictitious forces in S' are not *true* gravitational forces (i.e. associated with curvature).

---

* See Appendix 2.
† In practice where every measurement has limited accuracy, D is always finite (not just a point) within the error of measurement.

In general, the characterization of gravitational interaction by the curvature of space is an absolute one (an *observable*), because curvature is an *intrinsic* property. Gravitational fields are unambiguously characterized as *apparent* when the space is flat, so that a suitable coordinate transformation can "transform them away." True gravitational fields can never be transformed away.

It follows that apparent gravitational fields are entirely due to the choice of the coordinate system (motion of the observer); for a suitably moving observer they will not be present. Apparent gravitational fields are a characteristic feature of the motion of the observer (rather than of the observed physical system), while true gravitational fields are the same for all observers no matter what their motion. This distinction completely resolves the ambiguity which existed in Newtonian physics.

Einstein's statement of the equivalence principle as an equivalence between accelerated observers and gravitational fields is now seen to be restricted to apparent gravitational fields. True gravitational fields cannot be simulated by acceleration (i.e. by a coordinate transformation).

On the other hand, the statement (E) above concerning the equation of motion of a test particle in a gravitational field is meaningful also in general relativity where only *true* gravitational fields are recognized as observables. In fact, (E) plays an important role in general relativity because this equation of motion is the *geodesic equation*. A geodesic, being the shortest distance between two points, is a purely geometrical property and is consequently independent of the physical characteristics of the test particle.

3–11 EXPERIMENTAL CONFIRMATION

The principle of equivalence has been confirmed by a variety of experiments, some of which can claim extremely high accuracy. The most important of these involve a comparison of the ratio $m_{\text{inertial}}/m_{\text{grav}}$ for various bodies. This can be done by means of a pendulum, but a torsion balance, as used by Baron R. von Eötvös, yields much higher accuracy. These experiments, which he and his collaborators carried out since 1890 over a period of more than 30 years, confirmed that the ratio $m_{\text{inertial}}/m_{\text{grav}}$ is the same for a large variety of substances to an accuracy of the order of $\frac{1}{3} \times 10^9$. Similar experiments by J. Renner in 1935 improved this accuracy by a factor of about 2. Recent experiments by P. G. Roll, R. Krotkov, and R. H. Dicke* yield a further improvement by several orders of magnitude to about 10^{11}.

The so-called *gravitational red shift* is a direct consequence of the principle of equivalence. It was confirmed by the frequency shift of spectral lines from the sun as compared with the corresponding terrestrial line. Much greater accuracy was achieved in 1960 when two atomic clocks were compared in the laboratory

---

* *Ann. Phys.* (N.Y.), **26**, 442 (1964).

by means of the Mössbauer effect.* These clocks were located above each other so that they were in different potentials of the earth's gravitational field. The resultant frequency difference confirmed the principle of equivalence within the experimental error of ten percent. This experiment, however, represents only a weak confirmation of the principle because it takes place in an apparent gravitational field. It is the Doppler shift of special relativity as seen by a freely falling observer. The solar red shift as seen by a terrestrial observer involves a true gravitational field: no motion of the observer can transform away *both* the local solar and the local terrestrial field.

Finally, we have the fact that the principle of equivalence is an integral part of the general theory of relativity. While a confirmation of the principle does not confirm the theory, every confirmation of general relativity is also a confirmation of the equivalence principle. Thus, the deflection of light by the sun as well as the perihelion motion of the planets involves the use of geodesics as the orbits of test particles. These effects therefore add further experimental evidence to the above tests of the principle of equivalence.

D. The Relativity of Acceleration

The relativity of velocity which is expressed by Galilean or Lorentz invariance is based on kinematical considerations (Newton's first law). A deeper understanding of the nature of gravitational forces and their dynamics leads to the conclusion that these forces do not have absolute significance. More specifically, there are gravitational interactions which are of an intrinsic nature and others which are not. But the distinction between the two cannot be made by specifying the respective forces. For this reason, the (essentially Newtonian) concept of force loses its significance. Gravitational interactions must not be described by forces. This situation is very closely related to the relativity of acceleration.

3–12 LOCAL INERTIAL FRAMES

The principle of equivalence tells us how to dodge gravitational fields: the freely falling nonrotating observer will not see a gravitational field in his immediate vicinity.† More precisely, given a finite four-dimensional space-time region D (containing the observer and his laboratory), there always exists a

---

* Cranshaw, Schiffer, and Whitehead, *Phys. Rev. Letters* **4**, 163 (1960); R. V. Pound and G. A. Rebka, *Phys. Rev. Letters* **4**, 337 and 397 (1960).

† The principle of equivalence [(E) of Section 3–8] does not ensure the existence of a transformation to comoving coordinates. One must postulate it. The comoving frame is defined as S' such that $a'^\mu = 0$ in (3–11). In general relativity, (3–11) becomes (3–33) and the transformation always exists.

critical experimental accuracy, such that no true gravitational field can be observed inside D as long as this accuracy is not exceeded.

Mathematically, this statement can be understood as follows. Let R be a measure of the curvature of Riemannian space inside D; let d be a typical dimension of D. Then, if $Rd^2 \ll 1$ is too small to be measurable, the space in D is flat within experimental accuracy. It then follows that inside D only apparent gravitational fields can be present and that one can choose a coordinate system such that, inside D, $g_{\mu\nu} = \eta_{\mu\nu}$ (geodesic coordinates). The motion of the associated observer is necessarily that of free fall according to the principle of equivalence. This situation is sometimes expressed by saying that a freely falling observer sees locally flat space.

PROBLEM 3-5

A man is falling down a mine shaft. His pipe and his watch fall out of his pockets and are falling freely with him at a distance d (perpendicular to the mine shaft) from each other. He can observe a change in this distance only when that change is at least δ. How deep does the mine shaft have to be so that the man will know that he is not in a homogeneous field?

It follows that a freely falling nonrotating observer will find himself locally in a force-free region of space. We were thus led to the force-free situation which we tried to construct previously in order to specify an inertial reference system. While the previous construction of such systems (sheet of ice, system at rest with respect to the fixed stars) was rather artificial, not precise, and impossible to implement (in the case of the fixed stars), *the identification of an inertial system with a freely falling nonrotating observer* seems rather natural; in particular, such an observer *can* be produced "operationally," as the orbiting of astronauts in recent years has shown. This seems to indicate progress when compared with Einstein's thought experiment of an observer in an elevator with cut cable.

In adopting a new definition we must first ascertain that it does at least as well as the old one. To this end it is necessary that *all the physics of special relativity be valid* inside D, just as it is valid relative to the fixed stars. Experimental evidence confirms this directly, while the validity of special relativity relative to the fixed stars is confirmed only indirectly: the laws of special relativity are tested in laboratories on earth making due allowance for fictitious forces (homogeneous gravitational field, rotation of the earth, etc.). Thus, the 10-parameter family of inertial observers is to be taken locally relative to free fall.

It must be realized that the adoption of the new definition of inertial systems and the rejection of the old one is forced upon us to a large extent by the general theory of relativity. This theory *must* contain the special theory of relativity in some suitable limit. As is shown in Section A2-5, this limit is provided by the local features of curved Riemannian space. Thus, the new definition necessarily refers to a *local* situation. And more than that, the logical necessity for the higher-level theory to contain the lower-level theory requires the higher-level one to specify under what special conditions and approximations the older, lower-level theory is valid. This specification involves, first of all, the new definition of the inertial reference frames. But, in addition, it requires that *the laws of special relativity be valid in each local inertial frame.* This implies in particular that the "universal" constants are the same in every such frame, i.e. are truly universal.

3-13 GRAVITATIONAL FORCES AND SPACE CURVATURE

The most striking difference between the two definitions of inertial systems is the following. The 10-parameter family referred to the fixed stars is unique. There is only one such 10-parameter family. The 10-parameter family referred to a falling observer is not unique. For *every* falling observer there is such a family of reference frames, and we note that two falling observers are in general *accelerated* with respect to each other. An observer orbiting the earth and one in an elevator with cut cable are *both* freely falling. Thus, two inertial frames can be in relative *acceleration* if they belong to different families.

The law of inertia is valid, as in special relativity, relative to *every* family of inertial systems. This means that a mass point subjected to *no* forces according to every observer of one family of inertial systems *is* subject to forces as seen by an observer belonging to any one of the other families of inertial systems. The forces in question are gravitational forces. Since each family of inertial reference systems is as good as any other, it is clear that it will be impossible to decide for which family the true gravitational field is zero. In fact, this can be determined only by a measurement of the space curvature.

The complete equivalence of all families of inertial systems, i.e. the fact that at every point in a gravitational field a family of inertial systems can be constructed, is the content of the *general principle of relativity.* This principle means that the laws of physics are the same no matter which family of inertial systems is used as a reference.

It should be noted that in its present form general relativity exceeds this requirement. The structure of the theory is intimately connected with the invariance of the basic equation under the *group of all point transformations.* This suggests the requirement that the laws of physics have the same form not only locally relative to all inertial frames but also generally relative to *any* (not necessarily freely falling) observer. This requirement of *general covariance* exceeds the principle of relativity. The latter is ensured by it but does not require it. If in the future the group of all point transformations will be restricted to one of its subgroups, covariance will have to be correspondingly restricted.

From these considerations follows a relativity of gravitational forces. Via the equation of motion (3-11), this is equivalent to a *relativity of acceleration.* One arrives thus at an apparent contradiction: on one hand, gravitational forces are relative since acceleration is relative; on the other hand, they are absolute, i.e. are intrinsic properties of space (space curvature).

This paradox is easily resolved if one remembers that the gravitational forces which enter (3–11) are, according to general relativity, the *first* derivatives of the metric tensor $g_{\mu\nu}$ (corresponding to the gradient of a potential in Newtonian physics), whereas the space curvature involves the *second* derivatives of $g_{\mu\nu}$. In general relativity [cf. Eqs. (A2–39) and (A2–40)], Eq. (3–11) becomes the geodesic equation* for $x^{\mu}(\tau)$,

$$\frac{d^2 x^{\mu}}{d\tau^2} \equiv a^{\mu}(\tau) = f_G^{\mu}(\tau), \tag{3–33}$$

$$f_G^{\mu}(\tau) = -\Gamma_{\alpha\beta}^{\mu}(x) \frac{dx^{\alpha}}{d\tau} \frac{dx^{\beta}}{d\tau} = -\Gamma_{\alpha\beta}^{\mu} v^{\alpha} v^{\beta}, \tag{3–34}$$

$$\Gamma_{\alpha\beta}^{\mu}(x) = \tfrac{1}{2} g^{\mu\lambda}(\partial_{\alpha} g_{\beta\lambda} + \partial_{\beta} g_{\alpha\lambda} - \partial_{\lambda} g_{\alpha\beta}). \tag{3–35}$$

Equation (3–34) expresses the gravitational field strength in terms of the Christoffel symbol $\Gamma_{\alpha\beta}^{\mu}$ of (3–35), which involves no derivatives of $g_{\mu\nu}$ higher than the first. This is to be contrasted with the expression for the curvature tensor, (A2–23), which can be written

$$R_{\kappa\lambda\mu\nu} = \tfrac{1}{2}(\partial_{\kappa}\partial_{\mu} g_{\lambda\nu} + \partial_{\lambda}\partial_{\nu} g_{\kappa\mu} - \partial_{\kappa}\partial_{\nu} g_{\lambda\mu} - \partial_{\lambda}\partial_{\mu} g_{\kappa\nu}) + g_{\alpha\beta}(\Gamma_{\kappa\mu}^{\alpha}\Gamma_{\lambda\nu}^{\beta} - \Gamma_{\kappa\nu}^{\alpha}\Gamma_{\lambda\mu}^{\beta}) \tag{3–36}$$

by means of (3–35) and which involves the second derivative of $g_{\mu\nu}$.

The essential point which resolves the paradox is provided by the fact that $R_{\kappa\lambda\mu\nu}$ is a *tensor* and consequently an (intrinsic) observable, agreed upon by *all* observers, while a^{μ} and f_G^{μ} are *not* vectors, thereby expressing the relativity of acceleration: they are different for different observers and are zero for some of them. The law of inertia (3–11) therefore loses its importance as an absolute criterion of forces when more than one family of inertial systems is involved.

On the other hand, when the above equations are referred to the local family of inertial frames, $g_{\mu\nu} = \eta_{\mu\nu}$ so that $\Gamma_{\alpha\beta}^{\mu} = 0$. The equation of motion of a mass point falling with any member of this family is therefore $a^{\mu} = 0$, expressing by the law of inertia that $f^{\mu} = 0$. However, the curvature tensor does *not* vanish; only the second term of (3–36) vanishes.† In the *local approximation*, $g_{\mu\nu} = \eta_{\mu\nu}$ not only at a point but (within experimental accuracy) in a (small) domain D containing that point, so that $\Gamma_{\alpha\beta}^{\mu} = 0$ and $R_{\kappa\lambda\mu\nu} = 0$ in D in that approximation.

While a^{μ} and f_G^{μ} are not vectors and thus have no absolute meaning, the equation of motion as such is an observable, that is, $a^{\mu} - f_G^{\mu}$ *is* a vector: *all* observers will agree that $a^{\mu} - f_G^{\mu} = 0$. The quantity $A^{\mu} \equiv a^{\mu} - f_G^{\mu}$ is the covariant‡ time derivative of the velocity v^{μ}. Since $v^{\mu}v_{\mu} = -1$ in units where the velocity

---

* It is sometimes convenient to write ∂_{μ} for $\partial/\partial x^{\mu}$.

† At a given point in a curved space, one can choose the coordinate system such that $g_{\mu\nu} = \eta_{\mu\nu}$ and that the first derivatives of $g_{\mu\nu}$ vanish. But one cannot make the second derivatives vanish, too, unless the space is flat in the neighborhood of that point.

‡ See Section A2–2, esp. Eq. (A2–25).

of light $c = 1$, v^μ can be regarded as the unit tangent vector along the (timelike) world line; consequently, A^μ is the first curvature vector of this line, an intrinsic quantity. It should not be confused with the acceleration, a^μ, which has a quite different operational meaning.

One can summarize this situation as follows: the equation of motion (3-11) in the form (3-33) for motion in a gravitational field retains its fundamental importance for the determination of the motion of a mass point from the Newtonian initial values (position and velocity), but gravitational forces have no longer absolute meaning; they are not intrinsic quantities. They are replaced by the curvature tensor which does provide an absolute measure of the presence of gravitational interactions; it is an intrinsic observable.

If nongravitational forces are present or if the test body is spinning, the equation of motion will differ from the geodesic (3-33), (3-34) and will have the structure

$$a^\mu = f_G^\mu + f_P^\mu. \tag{3-37}$$

The force f_P^μ *will* be a vector, i.e. an intrinsic observable. Also, it will depend on the mass and possibly on other properties of the test body. We call it a *ponderomotive* force. Such a force retains its absolute significance.

The most important reference frame from the experimenters' point of view is the laboratory. It is *not* an inertial frame. Even if we disregard the spin of the earth and its motion in the field of the sun, the experimental physicist in his laboratory is an observer who is prevented from falling freely (toward the center of the earth) by being supported in the earth's gravitational field. This is accomplished by the intermolecular forces which are basically electromagnetic interactions and which ensure the solid nature of the earth's crust. Almost all experiments in physics are thus carried out in noninertial coordinate systems. To the extent that the curvature of the earth is negligible, these systems differ from inertial ones only by a constant acceleration. This acceleration must be taken into account when the laboratory results are converted to those valid in inertial systems.* That this conversion is successful and leads to the *same* laws of nature for two antipodes on earth is by itself strong evidence for the principle of relativity and the relativity of acceleration.

3-14 MACH'S PRINCIPLE

The discussion of the relativity of acceleration and of inertial reference frames would be incomplete without mentioning a problem which has been of great concern to many thinkers since the time of Newton, but which so far has escaped a satisfactory solution.

The problem is that of *the origin of inertia*. It is most vividly presented by an experiment first suggested by Newton. Consider an empty pail which is

---

* This conversion is the problem of the static homogeneous gravitational field and will be treated in Section 8-3.

hanging on a rope. The pail is rotated about its own symmetry axis, thereby twisting the rope. After the rope is sufficiently twisted, the pail is held at rest, filled with water, and then released. The required explanation of the ensuing events is a specific instance of the problem of the origin of inertia.

Before the pail is released, the water surface is plane. After release, the pail is spun by the rope; it reaches a maximum angular velocity, then proceeds into damped torsional vibrations, and eventually comes to rest. The water is more and more carried along by the rotation of the pail, thereby changing its surface shape to a paraboloid which reaches a maximum depth and eventually becomes plane again when the water comes to rest.

Firstly. it is clear that the shape of the water surface does not depend on its angular velocity ω relative to the *pail*. When water and pail rotate fastest, ω vanishes and the paraboloid is deepest. On the other hand, when they are both at rest, ω vanishes too, but the surface is plane.

Secondly, the angular velocity of the water relative to the earth is not the determining quantity either, because we can carry out this experiment on the North Pole: if we take into account the earth's rotation (but ignore its spherical shape), the water surface will not be plane when both pail and water are at rest relative to the earth but will be very slightly paraboloidal. The situation for which the surface would be plane is apparently the one in which *the water is at rest relative to a reference frame for which a Foucault pendulum remains in a given vertical plane.* Thus the angular velocity of the water relative to that frame characterizes the shape of the surface.

One is thus led to a conclusion first reached by Newton. According to Newton, the shape of the water surface is determined by the angular velocity of the water relative to absolute space. He suggested observations of the flatness of the water surface as a criterion for absolute rotation.

Apparently, this criterion is an alternative to the better known demonstration of the rotation of the earth by Foucault's pendulum. From the Newtonian point of view they differ in the type of fictitious force involved. In the case of the water it is the centrifugal force; in the case of the pendulum it is the Coriolis force.

The first criticism of Newton's interpretation of his pail experiment came from Bishop Berkeley (1685-1753). He felt that motion relative to absolute space is meaningless and that the essential point was the rotation of the pail relative to the rest of the universe and, in particular, relative to the fixed stars. Only motion relative to other matter is meaningful.

These ideas were later used and extended by Ernst Mach (1838-1916). His studies on this matter also led him to the definition of an inertial system as one which is in uniform motion relative to the fixed stars (cf. Section 3-2). Mach argued that no inertial effect would exist if "the rest of the universe" would not be present: the plane of the Foucault pendulum would not rotate because the earth would not rotate with respect to anything. Thus, the fixed stars and, more precisely, all the existent matter in the universe *are responsible* for the

inertial effect described in Newtonian mechanics by fictitious forces. The asser-
tions that the inertia of a body is *completely* determined by the universe around
it is known as *Mach's principle*, a name given by Einstein. According to this
principle, the paraboloidal shape of the water surface in Newton's pail is not a
manifestation of rotation relative to absolute space but a manifestation of the
gravitational interaction of the water with the rest of the universe, the *distant*
masses contributing most to the effect.

But is not the difference between "rotation relative to absolute space" and
"rotation relative to the fixed stars" a purely semantic one? Is not absolute
space characterized as the space in which the universe at large is at rest? The
answer is an emphatic "no," because Mach's principle is designed to ensure that
there is no difference between having the earth rotate relative to the fixed stars
and the fixed stars rotate relative to the earth. *Only their relative acceleration is
meaningful.*

PROBLEM 3–6

What is the difference between the identification of absolute space with the rest
system of the fixed stars and the identification of absolute space with the rest system
of the earth in Newton's mechanics? Can the rotation of a Foucault pendulum
relative to the earth be explained in the latter case? Contrast this with the Berkeley-
Mach ideas.

Clearly, from the point of view of Mach's principle, the inertial system which
was defined as moving uniformly relative to the universe at large (i.e. approximately relative to the fixed stars) receives a very different significance.
Originally, it was so defined because it presumably specified "no interaction."
Being far removed from other bodies meant not being influenced by forces.
However, it means just the opposite. The inertial system is a preferred system
because it is determined by the universe as a whole. And it is so determined by
the presence (rather than the absence) of interactions, the gravitational interactions with all the distant masses. Obviously, a new level of understanding
has been reached.

There seems to be no alternative: one must choose between absolute space
and absolute motion, and Mach's principle and relative motion. Needless to
say, Einstein chose Mach's principle and was in fact greatly influenced by
Mach. However, when the general theory of relativity was completed in its
present form, it was discovered that Mach's principle is contained in it only in
a token way. The theory indeed yields an inertial effect due to the presence
and acceleration of another body, and centrifugal and Coriolis forces are produced; but the gravitational field at a test body should be determined by the
mass distribution of the whole universe, and this is not the case in the present
formulation of the theory. There is no mechanism in the theory which would
provide for that. In fact, if it were present, a nearby mass would contribute a
completely negligible amount to the inertia of a body.

While various attempts at incorporating Mach's principle into general relativity exist, none seems satisfactory and none has been generally accepted. It is likely that this problem is linked in a fundamental manner to the structure of the universe at large and to its expansion. In particular, the question of a finite and closed or infinite and flat universe may have to be settled in this connection.* Thus, the problem of Mach's principle may not be completely separable from cosmological questions.

A lot of work went into the search for boundary conditions on the gravitational field equations which would be equivalent to the effect of distant masses. So far these attempts have not been successful.

Finally, it must be noted that in a theory which accounts for Mach's principle by explaining inertia as due to a gravitational interaction with the distant masses, the inertial mass of a body will be computed in terms of its gravitational mass. The ratio $m_{\text{inertial}}/m_{\text{grav}}$ will then depend on the characteristic properties of the universe at large and will thus be the same for all bodies, so long as the universe at large is the same as seen from any point. Thus, the principle of equivalence may then no longer be an independent assumption but may follow from the theory.

E. The Basic Problem of Classical Dynamics

3–15 CAUSALITY

The notion of causality has played an important role in philosophy since ancient times. Therefore, it is not surprising that it is used with a variety of denotative values. Even in physics "causality" has acquired several quite different meanings. The development of quantum mechanics and the great philosophical impact of the uncertainty principle has contributed greatly to this situation. However, we shall here be concerned only with causality in classical physics.

There are principally three different meanings of causality in classical physics: (a) predictability or Newtonian causality, (b) restriction of signal velocities to those not exceeding the velocity of light, and (c) the absence of "advanced" effects of fields with finite propagation velocity. While (b) and (c) will be of importance later in the discussion of electromagnetic interactions, we are presently concerned with Newtonian causality.

We require the equations of motion to enable us to compute the motion of a system from the knowledge of certain data given at a fixed time, t_0 say. As

---

* A. Einstein, *The Meaning of Relativity*, Princeton University Press, 1950.

mentioned earlier (Section 3-3), this equation must be of second order if the initial data are to involve position and velocity but not acceleration. Given this equation, one can of course compute *both* the future motion ($t > t_0$) and the past motion ($t < t_0$). This can be expressed as prediction and "retrodiction," respectively.

Our intuitive notion of causality involves a time ordering: what we call *cause* must precede the *effect*. Nevertheless, almost all the basic laws of physics* are known to be unchanged under a reversal of time (*time-reversal invariance*). The exceptions are certain statistical laws like the second law of thermodynamics (law of entropy). In particular, Newton's equations of motion know no *arrow of time*, a fact which was of some concern to Newton himself. Mathematically, there is no distinction between extrapolating into the future and extrapolating into the past by means of the equation of motion. It is pure convention to choose the parameter t for time in such a way that it is monotonically *increasing* in the direction which we associate with our intuitive experience of the passing of time. Having made this choice, future and past are defined. Consequently, the identification of *causality* with *prediction* rather than *retrodiction* in a time-symmetric system of equations is completely arbitrary. That this is indeed the case can be seen most easily from the fact that fundamental processes can also take place if their initial and final states are interchanged. This is just the statement of time-reversal invariance. The fact that certain statistical laws go only one way (entropy) may have to do with their statistical nature, but the mechanics on which they are based is time symmetrical.

There are two different cases to be distinguished, depending on whether the equation of motion is a differential equation of the usual Newtonian type, like Eq. (3-6), or whether it is an integrodifferential equation like Eq. (6-80), involving an *infinite* time interval. In the first case, the motion for $t_0 < t < t_1$ can be computed from the initial conditions and from the knowledge of the force for the same time intervals; in the second case, the force must be known for all times $t \geq t_0$. Does this second case violate causality?

Provided certain existence and uniqueness theorems hold, there is no doubt that both types of equations allow the prediction of the motion from $r(t_0)$ and $v(t_0)$, Newtonian causality is therefore satisfied. Nevertheless, the need to know the forces at times $t > t_1$ for the computation of the motion during $t_0 < t \leq t_1$ can violate causality, at least in principle, because it can admit advanced effects (third meaning of causality). This point will be studied in detail in connection with the equation of motion of a charged particle (see end of Section 6-7).

The two cases in question can be better understood when the nature of the forces involved is taken into account. The Newtonian forces (first case) are instantaneous. They correspond to an action-at-a-distance theory. Forces

---

* This refers to those which are *not* concerned with macroscopic phenomena, such as friction and other irreversible phenomena.

which give rise to integrodifferential equations (second case) are due to interactions that propagate with finite velocity, i.e. arise from field theory. The difference in the nature of the forces is reflected in the difference of the structure
of the equations of motion. The appearance of the limit $t \rightarrow \infty$ is a characteristic feature of field-theoretic interactions. It arises as a substitute for the
specification of an infinity of data at time t_0 or t_1, characterizing the fields over
all (three-dimensional) space. To appreciate this situation one has to study
the initial-value problem of field theory (Section 4-7).

3-16 THE INITIAL-VALUE PROBLEM FOR NEWTONIAN INTERACTIONS

We shall denote by "Newtonian interactions" those interactions that give
rise to equations of motion which are differential equations of second order. In
contradistinction to them are the interactions transmitted by a field (of finite
propagation velocity). They yield integrodifferential equations. However, in
certain approximations the latter reduce to differential equations. In electrodynamics this corresponds to the neglect of the effect of the fields produced by
the moving object itself on its own motion, i.e. of the self-forces. In this approximation the electromagnetic interactions reduce to the Newtonian type.
The present section shall be concerned only with Newtonian interactions, leaving the others for a later discussion (Sections 4-6).

The motion of a single mass point under the influence of a Newtonian force
is described nonrelativistically by

$$\frac{d^2\mathbf{r}}{dt^2} = \frac{1}{m} \mathbf{F}\left(\mathbf{r}, \frac{d\mathbf{r}}{dt}; t\right),$$ (3-38)

where the right-hand side is a given vector function. The mathematical prob-
lem is to find $\mathbf{r}(t)$ when $\mathbf{r}(t_0)$ and $\mathbf{v}(t_0) \equiv (d\mathbf{r}/dt)_{t=t_0}$ are given. This is an
initial-value problem for a set of three coupled ordinary second-order differential
equations. The existence and uniqueness of such a solution is ensured, provided
$\mathbf{F}$ satisfies certain conditions. These conditions are well known from the theory
of differential equations.

Physically, the problem posed must have a solution and this solution must
be unique. If it is not so, the problem is posed mathematically incorrectly.
Thus, for the physicist existence and uniqueness proofs are a test of his theory
rather than a statement of the physical behavior of systems.

If, instead of the one-body problem, one has a two-body problem,

$$\frac{d^2\mathbf{r}_1}{dt^2} = \frac{1}{m_1} \mathbf{F}_1\left(\mathbf{r}_1, \mathbf{r}_2, \frac{d\mathbf{r}_1}{dt}, \frac{d\mathbf{r}_2}{dt}; t\right),$$

$$\frac{d^2\mathbf{r}_2}{dt^2} = \frac{1}{m_2} \mathbf{F}_2\left(\mathbf{r}_1, \mathbf{r}_2, \frac{d\mathbf{r}_1}{dt}, \frac{d\mathbf{r}_2}{dt}; t\right),$$ (3-39)

the mathematical problem is qualitatively unchanged. The conditions on F_1 and F_2 follow again from well-known theorems on existence and uniqueness for given $r_i(t_0)$, $v_i(t_0)$ $(i = 1, 2)$. The physicist knows, furthermore, that the problem (3–39) can always be reduced to two independent one-body problems, one for the center of mass of the two-body system and one for its internal motion (relative coordinates).

It is obvious how these considerations can be extended to n mass points.

If we are working in special relativity, the equation for the one-body problem analogous to (3–38) is

$$\frac{d^2x^\mu}{d\tau^2} = \frac{1}{m} F^\mu\left(x, \frac{dx}{d\tau}, \tau\right). \tag{3–40}$$

Mathematically, this offers the same problem as (3–38), except that the restriction $v_\mu v^\mu = -c^2$ must also be satisfied. An alternative procedure is to reduce (3–40) to three-vector form by means of

$$x^\mu = (t, \mathbf{r}), \qquad v^\mu = (\gamma, \gamma\mathbf{v}), \qquad \mathbf{v} = d\mathbf{r}/dt,$$

$$a^\mu = (\gamma^4\mathbf{v}\cdot\mathbf{a}; \ \gamma^2\mathbf{a} + \gamma^4\mathbf{v}\cdot\mathbf{a}\,\mathbf{v}), \qquad \mathbf{a} = d^2\mathbf{r}/dt^2,$$

$$\tag{3–41}$$

where we chose units such that the velocity of light $c = 1$ (cf. Appendix 1). Another alternative will be discussed in Section 6–12.

The relativistic two-body problem analogous to (3–39) can be treated similarly. However, the reduction to center-of-mass coordinates is less trivial to carry out. The relativistic n-body problem is also mathematically well defined.

Unfortunately, this is not the end of the problem. Very often the system of mass points is governed by *constraints*, i.e. forces which do not enter the equation of motion explicitly. This leads to a restriction of the degrees of freedom involved. The ensuing problem can often be resolved by a suitable choice of the coordinate system. In any case, these constraints give rise to technical complications but they are not a difficulty in principle.

What constitutes the basic problem for the physicist, even after the mathematician has assured him of existence and uniqueness, is the actual solution of the initial-value problem. Only the simplest systems have exact solutions in closed form. It is for this reason that the more powerful formulations of classical mechanics are important—Lagrangian and Hamiltonian mechanics. They not only permit an easier elimination of constraints but also enable one to obtain more easily many characteristic features of the solution, if not the solution itself.

Because of the difficulty of actually solving the equations of motion, all possible means of simplification must be fully exploited. The most important of these are of two kinds: the specific symmetry properties of the system at hand and the general symmetry properties inherent in the theory applied. The former involve special coordinate systems and special constants of the motion. The latter involve the conservation laws characteristic of the theory and are closely related to the invariance properties of the theory.

3-17   CONSERVATION LAWS

The understanding of the time development of a physical system is obviously greatly aided by the knowledge of those physical quantities that do not change during this development. Such quantities are then said to be conserved. The laws which state under what conditions a given physical quantity is conserved are known as *conservation laws*. The determination of such laws is facilitated by the knowledge of their relationship to the symmetry properties of a system. By "symmetry properties" we refer here primarily to the invariance properties of a physical system under a group of transformations. This connection between invariance properties and conservation laws is provided by *Noether's theorem** in combination with *Hamilton's principle of least action.*

(a) Noether's theorem. In its simplest form Emmy Noether's theorem (1918) can be obtained as follows. Let $\{u_k(x)\}$ $(k = 1, 2, \ldots n)$ be a set of differentiable functions of the independent variable x; let $v_k(x) \equiv du_k/dx$ be their first derivatives; let $\mathcal{L}[x] \equiv \mathcal{L}(x;\{u_k\};\{v_k\})$ be a function of x and the $2n$ functions u_k and v_k. Consider an infinitesimal transformation T,

$$x \rightarrow x' = x + \delta x, \tag{3-42}$$

where $\delta x = \epsilon \xi(x)$ with ϵ infinitesimal and $\xi(x)$ arbitrary but differentiable. Under T,

$$u_k(x) \rightarrow u'_k(x') = u_k(x) + \delta u_k(x),$$
$$v_k(x) \rightarrow v'_k(x') = v_k(x) + \delta v_k(x). \tag{3-43}$$

We assume that $\mathcal{L}[x]$ is form invariant under T; i.e.

$$\mathcal{L}'[x'] = \mathcal{L}[x'] \equiv \mathcal{L}(x'; \{u'_k(x')\}; \{v'_k(x')\}). \tag{3-44}$$

Consider now the integral

$$I \equiv \int_X \mathcal{L}[x]\, dx. \tag{3-45}$$

By assumption, it is invariant under T so that the transformation (3-42) which maps the interval X into the interval X' does not change it:

$$\delta I \equiv \int_{X'} \mathcal{L}'[x']\, dx' - \int_X \mathcal{L}[x]\, dx = 0. \tag{3-46}$$

The quantity δI can be put into a more convenient form by means of a simple

---

* We shall always call *Noether's theorem* the first of two theorems by E. Noether. It refers to *finite* continuous groups. Her second theorem, referring to *infinite* continuous groups will be discussed briefly in Section 4-11.

calculation. Working only to first order in ϵ, and using (3–44), we have

$$\mathcal{L}'[x'] = \mathcal{L}[x] + \frac{\partial \mathcal{L}}{\partial x} \, \delta x + \frac{\partial \mathcal{L}}{\partial u_k} \, \delta u_k + \frac{\partial \mathcal{L}}{\partial v_k} \, \delta v_k$$

$$= \mathcal{L}[x] + \frac{d\mathcal{L}}{dx} \, \delta x + \frac{\partial \mathcal{L}}{\partial u_k} \, (\delta u_k - v_k \, \delta x) + \frac{\partial \mathcal{L}}{\partial v_k} \, (\delta v_k - w_k \, \delta x), \qquad (3\text{–}47)$$

where

$$\frac{d\mathcal{L}}{dx} \equiv \frac{\partial \mathcal{L}}{\partial x} + \frac{\partial \mathcal{L}}{\partial u_k} \, v_k + \frac{\partial \mathcal{L}}{\partial v_k} \, w_k, \qquad w_k \equiv \frac{dv_k}{dx},$$

and where we used the summation convention (all repeated indices in a product are summed). From (3–42) it follows that

$$dx' = \left(1 + \frac{d \, \delta x}{dx} \right) dx$$

so that (3–46) becomes, by means of (3–47),

$$\delta I = \int_X dx \left[\frac{d}{dx} \, (\mathcal{L} \, \delta x) + \frac{\partial \mathcal{L}}{\partial u_k} \, (\delta u_k - v_k \, \delta x) + \frac{\partial \mathcal{L}}{\partial v_k} \, (\delta v_k - w_k \, \delta x) \right].$$

The last term can be transformed by using the first order identity

$$\delta v_k = \frac{d}{dx} \, \delta u_k - v_k \, \frac{d \, \delta x}{dx}, \qquad\qquad (3\text{–}48)$$

which will be proven below. Integration by parts yields for this term

$$\frac{\partial \mathcal{L}}{\partial v_k} \frac{d}{dx} \, (\delta u_k - v_k \, \delta x) = \frac{d}{dx} \left[\frac{\partial \mathcal{L}}{\partial v_k} \, (\delta u_k - v_k \, \delta x) \right] - (\delta u_k - v_k \, \delta x) \, \frac{d}{dx} \frac{\partial \mathcal{L}}{\partial v_k}.$$

The final result for δI is therefore

$$\delta I = \int_X dx \left\{ \frac{d}{dx} \left[\mathcal{L} \, \delta x + \frac{\partial \mathcal{L}}{\partial v_k} \, (\delta u_k - v_k \, \delta x) \right] + \left(\frac{\partial \mathcal{L}}{\partial u_k} - \frac{d}{dx} \frac{\partial \mathcal{L}}{\partial v_k} \right) (\delta u_k - v_k \, \delta x) \right\}.$$
$$(3\text{–}49)$$

In order to prove (3–48) let us drop the subscript k for a moment and remember that we are interested only in the first-order terms. Then

$$v'(x) - v(x) = \frac{d}{dx} \, [u'(x) - u(x)] = \frac{d}{dx} \, [u'(x') - u(x')] + O(\epsilon^2)$$

$$= \frac{d}{dx} \, [u'(x') - u(x) + u(x) - u(x')] = \frac{d}{dx} \, (\delta u - v \, \delta x).$$

Therefore,

$$\delta v = v'(x') - v(x) = [v'(x) - v(x)] + [v(x') - v(x)] + O(\epsilon^2)$$

$$= \frac{d}{dx}(\delta u - v\,\delta x) + \frac{dv}{dx}\,\delta x = \frac{d\,\delta u}{dx} - v\,\frac{d\,\delta x}{dx},$$

which proves (3–48).

From (3–46) and (3–49) it follows that

$$\left(\frac{d}{dx}\frac{\partial \mathcal{L}}{\partial v_k} - \frac{\partial \mathcal{L}}{\partial u_k}\right)(\delta u_k - v_k\,\delta x) = \frac{d}{dx}\left[\mathcal{L}\,\delta x + \frac{\partial \mathcal{L}}{\partial v_k}(\delta u_k - v_k\,\delta x)\right]. \qquad (3\text{–}50)$$

This is Noether's theorem in its simplest form. It can be expressed in various ways. The most convenient form for us is as follows: "If $I[X]$ is invariant under the infinitesimal one-parameter group of transformations T, then the set of n equations

$$\frac{d}{dx}\frac{\partial \mathcal{L}}{\partial v_k} - \frac{\partial \mathcal{L}}{\partial u_k} = 0 \qquad (k = 1, 2, \ldots, n) \qquad (3\text{–}51)$$

implies that (summation convention!)

$$\mathcal{L}\,\delta x + \frac{\partial \mathcal{L}}{\partial v_k}(\delta u_k - v_k\,\delta x) = \text{const}, \qquad (3\text{–}52)$$

i.e. is independent of x."*

More general forms of Noether's theorem can be obtained by generalizing (a) from 1 to ν independent variables $x^1, x^2, \ldots x^\nu$, (b) from 1 to p parameters $\epsilon_1, \epsilon_2, \ldots \epsilon_p$, of the infinitesimal transformation group T, and (c) from 1 to m derivatives of the u_k occurring in $\mathcal{L}$.

While the generalization (c) is of minor interest in physics, the generalizations (a) and (b) are very important. The set $\{u_k\}$ then has derivatives† $v_k^\mu \equiv \partial^\mu u_k$ $(k = 1, 2, \ldots, n; \mu = 1, 2, \ldots, \nu)$ and the integral (3–45) reads

$$I = \int_{X_{(\nu)}} \mathcal{L}\,d^\nu x, \qquad (3\text{–}45')$$

where $d^\nu x$ is the invariant infinitesimal volume element of a ν-dimensional space and $X_{(\nu)}$ is a region in that space. For a p-parameter group T_p the infinitesimal transformations are

$$x^\mu \rightarrow x'^\mu = x^\mu + \delta x^\mu \qquad (3\text{–}53)$$

---

* The inverse of this theorem is also valid: given (3–52), the equation (3–51) ensures the existence of an infinitesimal transformation (3–42) which leaves I invariant.
† Note our convention $\partial_\mu \equiv \partial/\partial x^\mu$.

with

$$\delta x^\mu = \epsilon_i \xi_i^\mu(x) \equiv \sum_{i=1}^{p} \delta_i x^\mu. \tag{3-54}$$

Correspondingly,

$$\delta u_k = \sum_{i=1}^{p} \delta_i u_k; \quad \delta v_k^\mu = \sum_{i=1}^{p} \delta_i v_k^\mu. \tag{3-55}$$

Generalizations (a) and (b) lead to the following form of Noether's theorem: If $\mathfrak{L}(\{x^\mu\}; \{u_k\}; \{v_k^\mu\})$ with $\mu = 1, 2, \ldots, \nu$, $k = 1, 2, \ldots, n$, is invariant under the p-parameter infinitesimal group of transformations T_p, then the set of n equations*

$$\partial^\mu \frac{\partial \mathfrak{L}}{\partial v_k^\mu} - \frac{\partial \mathfrak{L}}{\partial u_k} = 0 \tag{3-56}$$

implies p *conservation laws*

$$\partial_\mu F_i^\mu = 0 \qquad (i = 1, 2, \ldots, p), \tag{3-57}$$

$$F_i^\mu \equiv \mathfrak{L}\, \delta_i x^\mu + \frac{\partial \mathfrak{L}}{\partial v_{k\mu}} (\delta_i u_k - v_{k\nu}\, \delta_i x^\nu). \tag{3-58}$$

The reason for this name for (3–57) will become apparent later on.

One more generalization is worth noting. The assumption (3–44) is not always valid and can be weakened to

$$\mathfrak{L}'[x'] = \mathfrak{L}[x'] + \frac{d\Omega[x']}{dx'} .$$

For several independent variables this equation reads†

$$\mathfrak{L}'[x'] = \mathfrak{L}[x'] + \partial_\mu \Omega^\mu[x'].$$

The derivation of Noether's theorem can be carried through as before and one obtains an additional term $\delta_i \Omega^\mu$ on the right-hand side of (3–58). It can be shown that $\partial_\mu \Omega^\mu$ satisfied the same equations as $\mathfrak{L}$, (3–56), provided Ω^μ does not depend on the v_k^μ (see p. 60, Hill, loc. cit.). This is desirable because it ensures the form invariance of (3–56) despite the added divergence in the relationship (3–44).

(b) Hamilton's principle. The *principle of least action* (1834) by Sir William Rowan Hamilton (1805–1865) is an application of the calculus of variations to the Lagrange equations of the second kind. It is an equivalent way of stating these equations.

---

* It must be understood that the divergence operation ∂_μ in these equations is meant as a total derivative, affecting the dependent variables $u_k(x)$, as well as the explicit occurrence of x.

† See preceding footnote.

Under certain conditions on the forces, especially when they are conservative (derivable from a potential energy), Newton's equations of motion for a system of mass points can be expressed in terms of a Lagrangian, i.e. a function $L(\{q_k\}\,;\{\dot{q}_k\})$ of the generalized coordinates q_k ($k = 1, 2, \ldots, n$) and their first time derivatives $\dot{q}_k$. The equations of motion then have the form

$$\frac{d}{dt}\frac{\partial L}{\partial \dot{q}_k} - \frac{\partial L}{\partial q_k} = 0. \tag{3-51'}$$

We owe these equations to Joseph Louis Lagrange (1736–1813). They are known as Lagrange's equations of the second kind. For a conservative system, L is the difference between kinetic and potential energy.

Hamilton showed that these equations are equivalent to the statement

$$\delta\int_{t_0}^{t_1} L\,dt = 0. \tag{3-59}$$

This is the principle of least action. It means the following: the solutions $\{q_k(t)\}$ of the system of mass points will be those functions for which the integral in (3–59) is an extremum when compared with neighboring functions also integrated between the *same* two instants of time t_0 and t_1. Two neighboring functions, $q'_k(t)$ and $q_k(t)$ are related by

$$q'_k(t) = q_k(t) + \delta q_k(t), \qquad \delta q_k(t) = \epsilon \xi_k(t), \tag{3-43'}$$

where ϵ is a small parameter.

The problem (3–59) is formally almost identical with (3–46), the essential difference being that the region of integration is not varied. In (3–49) this would mean $\delta x = 0$ and $\delta u_k = 0$ at the endpoints of the interval. It is then apparent from that equation that the principle of least action (3–59) and Lagrange's equations (3–51') are equivalent. In the calculus of variation, Eqs. (3–51') are known as the Euler conditions for the first variation, since Leonhard Euler (1707–1783) discovered them in 1744, many years before Lagrange was led to them in his study of analytical mechanics.

Hamilton's principle can obviously be extended to several independent variables. It then states that the *action integral* (3–45') is an extremum and leads to the Euler-Lagrange equations (3–56). This generalization, specifically to $\nu = 4$, is essential for the Lagrangian formulation of field theory in special relativity.

The combination of Noether's theorem and Hamilton's principle now clarifies the relation between conservation laws and symmetry groups. One must consider those groups which can be generated by infinitesimal transformations in the neighborhood of the identity transformation, so-called *Lie groups*. The following theorem is thus obtained.

If the Lagrangian is invariant under a p-parameter Lie group, the equations of motion ensure p conservation laws.

The Lie groups of special interest in classical mechanics are the four-dimensional translation group and the three-dimensional rotation group (forming the Euclidean group), the Galilean group, and the Lorentz group. The Euclidean group corresponds to the following seven conservation laws: conservation of energy (time-translation invariance), conservation of linear momentum (space-translation invariance), and conservation of angular momentum (rotation invariance). The additional three parameters of the Galilean and the Lorentz group yield the *center-of-mass theorem*, nonrelativistically and relativistically, respectively. This theorem states that the position of the center of mass of a system is a linear function of time (the velocity of the center of mass is conserved). In Eq. (4–172) we shall encounter a particular case of this theorem.

As an example, consider a nonrelativistic system of mass points in mutual interaction via forces which admit a Lagrangian. In the notation of (3–51′) the conservation law (3–52) reads

$$L\, \delta t + \frac{\partial L}{\partial \dot{q}_k}\, (\delta q_k - \dot{q}_k\, \delta t) = \text{const.} \tag{3–52′}$$

It states that the left-hand side is a *constant of motion*. Time translations are characterized by $\delta t \neq 0$ and $\delta q_k = 0$, yielding

$$H \equiv \frac{\partial L}{\partial \dot{q}_k}\, \dot{q}_k - L = \text{const.} \tag{3–60}$$

The Hamiltonian, defined by this equation, is therefore a constant of the motion. If L is (explicitly) time independent and does not contain linear terms in the $\dot{q}_k$, H equals the total energy of the system, i.e. the sum of kinetic and potential energy.

Space translations are characterized by a change of all coordinates by a fixed amount $\delta \mathbf{r}$, while $\delta t = 0$. Equation (3–52′) then yields

$$\frac{\partial L}{\partial \dot{q}_k}\, \delta \mathbf{r} \cdot \nabla q_k = \text{const,}$$

or, since $\delta \mathbf{r}$ is arbitrary,

$$\mathbf{P} \equiv \sum_k \frac{\partial L}{\partial \dot{q}_k}\, \nabla q_k = \text{const,} \tag{3–61}$$

where we have indicated the summation explicitly. This equation expresses the conservation of the total linear momentum of the system.

––––––––––

A generalization of translation invariance refers to the situation when L is invariant under translation along one of the generalized coordinates, q_k, say. It then follows from (3–52′) that this invariance yields

$$p_k \equiv \frac{\partial L}{\partial \dot{q}_k} = \text{const.} \tag{3–62}$$

This invariance will be ensured when L is independent of q_k (ignorable coordinate); in this case, the corresponding momentum, defined by (3–62), is a constant of motion.

PROBLEM 3–6

Show how the remaining conservation laws associated with Euclidean invariance follow for the system just described.

PROBLEM 3–7

Take the Lagrangian of N charged nonrelativistic mass points interacting with one another via Coulomb forces. Use Cartesian coordinates and obtain explicit expressions for all 10 conservation laws associated with Galilean invariance. Note that for the derivation of the center-of-mass theorem, (3–44) is not satisfied and an Ω-term must arise.

The basic problem of solving the equations of motion can thus be reduced: the solution is equivalent to finding as many independent integrals of motion as there are degrees of freedom; the knowledge of certain constants of the motion reduces the number of integrals which are still to be found. The remaining integrals of the motion, however, have no simple physical meaning and in general there is no short cut known for their determination.

CHAPTER 4

The Maxwell-Lorentz Field

There are only four types of fundamental interactions known in physics. Only two of them, the electromagnetic and the gravitational interactions, are of a classical nature, i.e. they can be understood on the level of classical physics. The other two cannot be understood on this level and require quantum mechanics; these are the "strong" interactions, which are about one hundred or more times stronger than the electromagnetic ones (responsible for nuclear forces), and the "weak" interactions, which are about 10^{10} times weaker than the electromagnetic ones (responsible for radioactivity). Of the two classical types of forces governing the details of our daily lives the electromagnetic force is by far the dominant one in variety and importance.

Maxwell's description of electromagnetic interactions is a field theory. This fact played a leading role in the development of theories of fundamental interactions since Maxwell. It is therefore of a very different nature from the mechanics of mass points which dominated Newtonian thinking. The clash between the conceptual foundations of these intrinsically different structures has been the source of some of the basic difficulties of elementary particle physics. It is symbolized by the appearance of point particles in a field theory. A comparison of the previous and the present chapter will exhibit the basic conceptual differences.

A. The Equations

4–1 LORENTZ'S MICROSCOPIC THEORY

The importance of electromagnetic interactions becomes apparent only when we realize that all interactions between the atomic nucleus and the atomic electrons, between the atoms and molecules of gases and of liquids, and between the atoms of a crystal or any solid are all, in the last analysis, electromagnetic interactions. All forces and energy sources involved in chemistry and technology are actually electromagnetic. The reactor technology of recent date and radioactivity are the only exceptions. While the understanding of most of this tremendous variety of manifestations of one and the same interaction requires the use of quantum mechanics, many features can be understood, at least approximately, on the classical level.

Maxwell's theory of electricity and magnetism was therefore a tremendous achievement. He reduced this large variety of phenomena to the knowledge of essentially two relations, the material relations between **B** and **H** and **D** and **E**, which must be known experimentally in order to apply his theory.

One can therefore not underestimate the ambitious program of Lorentz which was meant to do away even with these data and account for all electromagnetic phenomena in terms of the motion of electrons bound to atoms and their interaction with light. On the basis of our present knowledge of the quantum theory of matter we must say that this microscopic theory was extremely successful in view of the crudeness of the atomic description.

This success of the microscopic theory was not shared by Lorentz's attempt to account for the structure of the electron. The latter problem will be taken up in Chapter 6; in the present chapter we shall be concerned with Lorentz's equations for the electromagnetic field as determined by atoms and electrons, i.e. as determined by given charge and current distributions. The latter are assumed to be localized in space so that "outside" of them one is dealing with fields in the vacuum.

The fundamental electromagnetic field involves only two three-vectors, the electric field strength $\mathbf{E}$ and the magnetic induction or magnetic flux density $\mathbf{B}$. In a vacuum the microscopic and the macroscopic $\mathbf{E}$ and $\mathbf{B}$ are identical. Otherwise $\mathbf{E}_{macro}$ and $\mathbf{B}_{macro}$ are averages of $\mathbf{E}$ and $\mathbf{B}$; the latter symbols will refer to the microscopic fields from now on. The displacement $\mathbf{D}$ and the magnetic field strength $\mathbf{H}$ have no microscopic meaning and are by definition macroscopic quantities. They can be derived from $\mathbf{E}$ and $\mathbf{B}$ and the knowledge of the electric and magnetic multipole distributions in a given volume of matter by suitable averages. All averages are to be carried out according to the laws of statistical mechanics. We shall not be concerned with them,* but shall concentrate on the fundamental fields $\mathbf{E}$ and $\mathbf{B}$.

Lorentz's theory of electrons is based on the following set of equations (in Gaussian units)

$$\nabla \times \mathbf{B} - \frac{1}{c}\frac{\partial \mathbf{E}}{\partial t} = \frac{4\pi}{c}\,\mathbf{j}, \tag{4-1}$$

$$\nabla \cdot \mathbf{E} = 4\pi\rho, \tag{4-2}$$

$$\nabla \times \mathbf{E} + \frac{1}{c}\frac{\partial \mathbf{B}}{\partial t} = 0, \tag{4-3}$$

$$\nabla \cdot \mathbf{B} = 0. \tag{4-4}$$

The charge density ρ and the current density $\mathbf{j}$ are due to electrons and ions. The similarity of this set of equations to Maxwell's macroscopic equations is obvious:

$$\nabla \times \mathbf{H} - \frac{1}{c}\frac{\partial \mathbf{D}}{\partial t} = \frac{4\pi}{c}\,\mathbf{j}_{macro}, \tag{4-5}$$

$$\nabla \cdot \mathbf{D} = 4\pi\rho_{macro}, \tag{4-6}$$

$$\nabla \times \mathbf{E}_{macro} + \frac{1}{c}\frac{\partial \mathbf{B}_{macro}}{\partial t} = 0, \tag{4-7}$$

$$\nabla \cdot \mathbf{B}_{macro} = 0. \tag{4-8}$$

---

* The most satisfactory approach to the averaging problem can be found in P. Mazur and B.R.A. Nijboer, *Physica*, **19**, 971 (1953).

The essential feature of the averaging process lies in the first two equations, where ρ_{macro} and j_{macro} are obtained from the charge and current densities of the electrons and ions of the material, and H and D are determined as explained above.

In the following sections we shall study the problems posed by the *Maxwell-Lorentz equations* (4-1) through (4-4).

4-2 FREE ELECTROMAGNETIC FIELDS

In a region of space-time in which $\rho = 0$ and $j = 0$, the Maxwell-Lorentz equations (4-1) and (4-2) become

$$\nabla \times \mathbf{B} - \frac{1}{c}\frac{\partial \mathbf{E}}{\partial t} = 0, \tag{4-9}$$

$$\nabla \cdot \mathbf{E} = 0. \tag{4-10}$$

These two equations, together with (4-3) and (4-4), are the electromagnetic field equations in a matter-free region. They specify that in such a region both $\mathbf{E}$ and $\mathbf{B}$ are *solenoidal* (divergence-free) fields. Furthermore, they both satisfy the wave equation

$$\nabla^2 \mathbf{E} = \frac{1}{c^2}\frac{\partial^2 \mathbf{E}}{\partial t^2},$$

$$\tag{4-11}$$

$$\nabla^2 \mathbf{B} = \frac{1}{c^2}\frac{\partial^2 \mathbf{B}}{\partial t^2},$$

as can easily be seen by applying the curl operator to (4-3) and (4-9) and eliminating $\mathbf{B}$ and $\mathbf{E}$, respectively. If we assume that the fields have a Fourier transform, the general solution for $\mathbf{E}(\mathbf{r}, t)$ will be of the form

$$\mathbf{E}(\mathbf{r}, t) = \frac{1}{(2\pi)^{3/2}} \int [\widetilde{\mathbf{E}}_1(\mathbf{k})e^{i\mathbf{k}\cdot\mathbf{r}-i\omega t} + \widetilde{\mathbf{E}}_2(-\mathbf{k})e^{i\mathbf{k}\cdot\mathbf{r}+i\omega t}] \, d^3k \tag{4-12}$$

and similarly for $\mathbf{B}$, provided

$$\frac{\omega}{c} = k \equiv |\mathbf{k}|. \tag{4-13}$$

To prove this we write the four-dimensional Fourier transform

$$\mathbf{E}(\mathbf{r}, t) = \frac{1}{(2\pi)^2} \int \mathbf{F}(\mathbf{k}, \alpha)e^{i\mathbf{k}\cdot\mathbf{r}-i\alpha t} \, d^3k \, d\alpha.$$

In order that (4-11) be satisfied, the integral must vanish when $\mathbf{F}$ is replaced by $(k^2 - \alpha^2/c^2)\mathbf{F}$. Thus, $\mathbf{F}$ must vanish unless $\alpha = \pm\omega$ with ω defined by (4-13). This can be ensured by means of the Dirac δ-function. $\mathbf{F}$ must have the form

$$\mathbf{F}(\mathbf{k}, \alpha) = \sqrt{2\pi}\, [\delta(\alpha - \omega)\widetilde{\mathbf{E}}_1(\mathbf{k}) + \delta(\alpha + \omega)\widetilde{\mathbf{E}}_2(-\mathbf{k})].$$

After integration over α, Eq. (4-12) results.

The dummy variable **k** can be replaced by $-\mathbf{k}$ in either of the two terms of (4–12). This operation leads to the conclusion that $\mathbf{E}(\mathbf{r}, t)$ will be real provided the *reality condition*

$$\mathbf{E}_1^*(\mathbf{k}) = \mathbf{E}_2(\mathbf{k}) \tag{4–14}$$

is satisfied. Equation (4–12) can then also be written in the form

$$\mathbf{E}(\mathbf{r}, t) = \frac{1}{(2\pi)^{3/2}} \int \widetilde{\mathbf{E}}(\mathbf{k}) e^{i\mathbf{k}\cdot\mathbf{r} - i\omega t}\, d^3k + \text{complex conjugate}, \tag{4–12'}$$

where the index 1 has been dropped.

The integrand in (4–12) represents the coherent superposition of two plane waves moving in opposite directions with wave vectors **k** and $-\mathbf{k}$, circular frequency ω, and amplitudes $\widetilde{\mathbf{E}}_1$ and $\widetilde{\mathbf{E}}_2$. Because of the solenoidal nature of this field, (4–10), the amplitudes are orthogonal to the propagation directions,

$$\mathbf{k} \cdot \widetilde{\mathbf{E}}(\mathbf{k}) = 0 \tag{4–15}$$

representing *transverse* waves.

PROBLEM 4–1

Show that $\mathbf{E}(\mathbf{r}, t)$ is determined for all times when $\mathbf{E}(\mathbf{r}, 0)$ and $\partial\mathbf{E}/\partial t$ $(t = 0)$ are given. Find $\widetilde{\mathbf{E}}(\mathbf{k})$ in (4–12') in terms of the two given functions.

The Fourier expansion of $\mathbf{B}(\mathbf{r}, t)$ in terms of $\widetilde{\mathbf{B}}(\mathbf{k})$ is of course completely analogous to (4–12') and (4–15) since **B** and **E** satisfy the same equations (4–11) and are both solenoidal. The relationship of $\widetilde{\mathbf{B}}(\mathbf{k})$ to $\widetilde{\mathbf{E}}(\mathbf{k})$ can be obtained from (4–3) or (4–9) by substitution of the Fourier transforms:

$$\widetilde{\mathbf{B}}(\mathbf{k}) = \hat{\mathbf{k}} \times \widetilde{\mathbf{E}}(\mathbf{k}). \tag{4–16}$$

The vector $\hat{\mathbf{k}}$ is the unit vector along **k**. Equations (4–15) and (4–16) imply that **k**, $\widetilde{\mathbf{E}}(\mathbf{k})$, and $\widetilde{\mathbf{B}}(\mathbf{k})$ form a right-handed Cartesian coordinate system.

The Fourier decomposition does not require the use of a Cartesian spatial coordinate system. For example, we can use spherical coordinates. The well-known Rayleigh expansion*

$$e^{i\mathbf{k}\cdot\mathbf{r}} = \sum_{l=0}^{\infty} i^l(2l+1)j_l(kr)P_l(\cos\Theta) \tag{4–17}$$

in Legendre polynomials P_l and spherical Bessel functions j_l is convenient for this purpose. The angle between **r** and **k** is here denoted by Θ. If one uses an

---

* J. W. Strutt, Third Baron Rayleigh (1842–1919), expanded a plane wave in powers of cos Θ, to which (4–17) is closely related (*The Theory of Sound*, Dover, New York, 1945, §272 and §343). For a modern derivation see, e.g., P. M. Morse and H. Feshbach, *Methods of Theoretical Physics*, McGraw-Hill, New York, 1953, p. 1466.

arbitrary polar axis with respect to which the directions of **r** and **k** are given by ϑ, φ, and ϑ', φ', then

$$P_l(\cos\Theta) = \frac{4\pi}{2l+1} \sum_{m=-l}^{l} Y_l^{m*}(\vartheta', \varphi') Y_l^m(\vartheta, \varphi), \qquad (4\text{-}18)$$

according to the addition theorem for the spherical harmonics Y_l^m. Substitution of this theorem in (4-17) and use of the result in (4-12') yields

$$\mathbf{E}(\mathbf{r}, t) = \frac{2}{\sqrt{2\pi}} \int d^3k\, e^{-i\omega t} \tilde{\mathbf{E}}(\mathbf{k}) \sum_{l=0}^{\infty} \sum_{m=-l}^{l} i^l j_l(kr) Y_l^{m*}(\vartheta', \varphi') Y_l^m(\vartheta, \varphi)$$

$$+ \text{ complex conj.}$$

If the integration over the directions ϑ', φ' of **k** is carried out, using $d^3k = \sin\vartheta'\, d\vartheta'\, d\varphi' \omega^2\, d\omega/c^3$, one can define

$$\mathbf{E}_l^m(\omega) \equiv \frac{\omega^2}{c^3} \int \tilde{\mathbf{E}}(\mathbf{k}) Y_l^m(\vartheta', \varphi') \sin\vartheta'\, d\vartheta'\, d\varphi',$$

and one obtains

$$\mathbf{E}(\mathbf{r}, t) = \frac{1}{\sqrt{2\pi}} \int_0^{\infty} d\omega e^{-i\omega t} \sum_{l=0}^{\infty} \sum_{m=-l}^{l} 2i^l j_l\left(\frac{\omega r}{c}\right) \mathbf{E}_l^m(\omega) Y_l^m(\vartheta, \varphi)$$

$$+ \text{ complex conj.} \quad (4\text{-}19)$$

The conventional forms of the *multipole expansion* of the radiation field can be derived from this result by use of (4-10). It is to be noted that j_l is the sum of the spherical Hankel functions of first and second kind

$$2j_l(kr) = h_l^{(1)}(kr) + h_l^{(2)}(kr).$$

Since these functions have the asymptotic behavior

$$i^l h_l^{(1)}(kr) \sim \frac{e^{ikr}}{ikr} \qquad (kr \gg 1),$$

$$(-i)^l h_l^{(2)}(kr) = \frac{e^{-ikr}}{-ikr} \qquad (kr \gg 1),$$

the expression (4-19) is asymptotically a superposition of spherically outgoing and ingoing waves.

4-3 POTENTIALS AND GAUGES

The homogeneous Maxwell-Lorentz equations (4-3) and (4-4) permit the introduction of a vector function **A** and a scalar function ϕ in terms of which the electromagnetic field strengths **E** and **B** can be expressed. These functions, known as *potentials*, have no physical meaning and are introduced solely for the purpose of mathematical simplification of the equations. While the field strengths are directly measurable, the potentials are not measurable; certain combinations of their space and time derivatives, however, *are* measurable and are equivalent to the field strengths.

Equation (4–4) can be satisfied identically by making it the divergence of a *curl*,

$$\mathbf{B} = \nabla \times \mathbf{A}. \tag{4-20}$$

Equation (4–3) then becomes

$$\nabla \times \left(\mathbf{E} + \frac{1}{c} \frac{\partial \mathbf{A}}{\partial t} \right) = 0.$$

It, too, can be satisfied identically, this time by making it the curl of a gradient,

$$\mathbf{E} = -\nabla \phi - \frac{1}{c} \frac{\partial \mathbf{A}}{\partial t}. \tag{4-21}$$

These equations, (4–20) and (4–21), define the potentials. There is no loss of generality: there *always* exist functions $\mathbf{A}$ and ϕ such that $\mathbf{E}$ and $\mathbf{B}$ can be written in the forms (4–20) and (4–21). In fact, there exists a whole family of such functions for a given set $\mathbf{E}$ and $\mathbf{B}$: we can make the transformations

$$\mathbf{A} \rightarrow \mathbf{A}' = \mathbf{A} + \nabla \Lambda, \tag{4-22a}$$

$$\phi \rightarrow \phi' = \phi - \frac{1}{c} \frac{\partial \Lambda}{\partial t}, \tag{4-22b}$$

with Λ an arbitrary function, without changing the form of Eqs. (4–20) and (4–21). This can easily be verified. The invariance of $\mathbf{E}$ and $\mathbf{B}$ under (4–22) is known as *gauge invariance*, and (4–22) are the *gauge transformations*.

It follows immediately that, if one expresses the theory in terms of the potentials rather than in terms of the field strengths, the theory will have gauge invariance as an additional symmetry property. This property does not exist, i.e. cannot be defined, unless potentials are introduced. Consequently, gauge invariance has no physical meaning, but must be satisfied for all *observable* quantities in order to ensure that the arbitrariness in the choice of $\mathbf{A}$ and ϕ does not affect the field strengths.

It is often convenient to restrict the transformations (4–22) by imposing certain conditions on Λ. These *special gauges* are as follows.

(a) Covariant gauges. Λ is required to satisfy the wave equation

$$\nabla^2 \Lambda = \frac{1}{c^2} \frac{\partial^2 \Lambda}{\partial t^2}. \tag{4-23}$$

As a consequence, the expression

$$I \equiv \nabla \cdot \mathbf{A} + \frac{1}{c} \frac{\partial \phi}{\partial t} = \nabla \cdot A' + \frac{1}{c} \frac{\partial \phi'}{\partial t} \tag{4-24}$$

is gauge invariant. Conversely, the gauge invariance of I requires (4–23). The invariant I is arbitrary. It can be chosen zero:

$$\nabla \cdot \mathbf{A} + \frac{1}{c}\frac{\partial \phi}{\partial t} = 0. \tag{4-25}$$

This is the Lorenz *condition* on the potential.[*] It is arbitrarily imposed, strictly as a matter of computational convenience. It implies that the inhomogeneous Maxwell-Lorentz equations, (4–1) and (4–2), become

$$\left(\nabla^2 - \frac{1}{c^2}\frac{\partial^2}{\partial t^2}\right)\mathbf{A} = -\frac{4\pi}{c}\,\mathbf{j}, \tag{4-26a}$$

$$\left(\nabla^2 - \frac{1}{c^2}\frac{\partial^2}{\partial t^2}\right)\phi = -4\pi\rho. \tag{4-26b}$$

These equations, together with the Lorenz *condition* (4–25), are completely equivalent to the Maxwell-Lorentz equations (4–1) through (4–4), with which they are related by (4–20) and (4–21). Note that the homogeneous Maxwell-Lorentz equations are identically satisfied when (4–20) and (4–21) are substituted.

The gauge characterized by (4–25) is the Lorenz *gauge*. Equations (4–26) therefore express the Maxwell-Lorentz equations in the Lorentz gauge.

The advantage of the covariant gauges is that they lead to equations which are invariant under Lorentz transformations.

PROBLEM 4–2

Show that, if the Lorenz condition $I = 0$ is *not* assumed, the inhomogeneous Maxwell-Lorentz equations (4–1) and (4–2) can be written in terms of the potentials so that they differ from (4–26) only by terms involving derivatives of I.

(b) Noncovariant gauges. One can impose on Λ the condition

$$\nabla^2\Lambda = 0. \tag{4-27}$$

This ensures the gauge invariance of $\nabla \cdot \mathbf{A}$, namely

$$\nabla \cdot \mathbf{A} = \nabla \cdot \mathbf{A}'. \tag{4-28}$$

In particular, one often chooses
$$\nabla \cdot \mathbf{A} = 0. \tag{4-29}$$

This choice is obviously possible because the specification of $\nabla \times \mathbf{A}$ in (4–20) is independent of $\nabla \cdot \mathbf{A}$. The choice (4–29) characterizes the *Coulomb gauge*, also

---

[*]L. Lorenz (1829–1891) was older than H.A. Lorentz (1853–1928).

called *radiation gauge*. In this gauge (4–1) and (4–2) become

$$\left(\nabla^2 - \frac{1}{c^2}\frac{\partial^2}{\partial t^2}\right)\mathbf{A} = -\frac{4\pi}{c}\mathbf{j} + \nabla\frac{1}{c}\frac{\partial\phi}{\partial t} \tag{4-30a}$$

and

$$\nabla^2\phi = -4\pi\rho. \tag{4-30b}$$

The last equation is identical with the fundamental equation of electrostatics. For this reason this gauge is especially convenient for the purpose of distinguishing the static (or quasistatic) features of a problem from those of an essentially dynamical nature.

4-4 THE SOLENOIDAL FORM OF THE MAXWELL-LORENTZ EQUATIONS

In the special theory of relativity an essential distinction must be made between uniform and nonuniform motion relative to an inertial frame. A comoving observer would be inertial in the first case but noninertial in the second. Correspondingly, there exists a sharp distinction between the electromagnetic fields produced by charges that move uniformly and nonuniformly. The distinction appears in the form of *radiation fields* which are emitted if and only if a charge is moving nonuniformly, i.e. is accelerated. Radiation fields are therefore also called *acceleration fields*, while the other fields are sometimes denoted as *velocity fields*. This relationship will become apparent in Section 4–8.

A different separation of the fields will be studied in the present section. It is the less common separation into solenoidal and irrotational fields.

A *solenoidal* vector field $\mathbf{V}_S$ is defined as one which is divergence-free, and an *irrotational* field $\mathbf{V}_I$ as one which is curl-free,

$$\nabla\cdot\mathbf{V}_S = 0, \qquad \nabla\times\mathbf{V}_I = 0. \tag{4-31}$$

We now separate the electromagnetic fields accordingly, $\mathbf{E} = \mathbf{E}_S + \mathbf{E}_I$, $\mathbf{B} = \mathbf{B}_S + \mathbf{B}_I$. Looking at (4–4) we see that $\mathbf{B} = \mathbf{B}_S$. From (4–3) it follows that

$$\nabla\times\mathbf{E}_S + \frac{1}{c}\frac{\partial\mathbf{B}_S}{\partial t} = 0, \tag{4-32}$$

since $\nabla\times\mathbf{E}_I = 0$ by definition. Similarly, (4–2) does not involve a mixture of irrotational and solenoidal fields; it involves only $\mathbf{E}_I$:

$$\nabla\cdot\mathbf{E}_I = 4\pi\rho. \tag{4-33}$$

If we take the time derivative of this equation and compare it with the divergence of (4–1),

$$\nabla\cdot\left(\frac{\partial\mathbf{E}_I}{\partial t} + 4\pi j_I\right) = 0, \tag{4-34}$$

we find
$$\nabla \cdot j_I + \frac{\partial \rho}{\partial t} = 0, \tag{4-35}$$

which is the well-known differential form of the *law of charge conservation*.

A theorem on harmonic functions states that if $\nabla \cdot \mathbf{V}_I = 0$ everywhere and $\mathbf{V}_I$ vanishes asymptotically faster than $1/r^{3/2}$, then $\mathbf{V}_I = 0$. If we assume this asymptotic behavior for the irrotational vectors, Eq. (4-34) implies

$$-\frac{\partial \mathbf{E}_I}{\partial t} = 4\pi j_I. \tag{4-36}$$

This result can be combined with (4-1) and leads to the solenoidal equations

$$\nabla \times \mathbf{B}_S - \frac{1}{c} \frac{\partial \mathbf{E}_S}{\partial t} = \frac{4\pi}{c} \mathbf{j}_S. \tag{4-37}$$

We have thus succeeded in separating the Maxwell-Lorentz equations into two sets of equations, one involving only irrotational, the other only solenoidal quantities. These equations are (4-33) and (4-35) for the $\mathbf{V}_I$ and (4-32) and (4-37) for the $\mathbf{V}_S$. Equation (4-36) is a consequence of (4-33) and (4-35) and the asymptotic property of the $\mathbf{V}_I$.

We now return to the potentials and to Eqs. (4-20) and (4-21). The fact that $\mathbf{B} = \mathbf{B}_S$ is ensured by (4-20). The fact that $\mathbf{E} = \mathbf{E}_I + \mathbf{E}_S$ can be assured similarly by putting

$$\mathbf{E}_I = -\nabla \phi, \tag{4-38}$$

$$\mathbf{E}_S = -\frac{1}{c} \frac{\partial \mathbf{A}}{\partial t}. \tag{4-39}$$

Equations (4-31) will then be satisfied identically, provided

$$\nabla \cdot \mathbf{A} = 0, \tag{4-29}$$

i.e. provided we work in the Coulomb gauge. Thus, in this gauge *the vector potential describes the solenoidal fields and the scalar potential describes the irrotational ones.*

PROBLEM 4-3

Show that j_I and j_S are correctly represented by

$$\mathbf{j}_I(\mathbf{r}, t) = -\nabla \int \frac{\nabla' \cdot \mathbf{j}(\mathbf{r}', t) \, d^3 r'}{4\pi |\mathbf{r} - \mathbf{r}'|}, \tag{4-40a}$$

$$\mathbf{j}_S(\mathbf{r}, t) = \nabla \times \nabla \times \int \frac{\mathbf{j}(\mathbf{r}', t) \, d^3 r'}{4\pi |\mathbf{r} - \mathbf{r}'|}, \tag{4-40b}$$

i.e. that they satisfy $\mathbf{j} = \mathbf{j}_I + \mathbf{j}_S$ as well as Eqs. (4-31).

PROBLEM 4–4

Solve (4–30b), substitute the result into (4–30a), and show that (4–30a) becomes

$$\left(\nabla^2 - \frac{1}{c^2}\frac{\partial^2}{\partial t^2}\right)\mathbf{A} = -\frac{4\pi}{c}\,\mathbf{j}_S,$$

where $\mathbf{j}_S$ is given by (4–40b). Note that charge conservation must be used in this derivation.

The Coulomb gauge condition (4–29) assures the transversality (4–15), $\mathbf{k} \cdot \tilde{\mathbf{E}} = 0$, $\tilde{\mathbf{E}}\,\mathbf{k} \cdot \tilde{\mathbf{B}} = 0$, and the fact that for a given $\mathbf{k}$ there are only two degrees of freedom; the latter corresponds to the possibility of having only two linearly independent kinds of *polarization* for a given $\mathbf{k}$, for example, left and right. The two degrees of polarization are expressed by the *two* functions that determine the vector $\tilde{\mathbf{A}}(\mathbf{k})$, the amplitude of the plane wave expansion of $\mathbf{A}(x)$, as restricted by (4–29),

$$\mathbf{k} \cdot \tilde{\mathbf{A}}(\mathbf{k}) = 0. \tag{4–41}$$

This equation justifies the name *transverse gauge* sometimes used for the Coulomb gauge. For any other gauge $\tilde{\mathbf{A}}(\mathbf{k})$ would not be restricted in this way; the extra degrees of freedom which are then permitted have no physical meaning. They do not occur in observable phenomena. Their elimination is ensured by the gauge invariance of all measurable quantities.

4–5 THE COVARIANT FORM OF THE MAXWELL-LORENTZ EQUATIONS

The Maxwell-Lorentz equations are consistent with the Lorentz relativity of velocity; they are invariant under the Lorentz group of transformations. This is an experimental fact: two inertial observers can use the same form of these equations and find that their description of the observed phenomena is correct. This being so, it is easy to find the transformation properties of the electromagnetic fields.

For this purpose it is most convenient to start with the well-confirmed law of charge conservation (4–35), which must be invariant since it is a consequence of the field equations,

$$\nabla' \cdot \mathbf{j}' + \frac{\partial}{\partial t'}\rho' = \nabla \cdot \mathbf{j} + \frac{\partial}{\partial t}\rho.$$

Since we know the mappings $\nabla \to \nabla'$ and $(\partial/\partial t) \to (\partial/\partial t')$ under Lorentz transformations, the transformation properties of $\mathbf{j}$ and ρ can be found from the invariance of the above expression. The covariant four-vector

$$\partial_\mu \equiv \left(\frac{\partial}{c\,\partial t}, \nabla\right) \tag{4–42}$$

requires the contravariant four-vector

$$j^\mu \equiv (c\rho, \mathbf{j})\tag{4-43}$$

to ensure that

$$\partial_\mu j^\mu = 0\tag{4-44}$$

is an invariant. This is the covariant form of the law of charge conservation (4–35). The charge and current densities must therefore transform so that j^μ in (4–43) is a covariant four-vector, the four-current or relativistic current density.

In exactly the same way, one establishes from the invariance of the Lorenz gauge condition (4–25) that

$$A^\mu = (\phi, \mathbf{A})\tag{4-45}$$

is a contravariant four-vector. The Maxwell-Lorentz equations in this gauge (4–26) then become

$$\Box A^\mu = -\frac{4\pi}{c} j^\mu,\tag{4-46}$$

where the d'Alembertian $\Box \equiv \partial_\alpha \partial^\alpha$. This is a manifestly covariant equation. It is now obvious that the covariant gauges are Lorentz covariant provided Λ is a scalar.

The transformation properties of the field strengths follow trivially from those of the potential in the Lorenz gauge. Since the transformation properties of the fields cannot depend on the gauge at least as long as it is a relativistic gauge, this derivation leads to a generally valid result. In view of (4–45) it follows from (4–20) and (4–21) that the electromagnetic field strengths form an antisymmetric tensor

$$F^{\mu\nu} = \partial^\mu A^\nu - \partial^\nu A^\mu, \qquad F^{\nu\mu} = -F^{\mu\nu},\tag{4-47}$$

with the following Cartesian components,

$$F^{\mu\nu} = \begin{matrix} & \overset{\rightarrow \nu}{} \\ {\scriptstyle\downarrow\mu} & \begin{bmatrix} 0 & E_x & E_y & E_z \\ -E_x & 0 & B_z & -B_y \\ -E_y & -B_z & 0 & B_x \\ -E_z & B_y & -B_x & 0 \end{bmatrix} \end{matrix}.\tag{4-48}$$

This can be written more compactly by use of the Ricci symbol in three dimensions,

$$\epsilon_{ijk} = \begin{cases} 1 & \text{when } i\,j\,k \text{ is an even permutation of 1, 2, 3,} \\ -1 & \text{when } i\,j\,k \text{ is an odd permutation of 1, 2, 3,} \\ 0 & \text{otherwise.} \end{cases}\tag{4-49}$$

With subscripts 1, 2, 3 instead of x, y, z, Eq. (4–48) is equivalent to

$$E_k = F^{0k}; \qquad B_k = F^{ij} \epsilon_{ijk}. \tag{4-50}$$

The field equations (4–1) through (4–4) become

$$\partial_\mu F^{\mu\nu} = -\frac{4\pi}{c} j^\nu \tag{4-51}$$

and

$$\partial_\lambda F_{\mu\nu} + \partial_\mu F_{\nu\lambda} + \partial_\nu F_{\lambda\mu} = 0. \tag{4-52}$$

The first of these contains both of the inhomogeneous equations, (4–1) and (4–2), while the second contains both of the homogeneous equations, (4–3) and (4–4). The conciseness and beauty of this form make it obvious that the covariant notation is much more appropriate to the theory than is the three-vector notation.* This should not be surprising, considering the Lorentz invariance of the theory.

In solving the Maxwell-Lorentz equations one is soon led into rather complex expressions. Great simplifications arise, however, when one uses covariant notation throughout. We shall therefore work with (4–51) and (4–52) instead of (4–1) through (4–4).

We conclude this section with some remarks concerning the covariant notation and gauges. The gauge transformations (4–22) are covariant,

$$A'_\mu = A_\mu + \partial_\mu \Lambda. \tag{4-53}$$

Gauges lose their covariance when nonconvariant conditions are imposed on Λ or on A_μ. Thus, the Lorenz condition is covariant,

$$\partial_\mu A^\mu = 0, \tag{4-25'}$$

whereas the Coulomb gauge condition (4–29) is not; neither is the weaker condition (4–27) of the noncovariant gauges. Nevertheless, these equations are sometimes written in four-vector notation as follows.

Let n^μ be a timelike unit vector,

$$n_\mu n^\mu = -1, \qquad n^0 > 0; \tag{4-54}$$

then

$$(\partial_\mu + n_\mu n^\alpha \partial_\alpha) A^\mu = 0 \tag{4-55}$$

is apparently a covariant equation. In the system in which $n^\mu = (1; 0, 0, 0)$ it reduces to (4–29). If n^μ transforms like x^μ (see Eq. (3–14)),

$$n'^\mu = \alpha^\mu_\nu n^\nu, \tag{4-56}$$

---

* As an example, substitute (4–47) into (4–51) and observe how Problem 4-2 can be solved trivially in four-vector notation.

under a Lorentz transformation, (4–55) is indeed covariant. This requires that n^μ be a function of x. In calculations where expressions like (4–55) are used, however, the quantity n^μ is usually treated as a constant; for example,

$$\partial_\alpha(n^\mu f) = n^\mu \partial_\alpha f.$$

In that case n^μ *is not a four-vector*, because it does not transform according to (4–56). Consequently, (4–55) is *not* a covariant equation under these circumstances. We shall call it *formally covariant.** Nevertheless, the use of the quantities n^μ and of noncovariant expressions like these is often very convenient. When a covariant expression is derived from a noncovariant one (e.g., $F^{\mu\nu}$ from a noncovariant gauge), this fact is manifest because the dependence on the n^μ disappears. Therein lies the advantage of working with formally covariant expressions. If one uses three-vector notation [i.e. (4–29) instead of (4–55)], it is not obvious that an expression is actually covariant, i.e. independent of n^μ.

PROBLEM 4–5

Define the formally covariant expressions for **A** and ϕ in the Coulomb gauge,

$$A_S^\mu \equiv A^\mu + n^\mu A_n, \qquad A_I^\mu \equiv -n^\mu A_n, \qquad A_n \equiv n^\alpha A_\alpha,$$

and show that, with $\partial \equiv n^\alpha \partial_\alpha$,

$$\Box A_S^\mu = -\frac{4\pi}{c}\,[j^\mu + n^\mu j_n - (\Box + \partial^2)^{-1}(\partial^\mu + n^\mu \partial)\,\partial j_n] \equiv -\frac{4\pi}{c}\,j_S^\mu,$$

$$(\Box + \partial^2)A_I^\mu = \frac{4\pi}{c}\,n^\mu j_n \equiv -\frac{4\pi}{c}\,j_I^\mu,$$

are the Maxwell-Lorentz equations for the potentials, (4–30), in formally covariant notation. Compare the first equation also with the equation of Problem 4–4. Verify that A_S^μ is a transverse vector field, i.e. it satisfies

$$\partial_\mu A_S^\mu = 0, \qquad n_\mu A_S^\mu = 0.$$

Show that this condition is equivalent to the Coulomb gauge condition (4–29) in formally covariant notation.

PROBLEM 4–6

If A_S^μ and A_I^μ satisfy the field equations of Problem 4–5, find the equations satisfied by

$$F_S^{\mu\nu} \equiv \partial^\mu A_S^\nu - \partial^\nu A_S^\mu \qquad \text{and} \qquad F_I^{\mu\nu} \equiv \partial^\mu A_I^\nu - \partial^\nu A_I^\mu.$$

---

* If n^μ can be specified in an invariant way, for example as the direction of the four-velocity of the uniformly moving center of mass of the system, relativistic invariance is of course ensured.

Show that these equations are just the solenoidal and irrotational Maxwell-Lorentz equations for the field strengths, (4–37) and (4–33), written in formally covariant notation. Recover the (truly) covariant Eq. (4–51). Note that the above equation for $F_S^{\mu\nu}$ has an inverse,

$$A_S^\mu = \partial^{-1} n_\alpha F_S^{\alpha\mu} \equiv - \int^x n_\alpha F_S^{\alpha\mu} n_\beta \, dx^\beta.$$

Also note that $F_S^{\mu\nu}$ and $F_I^{\mu\nu}$ are not tensors, but that $F^{\mu\nu} = F_S^{\mu\nu} + F_I^{\mu\nu}$ is a tensor.

The results of Problems 4–5 and 4–6 have an important application in quantum electrodynamics. They permit one to formulate that theory in terms of $F_S^{\mu\nu}$ as the fundamental electromagnetic field, thereby avoiding difficulties of gauge invariance associated with formulations based on A^μ.

B. The Solutions

4-6 COVARIANT SOLUTION OF THE SOURCE-FREE EQUATIONS

We start with Green's theorem in Minkowski space. Let $u(x)$ and $v(x)$ be two functions of x such that their product vanishes faster than $1/r^2$ at spatial infinity; let V_4 be a four-dimensional volume in Minkowski space bounded by two spacelike planes, σ_0 and σ_1. The theorem states that

$$\int_{V_4} [u \Box v - v \Box u] \, d^4 x = - \left(\int_{\sigma_1} - \int_{\sigma_0} \right) [u \, \partial_\mu v - v \, \partial_\mu u] \, d\sigma^\mu. \qquad (4\text{–}57)$$

It is proven easily by integration by parts and use of Gauss's theorem [Appendix 1, Eq. (A1–60)],

$$\int_{V_4} \partial_\mu f^{\mu\nu\cdots} \, d^4 x = \epsilon_\sigma \int_\sigma f^{\mu\nu\cdots} \, d\sigma_\mu, \qquad (4\text{–}58)$$

where σ is a closed (three-dimensional) surface enclosing V_4. The contributions to (4–57) from the integral over the timelike surface at spatial infinity vanishes by assumption of the asymptotic properties of u and v.

Assume that one can find a function $D(x)$ with the following properties:

$$\Box D(x) = 0, \qquad (4\text{–}59a)$$

$$D(x) = 0 \qquad \text{for} \quad x^2 > 0, \qquad (4\text{–}59b)$$

and

$$\partial_\mu D(x)|_{x_n = 0} = -n_\mu \, \delta(\sigma, x), \qquad (x_n \equiv -n^\mu x_\mu). \qquad (4\text{–}59c)$$

The surface (plane) delta function $\delta(\sigma, x)$ is defined by

$$\int_\sigma f(x') \, \delta(\sigma, x' - x) \, d\sigma(x') = f(x) \qquad (x \text{ on } \sigma). \qquad (4\text{–}60)$$

In (4–59c), n^μ is a timelike unit vector, as defined by (4–54) orthogonal to σ. If the plane is such that $n^\mu = (1; 0, 0, 0)$, then $\delta(\sigma, x) = \delta(\mathbf{r})$, the three-dimensional δ-function.

The homogeneous wave equation

$$\Box A_\mu(x) = 0 \tag{4–61}$$

is a second-order differential equation in time and therefore poses a Cauchy problem: to find $A_\mu(x)$ at any space-time point in terms of the Cauchy data, i.e. in terms of A_μ and its normal derivative on a given spacelike surface. The surface will be a plane in Minkowski space. Its solution can now be obtained from Green's theorem (4–57) by the substitution

$$u(x') = A_\alpha(x'), \qquad v(x') = D(x' - x).$$

The left-hand side of (4–57) vanishes. The plane σ_1 is chosen to contain the point x so that

$$\int_{\sigma_1} [A_\alpha(x')\, \partial'_\mu D(x' - x) - D(x' - x)\, \partial'_\mu A_\alpha(x')]\, d\sigma^\mu(x')$$
$$= -\int_{\sigma_1} A_\alpha(x') n'_\mu\, \delta(\sigma_1, x' - x)\, d\sigma^\mu(x').$$

Since $d\sigma^\mu = n^\mu\, d\sigma$, it follows from (4–60) and (4–54) that the above expression is just $A_\alpha(x)$. Thus, (4–57) yields

$$A_\alpha(x) = \int_{\sigma_0} [A_\alpha(x')\, \partial'_\mu D(x' - x) - D(x' - x)\, \partial'_\mu A_\alpha(x')]\, d\sigma^\mu(x'). \tag{4–62}$$

This elegant solution of the Cauchy problem (4–61) determines $A_\alpha(x)$ in terms of the function $D(x)$ and the Cauchy data (initial values), i.e. the functions A_α and their proper time derivatives $cn^\mu\, \partial_\mu A_\alpha$ on σ_0. A spacelike plane is given by

$$n^\mu x_\mu + c\tau = 0, \tag{4–63}$$

where n^μ is the plane normal and τ is an invariant. Therefore,

$$n_\mu = -c\, \partial_\mu \tau \qquad \text{and} \qquad n^\mu = \frac{dx^\mu}{c\, d\tau}. \tag{4–64}$$

The solution (4–62) to the Cauchy problem becomes

$$A_\alpha(x) = \int_{\sigma_0} \left[A_\alpha(x')\, \frac{d}{c\, d\tau'}\, D(x' - x) - D(x' - x)\, \frac{d}{c\, d\tau'}\, A_\alpha(x') \right] d\sigma(x'). \tag{4–62'}$$

Note that (4–59c) can now be written as

$$\frac{dD(x)}{c\, d\tau}\bigg|_\sigma = \delta(\sigma, x), \tag{4–65}$$

but this equation cannot be applied to (4-62') because x and x' will not in general be spacelike to each other.

Finally the function $D(x)$ must be determined. To this end one uses again the Fourier representation. Clearly, if $f(k)$ is sufficiently well behaved,*

$$D(x) = \frac{1}{(2\pi)^2} \int e^{ik \cdot x} f(k) \, \delta(k^2) \, d^4k \qquad (4\text{-}66)$$

will be a solution of the homogeneous wave equation (4-59a), since $k^2 \, \delta(k^2) = 0$.

PROBLEM 4-7

Prove the identity

$$\delta(k^2) = \frac{c}{2\omega} \left[\delta\left(k^0 - \frac{\omega}{c}\right) + \delta\left(k^0 + \frac{\omega}{c}\right) \right], \qquad (4\text{-}67)$$

where $\omega/c \equiv |\mathbf{k}|$, as in (4-13).

By means of the identity (4-67), $D(x)$ can be written in the form

$$D(x) = \frac{1}{(2\pi)^2} \frac{1}{2} \int e^{ik \cdot x} f(k) \left[\delta\left(k^0 - \frac{\omega}{c}\right) + \delta\left(k^0 + \frac{\omega}{c}\right) \right] \frac{d^3k}{\omega/c} \, dk^0,$$

and the time derivative becomes at $t = 0$,

$$\frac{\partial D(x)}{\partial t} \bigg|_{t=0} = \frac{-i}{2(2\pi)^2} \int e^{ik \cdot \mathbf{r}} \frac{d^3k}{\omega/c} f(k) \left[k^0 \, \delta\left(k^0 - \frac{\omega}{c}\right) + k^0 \, \delta\left(k^0 + \frac{\omega}{c}\right) \right] dk^0.$$

For the special case $n^\mu = (1;0,0,0)$ the condition (4-59c) requires that this expression be equal to

$$\delta(\mathbf{r}) = \frac{1}{(2\pi)^3} \int e^{ik \cdot \mathbf{r}} \, d^3k.$$

Therefore,

$$-i\pi f(k) = \tfrac{1}{2}\epsilon(k^0) \equiv \frac{k^0}{2\omega/c},$$

and (4-66) becomes

$$D(x) = \frac{i}{(2\pi)^3} \int e^{ik \cdot x} \epsilon(k^0) \, \delta(k^2) \, d^4k. \qquad (4\text{-}68)$$

The generalization to (4-59c) for arbitrary n^μ is assured by generalizing $\epsilon(k^0)$ to

$$\epsilon(k) = -\frac{n^\mu k_\mu}{|n^\nu k_\nu|}. \qquad (4\text{-}69)$$

---

* In the following it will often be convenient to use the notation $A \cdot B$ for $A_\mu B^\mu$ and A^2 for $A_\mu A^\mu$.

The final result is

$$D(x) = \frac{1}{(2\pi)^4} \int 2\pi i \epsilon(k) \, \delta(k^2) e^{ik \cdot x} \, d^4 k. \tag{4-68'}$$

This function is known as the *invariant Jordan-Pauli function*.

PROBLEM 4-8

Prove that (4-68) can be integrated to yield the following alternative forms of $D(x)$,

$$D(x) = \frac{1}{(2\pi)^3} \int e^{i\mathbf{k}\cdot\mathbf{r}} \sin \omega t \, \frac{d^3 k}{\omega/c} = \frac{1}{2\pi} \epsilon(x) \, \delta(x^2). \tag{4-68''}$$

The latter is again a generalization analogous to (4-68'). This result makes it obvious that condition (4-59b) is satisfied irrespective of the choice of the normal n^μ. Note how this expression, when combined with (4-62'), describes the propagation of the electromagnetic potentials along the light cone.

As a final comment on the solution (4-62) of the homogeneous wave equation, we note that this solution does not depend on the choice of the initial-value plane because the integrand has vanishing divergence (see Appendix, Section A1-5).

4-7 THE POTENTIALS

The inhomogeneous wave equation

$$\Box A^\mu = -\frac{4\pi}{c} j^\mu \tag{4-46}$$

poses a very different problem than the homogeneous equation. The specification of A^μ and its normal derivative on a spacelike surface no longer suffices because the presence of the current generates new electromagnetic fields. To clarify this situation, let us assume that the current is due to a moving charge represented by a world line in Minkowski space (Fig. 4-1). At any point Q on this world line electromagnetic fields will be produced which move with the velocity of light, i.e. propagate along the future light cone. This fact is sometimes referred to as *causality* (see Section 3-15). As a consequence, the electromagnetic fields at the event P are due to the motion of the charge at the event Q (and no other event of that charge!) as well as to whatever other fields may be brought to bear on this system by currents not included as part of the physical system under consideration (*external* fields). The fields at P due to Q are known as *retarded* fields because they are determined by the behavior of the source at an earlier (retarded) time. Conversely, one can determine the retarded point or points (if there are more than one charged world line) for a given point P by drawing the past light cone with vertex at P. This cone will intersect all world lines at the retarded points Q. This construction is unique because world lines of charges always have timelike tangent vectors.

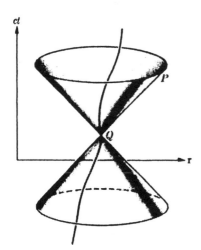

FIG. 4-1. Future and past light cones associated with a point Q on the world line of a charged particle.

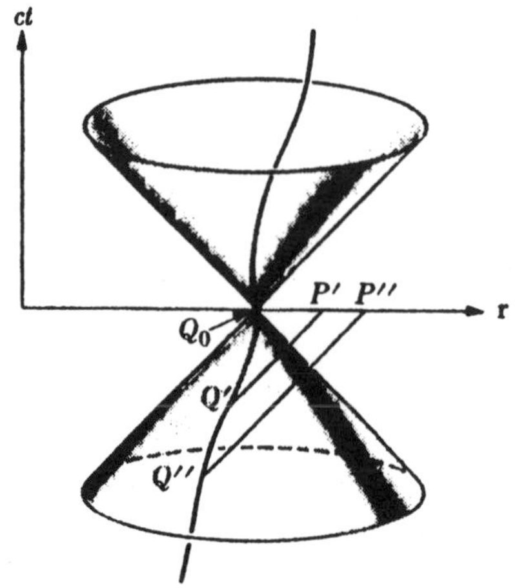

FIG. 4-2. Light cones with vertex Q_0 on the plane $x^0 = 0$.

Assume that there are no external fields present. Then the fields at $t = 0$ ($x^0 = 0$) are all entirely due to the charge shown in Fig. 4-2. The field at each point P' on $t = 0$ is due to a unique point Q' on the world line. This point Q' is responsible for all fields on a (hyper)circle in the (hyper)plane $t = 0$; or in three-dimensional space, all light spheres which started at Q' ($t_{Q'} < 0$) will have grown to a certain fixed radius by the time $t = 0$.

If one wants to specify a Cauchy problem at $t = 0$ together with the current for $t \geq 0$, the problem will separate into two problems: (a) the Cauchy problem with Cauchy data on $t = 0$; this will determine the fields for $t > 0$ *outside* the light cone whose vertex is Q_0 (Fig. 4-2); (b) the retarded field problem due to the current at $t \geq 0$; this will determine the fields *inside* and on the future light cone with vertex at Q_0. The Cauchy data for problem (a), however, are not known and must be found by solving a problem of type (b) for $t < 0$. Thus one simply has a retarded field problem [type (b)] *for all space-time.* It is very essential to realize that the finite propagation velocity of the field forces one into a problem posed for *all* space-time which would be very difficult (and physically awkward) to specify as (partially) a Cauchy problem. These considerations show clearly that field theory (i.e. interactions with finite propagation velocities) demands *asymptotic* problems, i.e. the study of field equations (and, as we shall see, equations of motion) involving *asymptotic conditions* specified at $t = \pm \infty$.

We now turn to the solution of the retarded field problem. The most convenient method for this purpose is *the Green function method.** Green functions

---

* The commonly used expressions "the Green's function" and "a Green's function" represent an atrocity to the English language. I doubt that those who use them ever refer to "a Shakespeare's sonnet."

are named after George Green (1793–1841) who introduced this type of function in potential theory. To solve (4–46) we first solve the auxiliary problem

$$\Box G(x - x') = -\delta(x - x'). \qquad (4\text{-}70)$$

The function $G(x - x')$ is a Green function. It must satisfy the same conditions that are imposed on A^{μ}. Thus, if we seek the retarded solution, we must impose the retardation condition also on G. As explained above, this condition specifies that

$$G(x_P - x_Q) = 0 \quad \text{for} \quad t_P < t_Q. \qquad (4\text{-}71)_{\text{ret}}$$

The Green function which satisfies (4–70) and (4–71) is denoted by $D_R(x - x')$, where the subscript R indicates "retarded." At this point it is clear that (4–71) is a necessary condition. That it is also sufficient to express the P-Q relationship of Fig. 4–1 will be seen presently.

For future reference it will be convenient to define another Green function by the condition

$$G(x_P - x_Q) = 0 \quad \text{for} \quad t_P > t_Q. \qquad (4\text{-}71)_{\text{adv}}$$

It ensures that the point P lies on the *past* light cone with vertex at Q. The function which satisfies (4–70) and (4–71)$_{\text{adv}}$ will be denoted by $D_A(x - x')$, where the subscript A indicates "advanced."

Once D_R is determined, the retarded solution of (4–46) is simply

$$A^{\mu}_{\text{ret}}(x) = \frac{4\pi}{c} \int D_R(x - x')j^{\mu}(x')\, d^4x'. \qquad (4\text{-}72)_{\text{ret}}$$

Similarly, an advanced solution can be obtained,

$$A^{\mu}_{\text{adv}}(x) = \frac{4\pi}{c} \int D_A(x - x')j^{\mu}(x')\, d^4x'. \qquad (4\text{-}72)_{\text{adv}}$$

Equations (4–72) are indeed solutions of the inhomogeneous equation. But so is

$$A^{\mu}(x) = \frac{4\pi}{c} \int G(x - x')j^{\mu}(x')\, d^4x', \qquad (4\text{-}73)$$

where

$$G(x - x') = aD_R(x - x') + (1 - a)D_A(x - x'). \qquad (4\text{-}74)$$

Obviously, it is the physical requirements, in particular the retardation condition (4–71)$_{\text{ret}}$, which make the solution unique.

Assuming a Fourier expansion of $G(x - x')$, Eq. (4–70) for the Green function becomes

$$-\frac{1}{(2\pi)^2} \int k^2 \widetilde{G}(k)e^{ik\cdot x}\, d^4k = -\frac{1}{(2\pi)^4} \int e^{ik\cdot x}\, d^4k.$$

Therefore,

$$\widetilde{G}(k) = \frac{1}{(2\pi)^2 k^2}$$

and

$$G(x) = \frac{1}{(2\pi)^4} \int \frac{1}{k^2} e^{ikx} d^4k. \tag{4-75}$$

This result shows that our assumption of the existence of a Fourier transform was incorrect, since the integrand in (4-75) has poles and the integral is meaningless. However, while a Fourier transform does not exist, an *integral representation* may exist if we are able to give meaning to the integral in (4-75) and verify that it satisfies the conditions.

The simplest way to give meaning to (4-75) is to take its Cauchy principal part. Thus, we define the function

$$D_P(x) \equiv \frac{1}{(2\pi)^4} P \int \frac{1}{k^2} e^{ik\cdot x} d^4k. \tag{4-76}$$

It is easy to verify that D_P satisfies the inhomogeneous equation (4-70). But so does any combination of (4-76) with a solution of the homogeneous equation. In fact, the general solution of (4-70) is of the form

$$D_{\text{inhom}}(x) = \alpha D_{\text{hom}}(x) + D_P(x), \tag{4-77}$$

where α is an arbitrary constant. The problem of finding D_R is thus reduced to finding D_{hom} and α which will satisfy the retardation condition $(4\text{-}71)_{\text{ret}}$. For this purpose it is most convenient to carry out the integration in (4-76). One finds

$$D_P(x) = \frac{1}{4\pi} \delta(x^2). \tag{4-78}$$

PROBLEM 4-9

Prove (4-78) from (4-76).

A comparison of the solution D, (4-68''), and D_P, (4-78), permits one to write down the correct linear combination by inspection. Clearly,

$$D_R(x) = D_P(x) + \tfrac{1}{2}D(x) = \frac{1}{4\pi} \delta(x^2)[1 + \epsilon(x)] = \frac{\Theta(x)}{2\pi} \delta(x^2) \tag{4-79}$$

satisfies the retardation condition $(4\text{-}71)_{\text{ret}}$, since it vanishes identically for $t < 0$ $(n^\mu x_\mu > 0)$.

In a completely analogous fashion one obtains

$$D_A(x) = D_P(x) - \tfrac{1}{2}D(x) = \frac{1}{4\pi} \delta(x^2)[1 - \epsilon(x)] = \frac{+\Theta(-x)}{2\pi} \delta(x^2) \tag{4-80}$$

These results can be written even more suggestively by employing the identity (4-67) for x^2 instead of k^2. One finds

$$D_R(x) = \frac{1}{4\pi r}\, \delta(r - ct) \tag{4-79'}$$

and

$$D_A(x) = \frac{1}{4\pi r}\, \delta(r + ct). \tag{4-80'}$$

With this result the solutions (4-72) can be written

$$
\begin{aligned}
A^\mu_{\substack{\text{ret}\\\text{adv}}}(x) &= \frac{1}{c} \int \frac{j^\mu(x')}{|\mathbf{r} - \mathbf{r}'|}\, \delta(|\mathbf{r} - \mathbf{r}'| \mp c(t - t'))\, d^4 x' \\
&= \frac{1}{c} \int \frac{j^\mu(\mathbf{r}', t \mp |\mathbf{r} - \mathbf{r}'|/c)}{|\mathbf{r} - \mathbf{r}'|}\, d^3 r'.
\end{aligned} \tag{4-81}
$$

These formulas express very clearly the retardation effect as discussed in connection with Fig. 4-1 and the corresponding advanced effects.

If the current density is due to a particle whose world line is given by $z^\mu(\tau)$, and whose charge is e, then

$$j^\mu(x) = ec \int_{-\infty}^{\infty} \delta(x - z(\tau))\, v^\mu(\tau)\, d\tau \tag{4-82}$$

with

$$v^\mu(\tau) \equiv \frac{dz^\mu(\tau)}{d\tau}. \tag{4-83}$$

The four-dimensional δ-function in (4-82) characterizes the current as due to a *point* charge, as will be shown below.

In order to compute the potential due to this current one can substitute it in (4-81) and carry out the integrations. However, we shall choose an alternative derivation starting from (4-72) and retaining the covariant notation.

We return to the solution (4-72) and assume that the current is due to a point charge as in (4-82). Then

$$
\begin{aligned}
A^\mu_{\substack{\text{ret}\\\text{adv}}}(x) &= 4\pi e \int_{-\infty}^{\infty} D_{\substack{R\\A}}(x - z) v^\mu\, d\tau \\
&= e \int_{-\infty}^{\infty} [1 \pm \epsilon(x - z)]\, \delta[(x - z)^2]\, v^\mu\, d\tau, \tag{4-84}
\end{aligned}
$$

or

$$A^\mu_{\substack{\text{ret}\\\text{adv}}} = \pm 2e \int_{\mp\infty}^{\tau_0} \delta[(x - z)^2] v^\mu\, d\tau. \tag{4-84'}$$

The integration of this expression requires the identity

$$\int f(\tau)\, \delta[g(\tau)]\, d\tau = \sum_i \frac{f(\tau_i)}{|\dot{g}(\tau_i)|}, \tag{4-85}$$

where τ_i are the zeros of $g(\tau) = 0$ which lie inside the integration interval. It can be proven as follows. Assume $g = g(\tau)$ can be inverted to $\tau = \tau(g)$. Then

$$\int f(\tau) \, \delta[g(\tau)] \, d\tau = \int f[\tau(g)] \, \delta(g) \, \frac{dg}{|\dot{g}[\tau(g)]|} .$$

The integral will contribute only when the interval of integration contains at least one zero of g. If it contains exactly one such zero, it gives

$$\frac{f[\tau(0)]}{|\dot{g}[\tau(0)]|} .$$

If it contains more than one, the result (4–85) ensues.

The first application of the identity (4–85) permits us to cast the covariant expression for the current density, (4–82), into a physically more obvious form. Applied to the factor $\delta(x^0 - z^0)$,

$$j^\mu(x) = ec \left. \frac{\delta(\mathbf{r} - \mathbf{z}) \, dz^\mu(\tau)/d\tau}{(dz^0/d\tau)} \right|_{x^0 = z^0} = e \, \delta(\mathbf{r} - \mathbf{z}) \, \frac{dz^\mu}{dt} = \rho \, \frac{dz^\mu}{dt} . \qquad (4\text{-}82')$$

The point character of the charge responsible for j^μ is obvious from the form of the charge density.

Now let

$$R^\mu \equiv x^\mu - z^\mu(\tau), \qquad R_\mu R^\mu = 0; \qquad (4\text{-}86)$$

then the derivative of the argument of the δ-function in (4–84) is $-2R^\mu v_\mu$ and the potentials are

$$A^\mu_{\substack{\text{ret} \\ \text{adv}}}(x) = \frac{ev^\mu}{|R^\alpha v_\alpha|} \cdot \frac{1 \pm \epsilon(R^0)}{2} . \qquad (4\text{-}87)$$

As was mentioned in connection with Fig. 4–1, the function $g = (x - z)^2$ has only two zeros: the future and the past light cones with vertices at x intersecting the world line $z(\tau)$ in exactly two points, Q_ret and Q_adv. The factor $1 + \epsilon(R^0)$ eliminates Q_adv and the factor $1 - \epsilon(R^0)$ eliminates Q_ret. This factor can be disposed of by defining R^μ in a suitable way. Thus,

$$R^0 = \pm |R| \quad \text{for} \quad \begin{cases} \text{ret} \\ \text{adv} \end{cases}, \qquad R^\mu_{\substack{\text{ret} \\ \text{adv}}} \equiv (\pm|R|, \, R) \qquad (4\text{-}88)$$

makes the last factor in (4–87) superfluous.

It is now convenient to define a spacelike unit vector u_μ orthogonal to v^μ,

$$u^\mu u_\mu = 1, \qquad u^\mu v_\mu = 0. \qquad (4\text{-}89)$$

The null-vector R^μ can now be written

$$R^\mu_{\substack{\text{ret} \\ \text{adv}}} = \rho_{\substack{\text{ret} \\ \text{adv}}} \left(u^\mu \pm \frac{v^\mu}{c} \right), \qquad \rho_{\substack{\text{ret} \\ \text{adv}}} = u_\mu R^\mu_{\substack{\text{ret} \\ \text{adv}}} = \mp v_\mu \, R^\mu_{\substack{\text{ret} \\ \text{adv}}}/c. \qquad (4\text{-}90)$$

This equation defines $\rho > 0$. The potentials (4-87) now take on the simple form

$$A^\mu_{\substack{\text{ret} \\ \text{adv}}} = \pm \frac{e}{c} \frac{v^\mu}{\rho} \qquad (\rho_{\text{ret}} = \rho_{\text{adv}} = \rho). \qquad (4\text{-}91)$$

Since $v^\mu = (\gamma c, \gamma \mathbf{v})$ and $\mathbf{R} = \mathbf{r} - \mathbf{z}$, this equation becomes in noncovariant form

$$\mathbf{A}_{\substack{\text{ret} \\ \text{adv}}}(x) = \frac{e\mathbf{v}/c}{|\mathbf{r} - \mathbf{z}|(1 \mp v_{||}/c)}, \qquad \phi_{\substack{\text{ret} \\ \text{adv}}}(x) = \frac{e}{|\mathbf{r} - \mathbf{z}|(1 \mp v_{||}/c)}, \qquad (4\text{-}92)$$

$$v_{||} \equiv \mathbf{v} \cdot \frac{\mathbf{r} - \mathbf{z}}{|\mathbf{r} - \mathbf{z}|}. \qquad (4\text{-}93)$$

Note that in these equations $\mathbf{z}$ and $\mathbf{v}$ refer to the retarded or advanced source points Q_{ret} or Q_{adv}, not to the argument of $\mathbf{A}$ and ϕ. The result (4-92) was first obtained by Liénard (1898) and Wiechert (1900).

The meaning of ρ and u^μ introduced in (4-90) and (4-89) becomes apparent when one goes into the rest system of the charge: $v^\mu = (1; 0, 0, 0)$, $u^\mu = (0, \hat{\mathbf{u}})$, $\mathbf{R} = \rho \hat{\mathbf{u}}$, so that $\mathbf{A} = 0$ and $\phi = e/\rho$. Thus, $\hat{\mathbf{u}}$ is the unit vector which points from the retarded (advanced) position $\mathbf{z}$ of the charge to the field point $\mathbf{r}$, while $|\mathbf{R}|$ is the magnitude of the corresponding distance; u^μ and ρ are the covariant generalizations of these concepts.

4-8 THE FIELD STRENGTHS

The covariant solution (4-91) for the retarded and advanced potentials permits us to obtain the field strengths in covariant form. According to (4-47) we need only to differentiate.

However, this differentiation is not so elementary if one recalls that the source point Q is uniquely related to the field point P (Fig. 4-1), so that an infinitesimal change of P entails a correspondingly infinitesimal change of Q. Thus, the instant τ (associated with Q) depends on x. The relation between P and Q, viz. (4-86) and (4-88), must be preserved in the process of differentiation. It determines the change of τ due to x.

Differentiation with respect to τ of (4-86) and use of (4-90) yields

$$R_\mu \left(\frac{dx^\mu}{d\tau} - v^\mu \right) = 0, \qquad \left(u^\mu \pm \frac{v^\mu}{c} \right) \left(\frac{dx_\mu}{d\tau} - v_\mu \right) = 0,$$

$$\left(u^\mu \pm \frac{v^\mu}{c} \right) \frac{dx^\mu}{c\, d\tau} = \mp 1. \qquad (4\text{-}94)$$

Now $\partial \tau / \partial x^\mu$ is the reciprocal of $dx^\mu/d\tau$, because of (4-64). Therefore, (4-94) implies

$$\partial^\mu \tau \equiv \frac{\partial \tau}{\partial x_\mu} = \mp \left(u^\mu \pm \frac{v^\mu}{c} \right). \qquad (4\text{-}95)$$

PROBLEM 4–10

Using (4–95), prove that

$$\partial^\mu \rho_{\substack{ret \\ adv}} = u^\mu + a_u R^\mu_{\substack{ret \\ adv}}/c^2 \qquad (4\text{–}96)$$

with

$$a_u \equiv a^\mu u_\mu \qquad (4\text{–}97)$$

and ρ given by (4–90).

The derivative of the potential can now be computed,

$$c\,\partial^\mu A^\nu(x) = -\frac{ev^\nu}{\rho^2}\,\partial^\mu\rho + \frac{e}{\rho}\,a^\nu\,\partial^\mu\tau.$$

Using the above results, (4–95) and (4–96), one finds

$$F^{\mu\nu}_{\substack{ret \\ adv}}(x) = \frac{e}{\rho^2 c}\,(v^\mu u^\nu - v^\nu u^\mu)$$

$$+ \frac{e}{\rho c^2}\left[(a^\mu v^\nu - a^\nu v^\mu)/c - u^\mu\left(\frac{v^\nu}{c}\,a_u \pm a^\nu\right) + u^\nu\left(\frac{v^\mu}{c}\,a_u \pm a^\mu\right)\right]. \qquad (4\text{–}98)$$

The quantities v^μ, a^μ, u^μ, and ρ must be referred to the retarded and advanced points, respectively.

The field strengths, therefore, consist of two terms, the *velocity fields* which are independent of a^μ and are proportional to $1/\rho^2$, and the *acceleration fields* which are linear homogeneous functions of a^μ and proportional to $1/\rho$. The latter fields are also called *radiation fields*. The expression (4–98) gives these fields in terms of the position, velocity, and acceleration of the (retarded or advanced) source point; the field point x enters only into ρ and the vector u^μ.

The *form* of the velocity fields in (4–98) is independent of the reference to retarded or advanced fields. But their numerical values are not, because $\rho_{ret} \neq \rho_{adv}$. The velocity fields are sometimes called generalized Coulomb fields. They differ from the Coulomb fields only by a Lorentz transformation. In fact, in the instantaneous rest frame of the charge where $\rho = |\mathbf{R}|$, the velocity fields of (4–98) become

$$\mathbf{E} = \frac{e}{R^2}\,\hat{u}, \qquad \mathbf{B} = 0, \qquad (4\text{–}99)$$

a pure Coulomb field.

Similarly, the acceleration fields in the instantaneous rest frame become

$$\mathbf{E}_{\substack{ret \\ adv}} = \frac{e}{|\mathbf{R}|}\,\hat{u} \times (\hat{u} \times \mathbf{a})/c^2,$$

$$\mathbf{B}_{\substack{ret \\ adv}} = \mp\,\frac{e}{|\mathbf{R}|}\,\hat{u} \times \mathbf{a}/c^2 = \pm\hat{u} \times \mathbf{E}_{\substack{ret \\ adv}} \qquad (4\text{–}100)$$

in agreement with (4–16).

PROBLEM 4-11

Verify (4-99) and (4-100).

For future reference it will be convenient to cast the expression (4-98) for $F^{\mu\nu}$ into a different form, viz.

$$F^{\mu\nu}_{\substack{\text{ret}\\\text{adv}}}(x) = \pm \frac{e}{\rho c^2} \frac{d}{d\tau} \left(\frac{v^\mu R^\nu - v^\nu R^\mu}{\rho} \right). \qquad (4\text{-}101)$$

This form of $F^{\mu\nu}$ does not exhibit the separation into velocity fields and acceleration fields, but its conciseness is often advantageous.

To prove (4-101) one must differentiate (4-90),

$$\dot{\rho}_{\substack{\text{ret}\\\text{adv}}} = \mp a_\mu R^\mu / c \mp v_\mu(-\dot{z}^\mu)/c = \mp c(1 + a_\mu\rho/c^2). \qquad (4\text{-}102)$$

From this, one obtains

$$\frac{d}{d\tau}\left[\frac{1}{\rho} (x^\mu - z^\mu) \right] = -\frac{\dot{\rho}}{\rho^2} R^\mu + \frac{1}{\rho}(-v^\mu) = -\frac{v^\mu}{\rho} \pm \frac{R^\mu c}{\rho^2} (1 + a_\mu\rho/c^2).$$

Use of these results and differentiation in (4-101) yields (4-98) and establishes the equivalence of these two expressions.

Alternatively, Eq. (4-101) could also be proven directly by differentiation of (4-84').

PROBLEM 4-12

Show that the current due to a point charge, Eq. (4-82), gives the following separation into solenoidal and irrotational fields, according to Problem 4-5, if one chooses $n^\mu \equiv v^\mu/c$:

$$j^\mu_I \equiv -n^\mu j_n = j^\mu, \qquad j^\mu_s = 0,$$
$$(\Box + \partial^2)A^\mu_I = -\frac{4\pi}{c} j^\mu, \qquad \Box A^\mu_s = 0.$$

Note that v^μ is treated as a constant vector because of the identification $n^\mu = v^\mu/c$. This means that the observer is comoving.

C. The Conservation Laws

In this chapter we have so far studied the field equations and their solutions, but we have not fully exploited the symmetry properties of the theory. According to Noether's theorem (Section 3-17), these properties can be exploited for the derivation of conservation laws. In electrodynamics these laws are less evident and less familiar than in mechanics. We shall devote the next three sections to them.

4–9 THE LAGRANGIAN AND INVARIANCE UNDER TRANSLATIONS

The largest transformation group under which electrodynamics is invariant is the Lorentz group.* It is a 10-parameter Lie group and should therefore furnish us 10 conservation laws. We have four independent variables, corresponding to the coordinates of quasi-Euclidean Minkowski space-time. Therefore, the basic integral invariant I of Noether's theorem is, according to (3–45'),

$$I = \int \mathcal{L} \, d^4x. \tag{4–103}$$

The functional $\mathcal{L}$ must be the Lagrangian which, via Hamilton's principle of least action, yields the electromagnetic field equations. It must be invariant under the 10 infinitesimal transformations which determine the Lorentz group. Our first task is therefore to find $\mathcal{L}$ so that the Maxwell-Lorentz field equations take on the form (3–56). Our second task is to determine the conservation laws (3–57) and (3–58) which result from this $\mathcal{L}$.

There is no general recipe for finding the Lagrangian which will provide a given set of equations. A few special cases are the exception; for a conservative system of mass points with holonomic constraints the Lagrangian is just the difference between kinetic and potential energy. In general, only certain structural features can be assured from the outset. In particular, the invariance of $\mathcal{L}$ under the same invariance group as that for the equations is of great help in finding the Lagrangian. Its exact determination is, in the last analysis, educated guesswork. A theory necessarily involves setting down a set of basic equations. Hamilton's principle and the associated Lagrangian are an afterthought, a *post facto* construct by which these equations can be reduced to a variational principle; the latter is often considered "nicer." Physically, it adds nothing that is not contained in the basic equations. But the Lagrangian is of great convenience in studying symmetry properties and conservation laws and in suggesting possible generalizations of the theory at hand.

The Lagrangian for the Maxwell-Lorentz field equations contains the potentials A^μ as the dependent variables. Since the equations for A^μ are of second order, $\mathcal{L}$ contains only the first derivatives of A^μ. Specifically, a possible choice is

$$\mathcal{L} = -\frac{1}{8\pi} \partial_\mu A_\nu \, \partial^\mu A^\nu + \frac{1}{c} j_\mu A^\mu. \tag{4–104}$$

The principle of least action based on the action integral (4–103) with $\mathcal{L}$ given by (4–104) yields the Euler-Lagrange equations [see Eq. (3–56)],

$$\partial_\alpha \frac{\partial \mathcal{L}}{\partial \, \partial_\alpha A_\mu} - \frac{\partial \mathcal{L}}{\partial A_\mu} = 0, \tag{4–105}$$

---

* If attention is limited to the electromagnetic field, a larger invariance group exists (cf. Section 4–12).

which become the field equations

$$\Box A^\mu = - \frac{4\pi}{c} j^\mu. \tag{4-46}$$

These equations must be supplemented by the Lorentz condition (4-25′) and the connection between the field strengths and the potentials, (4-47), in order to be equivalent to the Maxwell-Lorentz equations.

An alternative choice for $\mathcal{L}$ is*

$$\mathcal{L} = - \frac{1}{16\pi} F_{\mu\nu} F^{\mu\nu} + \frac{1}{c} j_\mu A^\mu. \tag{4-106}$$

It yields, via (4-105),

$$\partial_\mu F^{\mu\nu} = - \frac{4\pi}{c} j^\nu \tag{4-51}$$

and has the advantage that it does not require the additional specification of the Lorentz condition. The choice (4-106) is therefore more satisfactory.

When the Lorentz condition is taken into account, the two Lagrangians (4-104) and (4-106) differ only by a divergence. When inserted into the action integral (4-103), this term gives no contribution since the action integral is to be taken over all space-time. It can be transformed into a surface integral by Gauss's theorem and the integrand vanishes on that (infinitely distant) surface.

Having obtained a suitable action integral, we now return to Noether's theorem and the conservation laws. The field variables u_k become the potentials A_α, the v_k^μ become $\partial^\mu A_\alpha$, and the conservation laws (3-57), (3-58) become

$$\partial_\mu F^\mu_i = 0 \qquad (i = 1, 2, \ldots, 10), \tag{3-57′}$$

$$F^\mu_i = \frac{\partial \mathcal{L}}{\partial \, \partial_\mu A_\alpha} \delta_i A_\alpha - \left(\frac{\partial \mathcal{L}}{\partial \, \partial_\mu A_\alpha} \partial_\nu A_\alpha - \delta^\mu_\nu \mathcal{L} \right) \delta_i x^\nu. \tag{4-107}$$

The infinitesimal transformations under consideration are derived from the inhomogeneous Lorentz group (3-13) and (3-14),

$$x'^\mu = \alpha^\mu_{\ \nu} x^\nu + \alpha^\mu. \tag{4-108}$$

Expansion up to first order gives

$$\alpha^\mu = \epsilon^\mu + \ldots, \tag{4-109}$$

$$\alpha^\mu_\nu = \delta^\mu_\nu + \omega^\mu_\nu + \ldots \tag{4-110}$$

The four ϵ^μ and the six linearly independent ω^μ_ν are the 10 infinitesimal parameters which characterize the $\delta_i x^\mu$ ($i = 1, 2, \ldots, 10$) of (3-54). In this section we shall study the invariance under the infinitesimal translation group

---

* $F_{\mu\nu}$ is here *defined* by (4-47) so that the homogeneous Maxwell-Lorentz equations are identically satisfied.

(parameters ϵ^μ), while the infinitesimal proper homogeneous Lorentz transformations (parameter $\omega^\mu{}_\nu$) will be left for the next section.

Initially, the system of interest is the *free* electromagnetic field ($j^\mu = 0$). In this case, (4–106) reduces to

$$\mathcal{L}_{\text{elm}} = -\frac{1}{16\pi} F_{\mu\nu} F^{\mu\nu} = \frac{1}{8\pi} (\mathbf{E}^2 - \mathbf{B}^2). \qquad (4\text{--}111)$$

The Lagrangian is substituted in (4–107). For translations we have $\delta A^\alpha = 0$. Since the $\delta x^\mu = \epsilon^\mu$ are arbitrary and linearly independent, the conservation law (3–57′) yields

$$\partial_\mu T^{\mu\nu}_{\text{elm}} = 0, \qquad (4\text{--}112)$$

where the coefficient of δx^ν is

$$
\begin{aligned}
T^{\mu\nu}_{\text{elm}} &\equiv + \left(\frac{\partial \mathcal{L}_{\text{elm}}}{\partial \, \partial_\mu A_\alpha} \, \partial^\nu A_\alpha - \eta^{\mu\nu} \mathcal{L}_{\text{elm}} \right) \\
&= -\frac{1}{4\pi} (+F^{\mu\alpha} \partial^\nu A_\alpha - \tfrac{1}{4}\eta^{\mu\nu} F_{\alpha\beta} F^{\alpha\beta}) \qquad (4\text{--}113) \\
&= \frac{1}{4\pi} (F^{\mu\alpha} F_\alpha{}^\nu + \tfrac{1}{4}\eta^{\mu\nu} F_{\alpha\beta} F^{\alpha\beta}) - \frac{1}{4\pi} F^{\mu\alpha} \partial_\alpha A^\nu.
\end{aligned}
$$

The divergence of the last term can be transformed,

$$\partial_\mu(F^{\mu\alpha} \partial_\alpha A^\nu) = \partial_\mu \partial_\alpha(F^{\mu\alpha} A^\nu) - \partial_\mu(A^\nu \partial_\alpha F^{\mu\alpha}).$$

In this expression, the first term vanishes because it is a product of a symmetric tensor ($\partial_\mu \partial_\alpha$) and an antisymmetric one ($F^{\mu\alpha}$), while the second term vanishes because of the free field equations $\partial_\alpha F^{\alpha\mu} = 0$. Consequently, the *symmetric* tensor

$$\Theta^{\mu\nu}_{\text{elm}} = \frac{1}{4\pi} (F^{\mu\alpha} F_\alpha{}^\nu + \tfrac{1}{4}\eta^{\mu\nu} F_{\alpha\beta} F^{\alpha\beta}) \qquad (4\text{--}114)$$

also has vanishing divergence,

$$\partial_\mu \Theta^{\mu\nu}_{\text{elm}} = 0. \qquad (4\text{--}115)$$

The tensor $T^{\mu\nu}_{\text{elm}}$ is the *generator* of the infinitesimal translations (being the coefficient of the ϵ_ν). It is better known as the *canonical electromagnetic energy-momentum tensor;* $\Theta^{\mu\nu}_{\text{elm}}$ is the *symmetrical electromagnetic energy-momentum tensor.* Actually, this is a misnomer because 9 of the 16 components of these tensors refer to stress. When included, their names would become even longer. We shall therefore refer to $\Theta^{\mu\nu}_{\text{elm}}$ simply as the *electromagnetic energy tensor;* its dimensions are those of an energy density. It is gauge invariant, while the canonical tensor $T^{\mu\nu}_{\text{elm}}$ is not.

An important property of $\Theta^{\mu\nu}_{\text{elm}}$ is its vanishing trace,

$$\Theta^\mu{}_{\mu\,\text{elm}} = \frac{1}{4\pi} (F^{\mu\alpha} F_{\alpha\mu} + \tfrac{1}{4} \delta^\mu_\mu F^{\alpha\beta} F_{\alpha\beta}) = 0 \qquad (4\text{--}116)$$

since $\delta^\mu_\mu = 4$. The tensor can be expressed in terms of the three-vectors **E** and **B** as follows. The space part Θ^{ik}_{elm} ($i, k = 1, 2, 3$) becomes the matrix of a dyadic,

$$\mathsf{T} = \frac{1}{4\pi} [\mathbf{EE} + \mathbf{BB} - \tfrac{1}{2}(E^2 + B^2)\mathbf{1}].\tag{4-117}$$

This is the *Maxwell stress dyadic*; Θ^{ik}_{elm} is also called the Maxwell stress tensor. The time and mixed space-time components are

$$\Theta^{00} = -\frac{1}{8\pi}(E^2 + B^2) \equiv -U,\tag{4-118}$$

and

$$\Theta^{0k} = -\frac{1}{4\pi}(\mathbf{E} \times \mathbf{B})_k,\tag{4-119}$$

where U is known as the electromagnetic *energy density** and

$$\mathbf{S} \equiv \frac{c}{4\pi}\mathbf{E} \times \mathbf{B}\tag{4-120}$$

is the *Poynting vector*, an energy flux density.

The differential conservation law (4-115) can therefore also be written as

$$\nabla \cdot \mathsf{T} - \frac{1}{c^2}\frac{\partial \mathbf{S}}{\partial t} = 0,\tag{4-121}$$

$$\frac{\partial U}{\partial t} + \nabla \cdot \mathbf{S} = 0.\tag{4-122}$$

These equations express momentum and energy conservation of the electromagnetic field in differential form. The physical meaning of these conservation laws can be expressed much simpler, however, in terms of the corresponding integral conservation laws. We shall therefore derive the latter and then state their physical meaning.

One might be tempted to integrate (4-121) and (4-122) over a spatial volume V_3. However, care must be taken because the integral of a tensor over a (three-dimensional) plane is not, in general, a tensor. We might thus lose relativistic invariance; but this invariance is essential for the correct identification of energy and momentum. The easiest way to proceed correctly is to work with manifestly covariant expressions.

Consider the quantity

$$P^\mu_{\text{elm}} \equiv \frac{1}{c} \int \Theta^{\mu\nu}_{\text{elm}} \, d\sigma_\nu.\tag{4-123}$$

Since $\Theta^{\mu\nu}_{\text{elm}}$ is divergence free (4-115), this quantity is independent of the space-

---

* See the statements following Eq. (4-126).

like plane σ (cf. Al-58); P_{elm}^{μ} is therefore a vector. In the particular Lorentz frame S_0 in which the surface normal $n^{\mu} = (1; 0, 0, 0)$, the components of P_{elm}^{μ} are

$$P_{\text{elm}}^{\mu(0)} = \left(\frac{1}{c} W_{\text{elm}}^{(0)}, \mathbf{P}_{\text{elm}}^{(0)} \right) \tag{4-124$^{(0)}$}$$

with

$$W_{\text{elm}}^{(0)} = \int U^{(0)} \, d^3x \tag{4-125$^{(0)}$}$$

and

$$\mathbf{P}_{\text{elm}}^{(0)} = \frac{1}{c^2} \int \mathbf{S}^{(0)} \, d^3x. \tag{4-126$^{(0)}$}$$

In any other inertial system P_{elm}^{μ} can be obtained from $P_{\text{elm}}^{\mu(0)}$ by a Lorentz transformation. It will then no longer have the forms (4-125)$^{(0)}$ and (4-126)$^{(0)}$. We now identify P_{elm}^{μ} of (4-123) with the *momentum four-vector of the electromagnetic field*.

It would be incorrect to define this vector by (4-125)$^{(0)}$ and (4-126)$^{(0)}$, since these equations refer to a special coordinate system. This can be seen explicitly by choosing an arbitrary n^{μ},

$$n^{\mu} = (\gamma, \sqrt{\gamma^2 - 1} \, \hat{\mathbf{n}}), \tag{4-127}$$

with $\gamma \geq 1$ arbitrary and $\hat{\mathbf{n}}$ a unit three-vector. Equation (4-123) then takes on the general form*

$$P_{\text{elm}}^{\mu} = \left(\frac{1}{c} W_{\text{elm}}, \mathbf{P}_{\text{elm}} \right) \tag{4-124}$$

with

$$W_{\text{elm}} = \gamma \int U \, d\sigma - \frac{\sqrt{\gamma^2 - 1}}{c} \int \mathbf{S} \cdot \hat{\mathbf{n}} \, d\sigma, \tag{4-125}$$

$$\mathbf{P}_{\text{elm}} = \frac{\gamma}{c^2} \int \mathbf{S} \, d\sigma + \frac{\sqrt{\gamma^2 - 1}}{c} \int \mathsf{T} \cdot \hat{\mathbf{n}} \, d\sigma. \tag{4-126}$$

Only in the special coordinate system $\gamma = 1$ do these equations reduce to those for $P_{\mu}^{(0)}$. It also follows that the designations "energy density" for U and "momentum density" for $(1/c^2)\mathbf{S}$ are meaningful *only relative to* S_0.

PROBLEM 4-13

Show that (4-125) and (4-126) differ from (4-125)$^{(0)}$ and (4-126)$^{(0)}$ exactly by a Lorentz transformation with a velocity $\mathbf{v}$ whose direction is $\hat{\mathbf{n}}$ and whose magnitude is $v = c\sqrt{1 - 1/\gamma^2}$.

---

* Since $\Theta_{\text{elm}}^{\mu\nu}$ is a tensor, *both* indices must be transformed: for each choice of a Lorentz frame (σ_{ν}) $\Theta_{\text{elm}}^{\mu\nu}$ is identified with T, $\mathbf{S}$, and U in that frame; this ensures that P_{elm}^{μ}, computed with σ_{ν}, and $P_{\text{elm}}'^{\mu}$, computed with σ_{ν}', differ exactly by the corresponding Lorentz transformation.

The derivative of P^μ with respect to the proper time τ of an arbitrary inertial observer can be obtained as follows. The proper time τ is related to n^μ of that observer by (4–63) and (4–64). Therefore, we must choose σ in (4–123) accordingly [cf. Eq. (A1–62)],

$$\frac{dP^\mu_{\text{elm}}}{d\tau} = \frac{1}{c} \int \frac{d\Theta^{\mu\nu}}{d\tau} \, d\sigma_\nu,$$

$$= \frac{1}{c} \int \frac{d\Theta^{\mu\nu}}{d\tau} (-c \, \partial_\nu \tau) \, d\sigma \qquad (4\text{--}128)$$

$$= - \int \partial_\nu \Theta^{\mu\nu} \, d\sigma = 0.$$

The last equality follows from (4–115) and does not depend on the choice of σ. Equation (4–128) is the *integral covariant form* of the conservation law for energy and momentum of the field:

$$P^\mu_{\text{elm}} = \text{const}, \qquad (4\text{--}129)$$

i.e. is independent of τ.

The equalities (4–128) can be written in terms of three-vectors; using Eq. (4–124),

$$\frac{dW_{\text{elm}}}{d\tau} = \int \frac{\partial U}{\partial t} \, d\sigma + \int \nabla \cdot \mathbf{S} \, d\sigma = 0 \qquad (4\text{--}130)$$

and

$$\frac{d\mathbf{P}_{\text{elm}}}{d\tau} = \frac{1}{c^2} \int \frac{\partial \mathbf{S}}{\partial t} \, d\sigma - \int \nabla \cdot \mathbf{T} \, d\sigma = 0. \qquad (4\text{--}131)$$

This, of course, leads back to (4–121) and (4–122), but the identification of $\mathbf{P}_{\text{elm}}$ and W_{elm} has been made.

The physical meaning of these conservation laws is now easily stated. *The total electromagnetic energy and momentum of a free electromagnetic field are constants* (4–129).

The differential laws can also be integrated over an arbitrary, not necessarily infinite, three-dimensional spatial volume V_3 with $n^\mu = (1; 0, 0, 0)$. We define the total energy flux out of V_3 by

$$\Phi_W \equiv \int \nabla \cdot \mathbf{S} \, d^3x = \int_\Sigma \mathbf{S} \cdot d^2\sigma, \qquad (4\text{--}132)$$

where Σ is the surface enclosing V_3. Also,

$$\mathbf{F}_M \equiv \int \nabla \cdot \mathbf{T} \, d^3x = \int_\Sigma \mathbf{T} \cdot d^2\sigma \qquad (4\text{--}133)$$

is the total force acting on V_3 due to the Maxwell stresses. Then (4–121) and (4-122) can be interpreted in terms of the field energy and momentum* contained in V_3 *in the special system* S_0,

$$-\frac{dW^{(0)}_{\text{elm}}}{dt} = \Phi_W, \tag{4-134}$$

$$\frac{d\mathbf{P}^{(0)}_{\text{elm}}}{dt} = \mathbf{F}_M. \tag{4-135}$$

The field energy decreases at a rate equal to the energy flux out of the volume. The time rate of change of the electromagnetic momentum equals the total Maxwell force (result of Maxwell stresses).

When the fields are not free and $j^\mu \neq 0$, the conservation law (4-115) is no longer valid. We are no longer dealing with a closed system. The divergence of (4-114) no longer vanishes; using the inhomogeneous field equations (4-51), we have

$$\partial_\mu \Theta^{\mu\nu}_{\text{elm}} = -\frac{1}{c} j_\alpha F^{\alpha\nu} + \frac{1}{4\pi}(F^{\mu\alpha}\partial_\mu F^\nu_\alpha + \tfrac{1}{2}F^{\alpha\beta}\partial^\nu F_{\alpha\beta}). \tag{4-136}$$

The expression in parentheses can be combined with the homogeneous field equations;

$$F^{\beta\alpha}\partial_\beta F^\nu_\alpha + \tfrac{1}{2}F^{\alpha\beta}\partial^\nu F_{\alpha\beta} = F_{\alpha\beta}(-\partial^\beta F^{\alpha\nu} + \tfrac{1}{2}\partial^\nu F^{\alpha\beta})$$

$$= F_{\alpha\beta}(-\partial^\beta F^{\alpha\nu} - \tfrac{1}{2}\partial^\alpha F^{\beta\nu} - \tfrac{1}{2}\partial^\beta F^{\nu\alpha})$$

$$= \tfrac{1}{2}F_{\alpha\beta}(\partial^\beta F^{\nu\alpha} + \partial^\alpha F^{\nu\beta}).$$

This is a product of a symmetric and an antisymmetric tensor and therefore vanishes. Thus, (4-136) becomes

$$\partial_\nu \Theta^{\nu\mu}_{\text{elm}} = \frac{1}{c} F^{\mu\nu} j_\nu. \tag{4-137}$$

This generalization of (4-115) includes the effect of the impressed (or externally given) current distribution on the conservation of the field energy and momentum. Equations (4-121) and (4-122) now become

$$\nabla \cdot \mathbf{T} - \frac{\partial \mathbf{S}}{c^2\,\partial t} = \rho\mathbf{E} + \frac{1}{c}\mathbf{j}\times\mathbf{B}, \tag{4-138}$$

$$\frac{\partial U}{\partial t} + \nabla\cdot\mathbf{S} = -\mathbf{j}\cdot\mathbf{E}. \tag{4-139}$$

The vector [cf. Eq. (2-8)]

$$\mathbf{f} = \rho\mathbf{E} + \frac{1}{c}\mathbf{j}\times\mathbf{B} \tag{4-140}$$

---

* These quantities agree with those defined in (4-125)$^{(0)}$ and (4-126)$^{(0)}$ in the limit $V_3 \to \infty$; in that limit the surface integrals over Σ vanish and $W^{(0)}_{\text{elm}}$ and $\mathbf{P}^{(0)}_{\text{elm}}$ become constants.

is the *Lorentz force density*. It is the force density exerted by an electromagnetic field $(\mathbf{E}, \mathbf{B})$ on a charge-current density $(\rho, \mathbf{j})$. Correspondingly, $\mathbf{j} \cdot \mathbf{E}$ is the rate of work done by the field on that charge. Together, these quantities form a four-vector,

$$f^{\mu} \equiv \frac{1}{c} F^{\mu\nu} j_{\nu} = \left(\frac{1}{c} \mathbf{j} \cdot \mathbf{E}, \rho \mathbf{E} + \frac{1}{c} \mathbf{j} \times \mathbf{B} \right). \tag{4-141}$$

The meaning of (4-138) and (4-139) can again be most easily expressed after we integrate over a spatial volume, V_3, analogous to (4-128): the rate of decrease of the four-momentum of the electromagnetic field equals the four-force which it exerts on a given charge-current distribution; from (4-137),

$$-\frac{dP^{\mu}_{\text{elm}}}{d\tau} = \int f^{\mu} \, d\sigma \equiv F^{\mu}. \tag{4-142}$$

In terms of the three-vector, using (4-130) and (4-131), we have

$$-\frac{dW_{\text{elm}}}{d\tau} = \int \mathbf{j} \cdot \mathbf{E} \, d\sigma \tag{4-143}$$

and

$$-\frac{d\mathbf{P}_{\text{elm}}}{d\tau} = \int \mathbf{f} \, d\sigma \equiv \mathbf{F}. \tag{4-144}$$

The field energy diminishes exactly by the work it does on the current; the field three-momentum diminishes exactly at a rate equal to the force it exerts on the charges and currents.

In Eq. (4-137) we encounter the Lorentz force (4-141) for the first time. While we have been concerned exclusively with the fields produced *by* a given j^{μ} we are led, via the conservation laws, to the action of the fields *on* j^{μ}. This action includes the action of the fields due to one charge on another (mutual interaction) as well as the action of the fields due to a charge on itself (self-interaction). If j^{μ} is due to only one charge, only the self-interaction will occur in the above equations. This fundamental problem will be studied in detail in Chapter 6.

The only important point regarding these interactions in connection with the above conservation laws is the appreciation of the difference between *open and closed systems*. A closed system is an isolated one, so that no external influences (external fields or currents) affect it. Translation invariance will then lead to an energy tensor with vanishing divergence, as in (4-115). An open system is under the influence of external forces which, in general, destroy its invariance properties. In the case of translation invariance this means that we no longer have an energy tensor of vanishing divergence and that the external forces affect the energy balance, the momentum balance, or both. The quantity P^{μ}_{elm} in (4-142) is no longer a four-vector since it depends on the surface σ. The momentum and energy given to or received from the external fields or currents (in this case from the j^{μ}) must be included in the balance as, for example, in

(4–143) and (4–144).   If we want to form a *closed* system containing the fields produced by j^μ, as well as j^μ itself, full account must be taken of the dynamics of the charges constituting j^μ.   The problem then becomes one of mutual determination: the fields are determined by the charges and their motion, and the motion of the charges is determined by the fields.   This problem will occupy us in Chapters 6 and 7.

4–10   INVARIANCE UNDER PROPER HOMOGENEOUS LORENTZ TRANSFORMATIONS

We are now concerned with the invariance of the theory under the infinitesimal transformations

$$x'^\mu = x^\mu + \omega^\mu{}_\nu x^\nu$$

or
$$(4\text{--}145)$$
$$\delta x^\mu = \omega^\mu{}_\nu x^\nu,$$

which follows from (4–108) and (4–110).   This transformation also affects A^μ, since

$$A'^\mu(x') = \alpha^\mu{}_\nu A^\nu(x)$$

and therefore

$$\delta A^\mu(x) = \omega^\mu{}_\nu A^\nu(x). \tag{4--146}$$

For the free electromagnetic field with Lagrangian (4–111) the conserved quantity F^μ of (4–107) becomes

$$-\frac{1}{4\pi} F^{\mu\alpha}\omega_{\alpha\nu}A^\nu - T^{\mu\alpha}_{\text{elm}}\omega_{\alpha\nu}x^\nu \equiv \frac{c}{2} M^{\mu\alpha\nu}\omega_{\alpha\nu}.$$

This defines the tensor $M^{\mu\alpha\nu}$.   Relabeling μ and α with α and μ and exploiting the antisymmetry of $\omega_{\mu\nu}$, we have

$$M^{\alpha\mu\nu} = -\frac{1}{c}\left(T^{\alpha\mu}_{\text{elm}}x^\nu - T^{\alpha\nu}_{\text{elm}}x^\mu\right) - \frac{1}{4\pi c}\left(F^{\alpha\mu}A^\nu - F^{\alpha\nu}A^\mu\right). \tag{4--147}$$

The infinitesimal parameters $\omega_{\mu\nu}$ are arbitrary, giving six independent conservation laws (3–57′)

$$\partial_\alpha M^{\alpha\mu\nu} = 0. \tag{4--148}$$

Attempting to attach physical significance to this conservation law, we notice that the first parenthesis of (4–147),

$$L^{\alpha\mu\nu} \equiv -\frac{1}{c}\left(T^{\alpha\mu}_{\text{elm}}x^\nu - T^{\alpha\nu}_{\text{elm}}x^\mu\right) \tag{4--149}$$

has the structure of an orbital angular momentum.   If we could associate

$$S^{\alpha\mu\nu} \equiv -\frac{1}{4\pi c}\left(F^{\alpha\mu}A^\nu - F^{\alpha\nu}A^\mu\right) \tag{4--150}$$

with the spin angular momentum, $M^{\alpha\mu\nu}$ could presumably be interpreted as total angular momentum density. Equation (4–148) would then express the conservation of angular momentum. Since we are dealing with a field, the angular momentum densities (third-rank tensors) would have to be related to the corresponding angular momenta by integrals over spacelike planes,

$$M^{\mu\nu} \equiv \int M^{\alpha\mu\nu}\, d\sigma_\alpha = L^{\mu\nu} + S^{\mu\nu}, \qquad (4\text{–}151)$$

$$L^{\mu\nu} \equiv \int L^{\alpha\mu\nu}\, d\sigma_\alpha, \qquad (4\text{–}152)$$

and

$$S^{\mu\nu} \equiv \int S^{\alpha\mu\nu}\, d\sigma_\alpha. \qquad (4\text{–}153)$$

But the identification of these quantities with physical observables hinges on their invariance properties: only gauge-invariant quantities are observables, and only tensors have relativistic meaning.

Equation (4–148) guarantees that $M^{\mu\nu}$ is independent of σ and is therefore a tensor. On the other hand,

$$\partial_\alpha L^{\alpha\mu\nu} = T^{\nu\mu}_{\text{elm}} - T^{\mu\nu}_{\text{elm}} \qquad (4\text{–}154)$$

and does not vanish, because the canonical energy tensor is not symmetrical. Consequently, $L^{\mu\nu}$ is not independent of σ. The same must then be true of $S^{\mu\nu}$. Therefore, the separation of $M^{\mu\nu}$ into $L^{\mu\nu}$ and $S^{\mu\nu}$ is not relativistically invariant.

Worse than that, $T^{\mu\nu}_{\text{elm}}$ is not gauge invariant, so that $L^{\mu\nu}$ is not gauge invariant either. It is suggestive to define instead of $L^{\alpha\mu\nu}$, (4–149), the gauge-invariant tensor

$$J^{\alpha\mu\nu} \equiv -\frac{1}{c}\,(\Theta^{\alpha\mu}_{\text{elm}}x^\nu - \Theta^{\alpha\nu}_{\text{elm}}x^\mu). \qquad (4\text{–}155)$$

The corresponding angular momentum,

$$J^{\mu\nu} \equiv \int J^{\alpha\mu\nu}\, d\sigma_\alpha, \qquad (4\text{–}156)$$

is indeed a tensor because the conservation law (4–115) and the symmetry of $\Theta^{\mu\nu}_{\text{elm}}$ yields

$$\partial_\alpha J^{\alpha\mu\nu} = 0, \qquad (4\text{–}157)$$

which ensures the σ-independence of $J^{\mu\nu}$. We shall identify $J^{\mu\nu}$ with the *total angular momentum* tensor of the electromagnetic field. It is a gauge-invariant tensor and is therefore a relativistic observable.

To be sure, only the space part J^{ij} of the tensor $J^{\mu\nu}$ describes the total angular momentum. Its conservation is due to invariance under rotations [coefficients ω^{ij} in (4–145)]. The mixed space-time components, J^{0k}, have a different physical meaning; their conservation expresses the center-of-mass theorem [see Eqs. (4-170) and (4-172) below] and is due to invariance under Lorentz transformations (coefficients ω^{0k}).

How is $J^{\mu\nu}$ related to $M^{\mu\nu}$? The answer follows easily from the relation between $\Theta_{elm}^{\mu\nu}$ and $T_{elm}^{\mu\nu}$, (4-113) and (4-114),

$$M^{\alpha\mu\nu} = -\frac{1}{c}\left(\Theta_{elm}^{\alpha\mu} - \frac{1}{4\pi}F^{\alpha\lambda}\partial_\lambda A^\mu\right)x^\nu + \frac{1}{c}\left(\Theta_{elm}^{\alpha\nu} - \frac{1}{4\pi}F^{\alpha\lambda}\partial_\lambda A^\nu\right)x^\mu + S^{\alpha\mu\nu}$$

$$= J^{\alpha\mu\nu} + \frac{1}{4\pi c}F^{\alpha\lambda}(\partial_\lambda A^\mu x^\nu - \partial_\lambda A^\nu x^\mu) + S^{\alpha\mu\nu}$$

$$= J^{\alpha\mu\nu} + \frac{1}{4\pi c}F^{\alpha\lambda}\partial_\lambda(A^\mu x^\nu - A^\nu x^\mu) - \frac{1}{4\pi c}(F^{\alpha\nu}A^\mu - F^{\alpha\mu}A^\nu) + S^{\alpha\mu\nu}.$$

Because of (4-150) the last two terms cancel, and because of the field equations (4-51) with $j_\mu = 0$, the second term is a divergence,

$$M^{\alpha\mu\nu} = J^{\alpha\mu\nu} - \frac{1}{4\pi c}\partial_\lambda[F^{\lambda\alpha}(A^\mu x^\nu - A^\nu x^\mu)]. \tag{4-158}$$

The integrals over the spacelike planes are therefore related by

$$M^{\mu\nu} = J^{\mu\nu} - \frac{1}{4\pi c}\int \partial_\lambda[F^{\lambda\alpha}(A^\mu x^\nu - A^\nu x^\mu)]\,d\sigma_\alpha.$$

This integral is independent of σ because

$$\partial_\alpha\partial_\lambda[F^{\lambda\alpha}(A^\mu x^\nu - A^\nu x^\mu)] = 0$$

vanishes identically, being the product of a symmetric ($\partial_\alpha\partial_\lambda$) and an antisymmetric tensor ($F^{\lambda\alpha}$). Therefore if we choose $n^\alpha = (1;0,0,0)$, we find that this integral becomes

$$\frac{1}{4\pi c}\int \nabla\cdot[\mathbf{E}(A^\mu x^\nu - A^\nu x^\mu)]\,d^3x = \frac{1}{4\pi c}\int d^2\boldsymbol{\sigma}\cdot\mathbf{E}(A^\mu x^\nu - A^\nu x^\mu). \tag{4-159}$$

If the fields vanish sufficiently fast at infinity, this (two-dimensional) surface integral vanishes and

$$M^{\mu\nu} = J^{\mu\nu}. \tag{4-160}$$

In this case, therefore, $M^{\mu\nu}$ is both gauge invariant and relativistically covariant.

Let us study in further detail the case when the surface integral vanishes.*

While a covariant separation of $J^{\mu\nu}$ into orbital and spin angular momenta does not exist, it is of some interest to exhibit such a separation in a special

---

* At times, this surface integral was assumed to vanish when the conditions for this to happen were not satisfied. As a consequence, confusion and paradoxes arose.

coordinate system. Consider the system S_0 defined by $n^\mu = (1;0,0,0)$. Let us describe the six independent components of $J^{\mu\nu}$ by two three-vectors,

$$\mathbf{J} \equiv (J_k), \qquad J_k \equiv J^{ij}\epsilon_{ijk}, \tag{4-161}$$

with ϵ_{ijk} as defined in Eq. (4-49), and

$$\mathbf{J'} \equiv (J'_k), \qquad J'_k \equiv J^{k0}. \tag{4-162}$$

This separation of $J^{\mu\nu}$ into $\mathbf{J}$ and $\mathbf{J'}$ corresponds exactly to the separation of the proper homogeneous Lorentz transformations (4-145) into rotations and "pure" Lorentz transformations.

The vector $\mathbf{J}$ in S_0 follows from (4-155) and (4-156) (for convenience we shall omit the explicit reference to S_0, viz. the index $^{(0)}$),

$$J_k = \frac{1}{c}\,\epsilon_{ijk}\int (\Theta^{0i}_{\text{elm}}x^j - \Theta^{0j}_{\text{elm}}x^i)\,d^3x,$$

$$\mathbf{J} = \frac{1}{c^2}\int \mathbf{r} \times \mathbf{S}\,d^3x. \tag{4-163}$$

The integrand is exactly the (total) angular momentum density, since $\mathbf{S}/c^2$ was previously found to be the linear momentum density in the reference frame S_0, (4-126)$^{(0)}$. By means of the vector potential $\mathbf{A}$ the integral can be separated as follows. The identity

$$\frac{4\pi}{c}\,\mathbf{S} = \mathbf{E} \times (\nabla \times \mathbf{A}) = (\nabla\mathbf{A})\cdot\mathbf{E} - \mathbf{E}\cdot\nabla\mathbf{A}$$

is used in (4-163) to yield, by means of partial integration,

$$\mathbf{J} = \frac{1}{4\pi c}\int \mathbf{r} \times (\nabla\mathbf{A})\cdot\mathbf{E}\,d^3x + \frac{1}{4\pi c}\int \mathbf{E}\cdot(\nabla\mathbf{A}) \times \mathbf{r}\,d^3x$$

$$= \mathbf{L} + \frac{1}{4\pi c}\int \nabla\cdot(\mathbf{E}\mathbf{A} \times \mathbf{r})\,d^3x + \mathbf{\Sigma}. \tag{4-164}$$

Here we define

$$\mathbf{L} \doteq \frac{1}{4\pi c}\int \mathbf{r} \times (\nabla\mathbf{A})\cdot\mathbf{E}\,d^3x \tag{4-165}$$

and

$$\mathbf{\Sigma} \equiv \frac{1}{4\pi c}\int \mathbf{E}\cdot(\nabla\mathbf{r}) \times \mathbf{A}\,d^3x + \frac{1}{4\pi c}\int \mathbf{r} \times \mathbf{A}\nabla\cdot\mathbf{E}\,d^3x$$

$$= \frac{1}{4\pi c}\int \mathbf{E} \times \mathbf{A}\,d^3x. \tag{4-166}$$

The last equality makes use of the solenoidal character of the free field, (4-10), and of the dyadic identity $\nabla\mathbf{r} = \mathbf{1}$, the unit dyadic. The integral over the three-dimensional divergence in (4-164) is identical with the three-space components of (4-159) and vanishes by assumption. Thus, when the surface integral vanishes,

$$\mathbf{J} = \mathbf{L} + \boldsymbol{\Sigma} \tag{4-167}$$

in the reference frame S_0, with $\mathbf{L}$ and $\boldsymbol{\Sigma}$ defined by (4-165) and (4-166).* In general, $\mathbf{L}$ and $\boldsymbol{\Sigma}$ are not separately gauge invariant; only their vector sum, $\mathbf{J}$, depends on the fields alone and not on the potentials. Consequently, $\mathbf{L}$ and $\boldsymbol{\Sigma}$ are, except for special cases, not separately observable.

The vector $\mathbf{J}'$, (4-162), can also be expressed in three-vector notation in S_0.

$$J'_k = -\frac{1}{c}\int (\Theta^{00}_{\text{elm}}x^k - \Theta^{0k}_{\text{elm}}x^0)\,d^3x,$$

$$c\mathbf{J}' = \int U\mathbf{r}\,d^3x - t\int \mathbf{S}\,d^3x. \tag{4-168}$$

If we define the "position of the center of energy" by

$$\mathbf{R} \equiv \frac{\int U\mathbf{r}\,d^3x}{\int U\,d^3x}, \tag{4-169}$$

$\mathbf{J}'$ can be expressed in the notation of $(4\text{-}124)^{(0)}$ through $(4\text{-}126)^{(0)}$, as

$$c\mathbf{J}' = \mathbf{R}W^{(0)} - ct\cdot c\mathbf{P}^{(0)}. \tag{4-170}$$

The physical significance of $\mathbf{J}$ and $\mathbf{J}'$ for the conservation laws is the following. Exactly as in the integral conservation law for the linear four-momentum, (4-128), the differential conservation law, (4-157), for the angular momentum tensor, (4-156), yields

$$\frac{dJ^{\mu\nu}}{d\tau} = 0. \tag{4-171}$$

The total angular momentum tensor of the free field is a constant of the motion. In the particular system S_0 this statement reads

$$\frac{d\mathbf{J}}{dt} = 0 \quad \text{and} \quad \frac{d\mathbf{J}'}{dt} = 0. \tag{4-171}^{(0)}$$

The latter gives, with energy-momentum conservation, (4-129),

$$\frac{d\mathbf{R}}{dt} = \frac{\mathbf{P}^{(0)}_{\text{elm}}}{W^{(0)}_{\text{elm}}/c^2}. \tag{4-172}$$

---

* This three-vector decomposition is completely equivalent to the space-part of the tensor decomposition (4-151), combined with the equality (4-160), as will become apparent from Eqs. (4-173), ff.

This is the *law of the center of gravity* or the *center-of-mass theorem* for the free electromagnetic field: the velocity of the center of gravity is a constant and equal to the total field momentum divided by the total field mass. This law is meaningful because the vanishing of the surface integral (4–159) ensures a certain confinement of the field. Furthermore, the equivalence between energy and gravitational mass (3–29) completely justifies the name "center of gravity" or "center of mass" for $\mathbf{R}$ in (4–169). The total mass of the electromagnetic field can be defined by $M_{\mathrm{elm}} \equiv W_{\mathrm{elm}}^{(0)}/c^2$, so that (4–173) becomes

$$\frac{d\mathbf{R}}{dt} = \frac{\mathbf{P}_{\mathrm{elm}}^{(0)}}{M_{\mathrm{elm}}} . \tag{4-172'}$$

Let us now consider the case when the surface integral (4–159) does *not* vanish. This case might be called "pathological" because it is due to an over-idealization of the actual physical situation: for an actual physical system one can always find a surface for which (4–159) vanishes. In particular, the example *par excellence* of a system with nonvanishing surface integral, the plane wave, is no doubt an oversimplification of reality, though a very useful one. This can also be seen in quantum mechanics where plane waves cannot be described by vectors in Hilbert space, and where wave packets are the correct description.

When the surface integral does not vanish, the equality (4–160) no longer holds. We must therefore return to $M^{\mu\nu}$ and its decomposition (4–151). In the system S_0 the space components of $L^{\mu\nu}$ become

$$L^{ij} = \frac{1}{c} \int (T^{0i} x^j - T^{0j} x^i)\, d^3x = -\frac{1}{4\pi c} \int (F^{0\alpha}\, \partial^i A_\alpha x^j - F^{0\alpha}\, \partial^j A_\alpha x^i)\, d^3x$$

because the off-diagonal elements of $\eta^{\mu\nu}$ vanish. Comparison with (4–165) shows that

$$\mathbf{L} \equiv (L_k), \qquad L_k = L^{ij}\epsilon_{ijk}. \tag{4-173}$$

Similarly, from (4–150),

$$S^{ij} = \frac{1}{4\pi c} \int (F^{0i} A^j - F^{0j} A^i)\, d^3x,$$

and comparison with (4–166) shows that

$$\mathbf{\Sigma} \equiv (\Sigma_k), \qquad \Sigma_k \equiv S^{ij}\epsilon_{ijk}. \tag{4-174}$$

Therefore, we can define

$$\mathbf{M} \equiv (M_k), \qquad M_k = M^{ij}\epsilon_{ijk} \tag{4-175}$$

and find

$$\mathbf{M} = \mathbf{L} + \mathbf{\Sigma} \tag{4-176}$$

in the coordinate system S_0. This equation replaces (4–167) when the surface integral does not vanish. However, it is also valid when (4–167) holds, since its derivation makes no assumptions concerning (4–159). But we now no longer know whether **M** is gauge invariant. This depends on the system under consideration.

The other three components of $L^{\mu\nu}$ and $S^{\mu\nu}$ can be expressed in a way completely analogous to the above. One defines

$$
\begin{aligned}
\mathbf{M}' &\equiv (M_k'), & M_k' &= M^{k0}, \\
\mathbf{L}' &\equiv (L_k'), & L_k' &= L^{k0}, \\
\mathbf{\Sigma}' &\equiv (\Sigma_k'), & \Sigma_k' &= S^{k0},
\end{aligned}
\tag{4–177}
$$

and finds

$$
\mathbf{M}' = \mathbf{L}' + \mathbf{\Sigma}'. \tag{4–178}
$$

$$
c\mathbf{L}' = RW_{elm}^{(0)} - c^2 t P_{elm}^{(0)} - \frac{ct}{4\pi} \int d^2\sigma \cdot \mathbf{EA} + \frac{1}{4\pi} \int d^2\sigma \cdot \mathbf{E}x\varphi, \tag{4–179}
$$

$$
c\mathbf{\Sigma}' = \frac{1}{4\pi} \int \mathbf{E}\phi \, d^3x. \tag{4–180}
$$

PROBLEM 4–14

Prove Eqs. (4–179) and (4–180).

While these equations are gauge dependent and not meaningful in general, they become meaningful in certain special cases. In particular, we shall apply them to a plane wave. For such a wave, $\phi = 0$, $\nabla \cdot \mathbf{A} = 0$ and, since **B** satisfies the wave equation (4–11) and is monochromatic with angular frequency $\omega = kc$,

$$
[\nabla^2 + k^2] \mathbf{B} = 0, \tag{4–181}
$$

it follows that

$$
\mathbf{A} = \frac{1}{k^2} \nabla \times \mathbf{B}. \tag{4–182}
$$

One easily verifies that this is indeed the inverse of (4–20) for a plane wave. With these relations, **L**, **L**′, and **Σ** *become gauge invariant*, and $\mathbf{\Sigma}' = 0$.

Let the plane wave be moving in the positive z-direction and let it be completely polarized. Then

$$
A_x = a \sin(kz - \omega t)/k, \qquad A_y = b \sin(kz - \omega t + \delta)/k. \tag{4–183}
$$

Consequently, $\mathbf{E} = -(1/c)\dot{\mathbf{A}}$ and $\mathbf{B} = \nabla \times \mathbf{A}$ yield the fields

$$
\begin{aligned}
E_x &= a \cos(kz - \omega t), & B_x &= -b \cos(kz - \omega t + \delta), \\
E_y &= b \cos(kz - \omega t + \delta), & B_y &= a \cos(kz - \omega t).
\end{aligned}
\tag{4–184}
$$

The orbital angular momentum **L** vanishes for reasons of symmetry: since **A** is independent of x and y, $(\nabla \mathbf{A}) \cdot \mathbf{E}$ has only a z-component; therefore, the integrand of (4–165) is a vector in the xy-plane. There being no preferred direction in that plane, $\mathbf{L} = 0$.

The surface integral in (4–178) is a component of the surface integral (4–159). One easily sees that this component vanishes. Furthermore, since $\boldsymbol{\Sigma}' = 0$,

$$\frac{d\mathbf{M}'}{dt} = 0 \quad \text{implies} \quad \frac{d\mathbf{L}'}{dt} = 0, \tag{4-185}$$

so that for a plane wave, (4–179) leads again to the center-of-mass theorem (4–172).

Finally, the total angular momentum $\mathbf{M} = \boldsymbol{\Sigma}$, the spin angular momentum, since $\mathbf{L} = 0$. For this, we find

$$\boldsymbol{\Sigma} = \frac{1}{4\pi c} \int \mathbf{E} \times \mathbf{A} \, d^3x = \frac{1}{4\pi\omega} \, \hat{\mathbf{k}} \int ab \sin \delta \, d^3x. \tag{4-186}$$

Since the time average of the total energy of the plane wave is

$$\overline{W} = \frac{1}{8\pi} \int (\overline{\mathbf{E}^2} + \overline{\mathbf{B}^2}) \, d^3x = \frac{1}{8\pi} \int (a^2 + b^2) \, d^3x, \tag{4-187}$$

we find the following ratio:

$$\frac{\boldsymbol{\Sigma}}{\overline{W}} = \hat{\mathbf{k}} \, \frac{2ab}{a^2 + b^2} \, \frac{\sin \delta}{\omega}. \tag{4-188}$$

This formula was first derived for the special case of circular polarization by Max Abraham; it was generalized to spherical waves by Arnold Sommerfeld.

The type of polarization is determined by the relative phase δ. For $\delta = 0$ we have a linearly polarized wave; its spin vanishes. For $\delta = \pi/2$ and $a = b$ we have right-circular polarization (as seen *in* the direction of propagation). The ratio (4–188) then reaches its maximum, viz. $1/\omega$. For $a = b$ and $\delta = -\pi/2$, the ratio is $-1/\omega$, describing left-circular polarization. The other cases are elliptically polarized waves.

The classical result (4–188) can be combined with Max Planck's light quantum hypothesis of the year 1900. If $\overline{W} = \hbar\omega$ where $\hbar$ is $1/(2\pi)$ times Planck's constant h, then right and left circularly polarized light must have a spin angular momentum of $\boldsymbol{\Sigma} = \hbar\hat{\mathbf{k}}$ and $-\hbar\hat{\mathbf{k}}$, respectively. The quantum field theoretical description of electromagnetic waves confirms this conclusion.*

PROBLEM 4–15

Show that when $j^\mu \neq 0$ and the electromagnetic field is not free, Eqs. (4–137) and (4–141) yield

$$-\frac{dJ^{\mu\nu}}{d\tau} = N^{\mu\nu} \equiv \int (x^\mu f^\nu - x^\nu f^\mu) \, d\sigma. \tag{4-189}$$

---

* The classical spin $\boldsymbol{\Sigma}$ corresponds to the expectation value of the quantum-mechanical vector operator for spin.

Give a physical interpretation of this result for the space components as well as for the mixed components.

4-11 GAUGE INVARIANCE

We shall restrict ourselves here to covariant gauges, in particular to the Lorentz gauge. The Lagrangian (4-106) is not invariant under the gauge transformations (4-53) because the interaction term $(1/c)j_\mu A^\mu$ yields an extra term, so that

$$\mathcal{L}' = \mathcal{L} + \frac{1}{c}\, j^\mu \,\partial_\mu \Lambda.$$

However, even if $\mathcal{L}$ were invariant under (4-53), Noether's theorem of Section 3-17 would not apply because $\partial_\mu \Lambda$ is not a set of four parameters but is a set of functions.*

The action integral I, (3-45'), however, can be invariant under these gauge transformations, provided Λ has the appropriate asymptotic behavior. The difference between the transformed integral and the original one is

$$I' - I = \frac{1}{c}\int j^\mu\, \partial_\mu \Lambda\, d^4x = \frac{1}{c}\int \partial_\mu (j^\mu \Lambda)\, d^4x - \frac{1}{c}\int \Lambda\, \partial_\mu j^\mu\, d^4x.$$

The integral over the divergence becomes a surface integral by Gauss's theorem and vanishes if Λ vanishes fast enough asymptotically. Under this condition, the action integral is gauge invariant exactly when

$$\partial_\mu j^\mu = 0, \tag{4-190}$$

since Λ is an arbitrary function within the restriction

$$\Box \Lambda = 0$$

and the above asymptotic condition. Since (4-190) is just the differential law of charge conservation, (4-35), we conclude that under the above conditions for Λ, *the gauge invariance of the action integral is equivalent to charge conservation, and this equivalence holds irrespective of whether the field equations are satisfied.*

When the action integral is gauge invariant, the corresponding field equations will also be gauge invariant. This is trivially satisfied for the Lagrangian

---

* A second theorem by Noether refers to invariance of $\mathcal{L}$ under infinitesimal transformations involving *functions* $\epsilon_1(x), \ldots, \epsilon_p(x)$ rather than p parameters in (3-55) (infinite continuous group). In that case one obtains p identities which involve linear homogeneous functionals of $\mathcal{L}$; an example of this is the left-hand side of (3-56). No conservation laws are obtained when the equations of motion are satisfied. As an example, this theorem can be applied to the gauge invariance of the *free* Lagrangian with the resultant identity $\partial_\mu \partial_\nu F^{\mu\nu} \equiv 0$.

(4-106), which leads to the field equations

$$\partial_\mu F^{\mu\nu} = -\frac{4\pi}{c} j^\nu. \tag{4-51}$$

The consistency of this equation with (4-190) is also trivial since charge conservation is a consequence of (4-51).

The situation is different for the Lagrangian (4-104), which leads to the field equations

$$\Box A^\nu = -\frac{4\pi}{c} j^\nu. \tag{4-46}$$

With the above asymptotic conditions on Λ, the action integral is gauge invariant provided the field equations (4-46) are satisfied. But these field equations require that

$$\partial_\nu A^\nu = 0, \tag{4-25'}$$

in order that they become equivalent to the Maxwell-Lorentz field equations. Thus, with the Lagrangian (4-104), the conservation law (4-190) emerges as a consequence of the field equations together with the Lorenz condition.

PROBLEM 4-16

Prove the above statement that the gauge invariance of the action integral with the Lagrangian (4-104) is assured by the field equations (4-46). The function Λ is assumed to vanish asymptotically so that all surface integrals vanish.

The differential law of charge conservation, (4-190), has an equivalent integral form. To this end, one defines the invariant Q as

$$Q \equiv -\frac{1}{c} \int j^\mu \, d\sigma_\mu. \tag{4-191}$$

It is an invariant because (4-190) assures that the integral is independent of the surface, [cf. Appendix 1 Eqs. (A 1-57) and (A 1-58)]. Thus,

$$\frac{dQ}{d\tau} = +\int \partial_\mu j^\mu \, d\sigma = 0, \tag{4-192}$$

and Q is a constant of the motion, i.e. independent of the proper time τ. In the particular Lorentz frame characterized by $n^\mu = (1; 0, 0, 0)$,

$$Q = \int \rho \, d^3 x, \tag{4-193}$$

in view of (4–43). The invariant Q can therefore be interpreted as the total charge of the system. Equation (4–192) is the (integral) *law of charge conservation.*

4–12 CONFORMAL INVARIANCE

As was first observed by Cunningham (1909) and by Bateman (1910), the Maxwell equations are invariant under a larger group of transformations than the Lorentz group, viz. under the 15-parameter *conformal group.* It was later found that this invariance property is in fact characteristic of all fields which propagate along light cones and of all particles of vanishing rest mass. Conformal invariance does not hold for particles of nonvanishing rest mass in Minkowski space.* This prevented the adoption of conformal invariance as the fundamental space-time symmetry property in place of Lorentz invariance.

The conformal group contains the Lorentz group as a subgroup. The additional five parameters can be interpreted as one scaling parameter and four parameters which constitute a four-vector of acceleration. The latter occur in the characteristic *acceleration transformation* of the conformal group; they transform to uniformly accelerated motion (Sections 5–3 and 6–11).

The conformal invariance of the Lagrangian gives rise to 15 conservation laws in accordance with Noether's theorem, 10 of which are identical with those derived above on the basis of Lorentz invariance. The other 5 have no simple physical interpretation.

Since general relativity admits all point transformations, the conformal transformations are also included in them, although in somewhat disguised form. One can show that the acceleration transformations correspond to a transformation from an inertial system to a nonstatic apparent gravitational field which, in the nonrelativistic limit, is a constant homogeneous gravitational field (Section 8–3).

While the physical contents of conformal invariance is thus contained in general relativity as a special case, the physical situation singled out by this invariance property does not seem to be of sufficient simplicity or importance to warrant special attention in a theory with nonvanishing rest masses. The interested reader is referred to the original literature.‡

---

* In a more general space, (Weyl space), one can have a conformal theory also for particles with $m \neq 0$. The rest mass is then no longer an invariant.

‡ Conformal invariance in physics was reviewed by T. Fulton, F. Rohrlich, and L. Witten, *Rev. Mod. Phys.* **34**, 442 (1962), in which references to other papers can be found. The physical meaning of the acceleration transformation is discussed by the same authors in *Nuovo Cim.* **26**, 652 (1962).

CHAPTER **5**

Electromagnetic Radiation

While in principle it is possible to consider all electromagnetic fields simply as intermediaries, transmitting interactions between charges, they are no doubt useful concepts to be studied in their own right, irrespective of the charges involved in their production or absorption. The *radiation* fields, however, have importance which goes beyond that. Radiation has so many of the physical attributes associated with matter that it receives a status almost *at par* with it in classical physics. In quantum physics this status is elevated to full emancipation when the photon emerges as a full-fledged particle, just as "fundamental" as any charged particle of matter. The present chapter takes account of this situation and is devoted to electromagnetic radiation in its own right.

5–1 MOMENTUM AND ENERGY OF RADIATION

Given the world line of a point charge $z^\mu(\tau)$, the electromagnetic field strengths $F^{\mu\nu}(x)$ due to that charge, at an arbitrary point x in spacetime, are determined by the Maxwell-Lorentz equations and the retardation condition.* They are given by Eq. (4–98) in terms of the four-velocity v^μ and the four-acceleration a^μ,

$$F^{\mu\nu}(x) = \frac{2e}{\rho^2} v^{[\mu}u^{\nu]} + \frac{2e}{\rho} \{a^{[\mu}v^{\nu]} - u^{[\mu}(v^{\nu]}a_u + a^{\nu]})\}. \qquad (5-1)$$

Here we use units such that $c = 1$ and the notation

$$a^{[\mu}b^{\nu]} \equiv \tfrac{1}{2}(a^\mu b^\nu - a^\nu b^\mu), \qquad (5-2)$$

which often helps to prevent certain expressions from becoming unwieldy.

Our aim is to express the energy and momentum carried by the radiation in terms of the dynamical variables of the moving source charge. We start, therefore, with the electromagnetic energy-momentum four-vector P^μ_{elm} defined in (4–123),

$$P^\mu_{\text{elm}} = \frac{1}{c} \int_{\text{spacelike } \sigma} \Theta^{\mu\nu}_{\text{elm}} \, d^3\sigma_\nu, \qquad (4-123')$$

where the integration extends over a spacelike plane. The notation $d^3\sigma^\mu$ indicates the three-dimensional nature of this plane explicitly. As was discussed

---

* In the present chapter all fields are retarded, so that the subscript "ret" used in Chapter 4 can be dropped without causing ambiguity.

in Section 4–7, this integral describes the resultant effect of fields emitted at different times from the same charge (see Fig. 4–2). If we want the energy and momentum of the fields produced at one given instant τ_Q, we must restrict the integral to the intersection of the future light cone from Q with the plane σ. The energy and momentum produced by the charge in the form of electromagnetic fields during the proper time interval $\Delta\tau$ will be given by the integral (4.123′) restricted* to the annulus between two hyperellipses on the spacelike plane σ. They are determined by the intersection of σ with the light cones at $Q(\tau)$ and at $Q(\tau + \Delta\tau)$ (see Fig. 5–1). We shall denote the result of this integration by ΔP^μ.

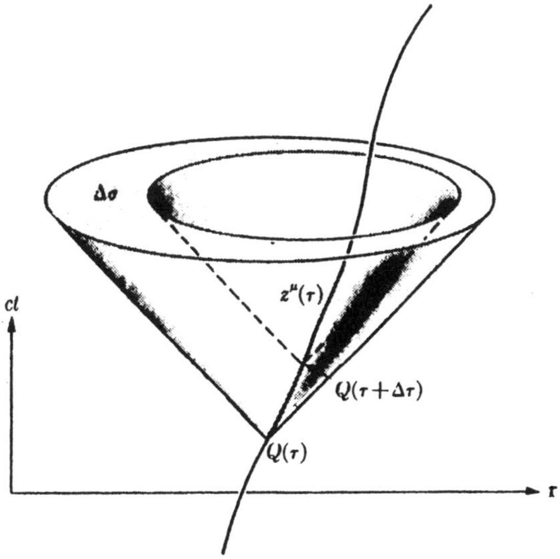

FIG. 5–1. Intersection of the spacelike plane σ with the future light cones from $Q(\tau)$ and $Q(\tau + \Delta\tau)$.

In order that ΔP^μ be a physically meaningful quantity, it must be the same (apart from a Lorentz transformation) for two inertial observers characterized by two different planes σ and σ'. Thus, if it is possible to show that

$$\Delta P^\mu = \frac{1}{c} \int_{\Delta\sigma} \Theta^{\mu\nu}_{\text{elm}} \, d^3\sigma_\nu \tag{5–3}$$

is independent of the choice of $\Delta\sigma$, then P^μ will be a four-vector. To this end Eqs. (A1–57) and (A1–58) are of no help to us, since we are not dealing with an integral over the entire plane σ.

Consider two surfaces σ and σ' which intersect the world line at S and S' and

---

* Equation (4–123′) is meaningful for free electromagnetic fields. In the presence of a charge the integral must be restricted since it diverges when the world line of the charge crosses σ.

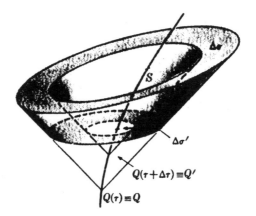

FIG. 5–2. Intersection of the future light cones from $Q(\tau)$ and $Q(\tau + \Delta\tau)$ with the spacelike planes orthogonal to the tangent vectors at S and S'. (Recall the definition of orthogonality in Minkowski space according to Fig. A1–1.)

which intersect the light cones so as to form (three-dimensional) annuli $\Delta\sigma$ and $\Delta\sigma'$ (Fig. 5–2). These two annuli, together with the two light cones, form the boundary surface Σ of a four-dimensional doubly connected volume V_4. If ΔP^μ is independent of $\Delta\sigma$, then

$$\Delta P^\mu(\Delta\sigma) = \Delta P^\mu(\Delta\sigma'). \tag{5–4}$$

In order to test this equality we use Gauss's theorem on V_4:

$$-\int_{V_4} \partial_\nu \Theta^{\mu\nu}_{\mathrm{elm}} \, d^4x = \int_\Sigma \Theta^{\mu\nu}_{\mathrm{elm}} \, d^3\sigma_\nu. \tag{5–5}$$

The differential energy-momentum conservation law, (4–115), ensures the vanishing of the left side of this equality. Therefore

$$\Delta P^\mu(\Delta\sigma) = \Delta P^\mu(\Delta\sigma') + \frac{1}{c}\int_{\Delta(\text{light cone } Q')} \Theta^{\mu\nu}_{\mathrm{elm}} \, d^3\sigma_\nu - \frac{1}{c}\int_{\Delta(\text{light cone } Q)} \Theta^{\mu\nu}_{\mathrm{elm}} \, d^3\sigma_\nu. \tag{5–6}$$

The integrals refer to those parts of Σ which consist of portions of the light cones emerging from Q and Q'. The equality (5–4) will hold exactly when

$$\int_{\Delta(\text{light cone } Q')} \Theta^{\mu\nu}_{\mathrm{elm}} \, d^3\sigma_\nu = \int_{\Delta(\text{light cone } Q)} \Theta^{\mu\nu}_{\mathrm{elm}} \, d^3\sigma_\nu. \tag{5–7}$$

As first shown by Schild,* Eq. (5–7) can be proven to hold in the limit $\Delta\sigma \to \infty$, $\Delta\sigma' \to \infty$, so that (5–4) is also valid in this limit. By $\Delta\sigma \to \infty$ we

---

* A. Schild, *J. Math. Anal. and Appl.* **1**, 127 (1960).

mean that $\Delta\sigma$ is to be found as the intersection of the (fixed) light cones and the surface $\sigma(\tau)$ in the limit $\tau \to +\infty$. The proof consists in showing that each side of (5–7) vanishes in this limit. To see this we express $\Theta_{\text{elm}}^{\mu\nu}$ in terms of v^μ and a^μ by substituting (5–1) into (4–114). A simple calculation yields

$$\Theta_{\text{elm}}^{\mu\nu} = \frac{e^2}{4\pi\rho^4}\left(u^\mu u^\nu - \frac{v^\mu v^\nu}{c^2} - \tfrac{1}{2}\eta^{\mu\nu}\right) + \frac{e^2}{2\pi\rho^3 c^2}\left[a_u \frac{R^\mu R^\nu}{\rho^2} - \left(\frac{v^{(\mu}a_u}{c} + a^{(\mu}\right)\frac{R^{\nu)}}{\rho}\right]$$

$$+ \frac{e^2}{4\pi\rho^2 c^4}(a_u^2 - a_\lambda a^\lambda)\frac{R^\mu R^\nu}{\rho^2}, \tag{5–8}$$

where

$$a^{(\mu}b^{\nu)} \equiv \tfrac{1}{2}(a^\mu b^\nu + a^\nu b^\mu),$$

in analogy with (5–2).

PROBLEM 5–1

Prove (5–8).

The three terms in (5–8) behave like ρ^{-4}, ρ^{-3}, and ρ^{-2}, respectively, in the limit $\Delta\sigma \to \infty$. The volume element $d^3\sigma^\mu$ on the light cone is

$$d^3\sigma^\mu = R^\mu\, d^2\omega, \tag{5–9}$$

where $d^2\omega$ is an invariant and the null vector R^μ is defined by (4–86). This is proven in Appendix 1 [(cf. (A1–66)]. Since $d^2\omega$ does not depend on ρ, and R^μ is given by (4–90), it follows that the integrals in (5–7) of the above three terms in $\Theta_{\text{elm}}^{\mu\nu}$ behave like ρ^{-3}, ρ^{-2}, and ρ^{-1}, respectively. Since the limit $\Delta\sigma \to \infty$ requires $\rho \to \infty$, both sides of (5–7) indeed vanish.*

It is clear that this proof does *not* depend on the spacelike character of $\Delta\sigma$. If the surfaces $\Delta\sigma$ and $\Delta\sigma'$ were, as in Fig. 5–3, intersections of the light cones with (timelike!) cylindrical surfaces surrounding the world line of the charge, the proof would still carry through. Thus, what is really proven here is that as one travels with the fields along the light cone, the flux of energy and momentum of the electromagnetic field which is produced between Q and Q' is more and more confined to the region between the light cones until, in the limit $\rho_{\min} \to \infty$, there is no flux at all in any direction which crosses the bounding light cones: the flux through $\Delta\sigma'$ is the same as the flux through $\Delta\sigma$ in the limit $\rho_{\min} \to \infty$.

Applying this result to (5–4), we find that

$$dP_{\text{rad}}^\mu \equiv \lim_{\rho_{\min}\to\infty} \frac{1}{c}\int \Theta_{\text{elm}}^{\mu\nu}\, d^3\sigma_\nu \tag{5–10}$$

$$d\sigma \text{ (timelike or spacelike)}$$

---

* More precisely, since ρ has different values for different points on $\Delta\sigma$, $\Delta\sigma \to \infty$ means $\rho_{\min} \to \infty$.

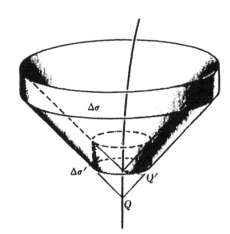

FIG. 5-3.   Intersection of the
future light cones from Q and Q'
with timelike surfaces.

is a four-vector. This expression is to be compared with (4-123′) on page 106
and the different physical meaning is to be noted.

In order to evaluate (5-10) we choose $d\sigma$ to be the (timelike) surface which is
a straight circular cylinder in the rest system of the charge,

$$d^3\sigma^\mu = u^\mu \rho^2 \, d\Omega \, c \, d\tau. \tag{5-11}$$

The element of solid angle is here denoted by $d\Omega$. If we further refer (5-10) to
the instantaneous rest frame of the charge, S_0, it can be expressed by means of
$u^\mu = (0, \hat{u})$ and Eqs. (4-117) and (4-119),*

$$\frac{dP_{rad}^{\mu(0)}}{d\tau} = \lim_{R\to\infty} \left(\frac{1}{c} \int \mathbf{S}^{(0)} \cdot \hat{u} R^2 \, d\Omega; \ -\int \mathbf{T}^{(0)} \cdot \hat{u} R^2 \, d\Omega \right). \tag{5-10$^{(0)}$}$$

The time component is just the integral of the normal component of the Poynting
vector over the surface of the three-dimensional light sphere of radius R, as given
in elementary discussions of radiation rates. The above equation shows clearly
that this is valid only in the system S_0.

Returning to an arbitrary reference frame but still integrating over the special
surface (5-11), we see that the energy tensor can be expressed in terms of the
kinematics of the charge, as in (5-8), yielding

$$\frac{dP_{rad}^\mu}{d\tau} = \lim_{\rho\to\infty} \frac{e^2}{4\pi c^4} \int (a_\lambda a^\lambda - a_u^2) \frac{R^\mu R^\nu u_\nu}{\rho^2} \, d\Omega. \tag{5-12}$$

The integral is *independent* of ρ, so that the limit symbol can be omitted. Using
Eqs. (4-90), we find the integral

$$\int (a_\lambda a^\lambda - a_u^2) \left(u^\mu + \frac{v^\mu}{c} \right) d\Omega.$$

---

* As usual, the semicolon separates the ($\mu = 0$)-component from the others ($\mu =$
1, 2, 3).

Since we have proven it to be a vector, it must differ exactly by a Lorentz transformation from its value in S_0, the rest system of Q. In that system it is easily computed:

$$\left(\int [\mathbf{a}^2 - (\mathbf{a} \cdot \hat{\mathbf{u}})^2] \, d\Omega; \int [\mathbf{a}^2 - (\mathbf{a} \cdot \hat{\mathbf{u}})^2] \hat{\mathbf{u}} \, d\Omega \right) = (\tfrac{2}{3} a^2 4\pi; 0).$$

Therefore, in the original system, S, the integral is

$$\frac{2}{3} \frac{v^\mu}{c} a_\lambda a^\lambda 4\pi,$$

and we have the final result*

$$\frac{dP^\mu_{\text{rad}}}{d\tau} = \frac{2}{3} \frac{e^2}{c^5} a_\lambda a^\lambda v^\mu. \tag{5-13}$$

This is the momentum four-vector *rate* (relative to the proper time of the charge) at which radiation is leaving the charge. It is a timelike vector of dimensions "momentum per unit time." That Eq. (5-12) involves only radiation fields follows from the fact that only the $(1/\rho)$-terms in $F^{\mu\nu}$ contribute to the integral in the limit. At the same time, it is seen that only for the radiation fields does $dP^\mu/d\tau$ have a nonvanishing value in the limit $\rho \to \infty$. The radiation field *detaches* itself from the charge which is its source and leads an independent existence; it is endowed with energy, momentum, and, as we saw in Section 4–10, at times also with angular momentum. In contradistinction, the velocity fields are permanently "attached" to the charge and are carried along with it; they have neither energy nor momentum at large distances from their source.

The momentum four-vector rate of radiation (5–13) is closely related to the energy rate of radiation emitted by the charge. This quantity is defined by

$$\mathfrak{R} \equiv -v_\mu \frac{dP^\mu_{\text{rad}}}{d\tau} = \frac{2}{3} \frac{e^2}{c^3} a_\lambda a^\lambda. \tag{5-14}$$

It is a Lorentz invariant and constitutes the relativistic generalization of the famous nonrelativistic *Larmor formula*

$$\frac{dW_{\text{rad}}}{dt} = \frac{2}{3} \frac{e^2}{c^3} \mathbf{a}^2.$$

As is obvious from (5–12), Eq. (5–14) can also be expressed by the zero component of $P^\mu_{\text{rad}} = (W_{\text{rad}}/c, \mathbf{P}_{\text{rad}})$,

$$\frac{dW_{\text{rad}}}{dt} = \frac{dP^0_{\text{rad}} \, c}{\gamma \, d\tau} = \frac{2}{3} \frac{e^2}{c^5} a_\lambda a^\lambda \frac{v^0 c}{\gamma} = \mathfrak{R}. \tag{5-15}$$

The physical meaning of the *invariant radiation rate* $\mathfrak{R}$ follows from this equation.

---

* It should be evident that despite its appearance, the left-hand side is in general *not* a total derivative.

The rate at which momentum and energy leave the charge in the form of radiation, as given by (5–13) and (5–15), can be used as a criterion for the presence or absence of radiation emission. Since $a_\lambda a^\lambda$ is a positive definite invariant (it vanishes if and only if $a^\lambda = 0$), $\mathcal{R}$ is positive definite, and one concludes that *a charge emits radiation relative to a Lorentz observer if and only if it is accelerated* ($a_\lambda a^\lambda \neq 0$). This is a Lorentz invariant criterion.

The integrand of (5–12) is of considerable interest. Since it is independent of ρ, the momentum rate per unit solid angle emitted in a given direction characterized by the unit vector u^μ is

$$\frac{dP^\mu_{\text{rad}}}{d\tau\, d\Omega} = \frac{e^2}{4\pi c^4}\,(a_\lambda a^\lambda - a_u^2)\left(u^\mu + \frac{v^\mu}{c}\right) = \frac{e^2}{4\pi c^4}\,a_\lambda^\perp a_\perp^\lambda\left(u^\mu + \frac{v^\mu}{c}\right). \tag{5–16}$$

The vector

$$a_\perp^\mu \equiv a^\mu - a_u u^\mu, \qquad a_\perp^\mu u_\mu = 0, \tag{5–17}$$

is the component of a^μ which is orthogonal to the direction u^μ.

The expression (5–16) is a four-vector. In fact, it is a *null vector*, as follows from (4–89). In the particular coordinate system S_0 in which $u^\mu = (0, \hat{u})$, the charge is at rest, and $a_\perp^\mu = (0; \mathbf{a}_\perp)$,

$$\frac{dP^{\mu(0)}_{\text{rad}}}{d\tau\, d\Omega} = \left(\frac{dW^{(0)}_{\text{rad}}}{c\,dt\,d\Omega}, \frac{d\mathbf{P}^{(0)}_{\text{rad}}}{dt\,d\Omega}\right) = \left(\frac{e^2}{4\pi c^4}\,\mathbf{a}_\perp^2;\ \frac{e^2}{4\pi c^4}\,\mathbf{a}_\perp^2\hat{u}\right), \tag{5–16}$$

$$\mathbf{a}_\perp = \hat{u} \times (\mathbf{a} \times \hat{u}).$$

From these results we learn that no radiation is *emitted in the direction of acceleration;* also, the three-momentum flux into a given solid angle in a given direction is just $1/c$ times the energy flux into that angle and in that direction. If the latter is given by Planck's quantum $\hbar\omega$, the momentum flux is $(\hbar\omega/c)\hat{u}$. Thus, if one associates a particle with this energy and this momentum, the fundamental relation of special relativity,

$$E = [(mc^2)^2 + (\mathbf{p}c)^2]^{1/2}$$

for a free particle of mass m, requires that the quantum have vanishing rest mass and that it move with the velocity of light,

$$v \equiv \frac{|\mathbf{p}|c}{E} = c. \tag{5–18}$$

The existence of the photon can therefore be deduced from Planck's hypothesis of 1900 combined with the special theory of relativity.*

---

* Note that the contribution of the *velocity* fields to the integrand of (5–10) yields spacelike four-vectors in a given direction and these cannot be interpreted as particle momenta.

5-2 LOCAL CRITERION FOR RADIATION

In the previous section we established a simple criterion which is necessary and sufficient for the emission of radiation by a charge as seen by a Lorentz observer: the presence of acceleration $(a_\lambda a^\lambda \neq 0)$ at the instant τ implies the emission of radiation at that instant.

When only the velocity of the charge at the instant τ is known, and not its acceleration, emission of radiation can be ascertained by a measurement of the field strengths. One follows the future light cone from the event of emission; the field strengths will asymptotically have a $(1/\rho)$-dependence if and only if they are radiation fields.

This criterion is not always convenient. The asymptotic region (wave zone) is reached when $\rho \gg \lambda$, the wave length of the radiation. Therefore, for any finite distance from the source there are wavelengths $\lambda \gtrsim \lambda_c$ for which a measurement at that distance is insufficient to test the presence of radiation fields.

However, a *local* criterion is possible in the following way.[*] Consider the expression $v_\mu \Theta_{elm}^{\mu\nu} \, d\sigma_\nu$ with $\Theta_{elm}^{\mu\nu}$ given by (5-8) and $d\sigma_\nu$ by (5-11). Now

$$v_\mu \Theta_{elm}^{\mu\nu} u_\nu = \frac{e^2}{4\pi\rho^2 c^4} (a_\lambda a^\lambda - a_u^2), \qquad (5\text{-}19)$$

and all the terms proportional to ρ^{-4} and ρ^{-3} vanish. Thus, the integral

$$\int v_\mu \Theta_{elm}^{\mu\nu} u_\nu \rho^2 \, d\Omega = \frac{e^2}{4\pi c^4} \int (a_\lambda a^\lambda - a_u^2) \, d\Omega$$

is independent of ρ. On the other hand, the invariant radiation rate $\mathfrak{R}$, defined in (5-14), is given, according to (5-10), exactly by the limit $\rho \to \infty$ of this integral. Therefore,

$$\mathfrak{R} = \int v_\mu \Theta_{elm}^{\mu\nu} u_\nu \rho^2 \, d\Omega, \qquad (5\text{-}20)$$

where the integral extends over the surface of the sphere of *arbitrary* radius ρ. In particular, ρ can be chosen arbitrarily small; the sphere can lie in the near zone $\rho \ll \lambda$. We also note that (5-20) reduces to the zero component of (5-10)[(0)] (times a factor c) when referred to the frame S_0.

This result was obtained by use of the special surface element (5-11) of a sphere. It must therefore be interpreted as follows. If at time t_0 a charge passes through the center of a sphere of radius ρ which is at rest with respect to an inertial observer S, this observer can make a measurement of the $F_{\mu\nu}$ on the surface of that sphere at the time $t = t_0 + \rho/c$. He can then compute the expression (5-20) and he will find the radiation rate $\mathfrak{R}$ which has exactly the

---

[*] This criterion was discussed by the author in *Nuovo Cimento* **21**, 811 (1961) and also in his Boulder Lectures: *Lectures in Theoretical Physics*, Vol. II, W. E. Brittin and B. W. Downs, Interscience, New York, 1960.

value of the *invariant* radiation rate as defined in (5–14). The vanishing of $\mathfrak{R}$ is a necessary and sufficient condition that no radiation was emitted at time t_0 and that, consequently, the charge was not accelerated at that time.

While this experiment is perhaps not easy to carry out in the laboratory, it does give an *operational criterion* for radiation which is *local* and does not involve the wave zone. It is important that the radius of the sphere in (5–20) is arbitrary and that the result is the invariant rate.

As a general comment on the above discussion of radiation we note that we have two different points of view, both contained in the same equation (5–10); in one case, the energy tensor is expressed in terms of the field strengths, (4–114), as in Eq. (5–10)$^{(0)}$; in the other, it is expressed in terms of the velocity and acceleration of the source, (5–8), as in Eqs. (5–13), (5–14), and (5–16). Correspondingly, $dP^{\mu}_{\rm rad}/d\tau$ can be interpreted as the momentum rate of the radiation field crossing a distant surface [or a spherical surface at *any* distance, in the special case of Eq. (5–20)]; or it can be interpreted as the rate at which the source gives off momentum in terms of radiation. These two points of view are equivalent. However, it is the latter point of view, especially as expressed by (5–13), which is of crucial dynamical significance for the motion of the source. It represents the reaction of the emitted radiation on the moving charge and will be of major importance for the equations of motion (Chapter 6, Section B).

In the remainder of the present chapter we shall give a few examples of radiation emitted by a charge in motion. These motions are of very special nature and deserve our attention for various reasons. The discussions will be brief, emphasizing certain characteristic features. No attempt will be made to be exhaustive, because it is our purpose here to discuss fundamental points of the theory and not its applications. The latter are legion and many of them have found excellent presentations in standard references.

5–3 RADIATION FROM UNIFORMLY ACCELERATED CHARGES

This problem is of interest because it has received considerable attention in the original literature with many contradictory results. For a more complete discussion of the history of this problem we refer to a recent paper.* Here we shall give only a brief outline with emphasis on the pitfalls which proved to be the source of the longstanding confusion.

Uniformly accelerated motion is defined nonrelativistically as motion of a point mass under a force which produces a constant acceleration, $\mathbf{a} = $ const. According to Newton's second law, this requires a constant force, $\mathbf{F} = $ const. As an example of this very simple situation, one thinks of the gravitational field which provides bodies with a gravitational acceleration, g, that is approximately constant (inside a laboratory, say). But here is the first pitfall: Newton's laws are valid relative to an inertial frame of reference and a laboratory is *not* an

---

* T. Fulton and F. Rohrlich, *Ann. Phys.* (N.Y.) **9**, 499 (1960).

inertial frame of reference; it is supported in a gravitational field. The *falling* object is (within the dimensions of the laboratory) an inertial system, so that an object at rest in the laboratory will be seen by a freely falling observer to undergo uniform acceleration. The inverse is not generally valid.*

The realization of this situation becomes crucial when one wants to apply electromagnetic theory. Maxwell's equations *must* be referred to an inertial system when written in their usual form of special relativity. Otherwise, the general theory of relativity must be used.

In special relativity, the definition of uniform acceleration is more subtle than the one given above. Here one must refer to the instantaneous Lorentz frames. The acceleration four-vector must be constant relative to those frames. This requirement specifies uniform acceleration unambiguously. Consider the world line of a point mass, $z^\mu(\tau)$. Let τ_1 and τ_2 be two instants of time on the interval $0 < \tau < T$. Let S_1 and S_2 be the Lorentz frames in which the point mass is at rest at τ_1 and τ_2, respectively. Let $a_1^\mu(\tau_1)$ be $a^\mu(\tau)$ relative to S_1 at $\tau = \tau_1$; let $a_2^\mu(\tau_2)$ be $a^\mu(\tau)$ relative to S_2 at $\tau = \tau_2$. If $a_1^\mu(\tau_1) = a_2^\mu(\tau_2)$ for all $0 < (\tau_1, \tau_2) < T$, the motion for $0 < \tau < T$ is uniformly accelerated. These statements remain valid if, instead of S_1 and S_2, one uses S_1' and S_2' which differ from the unprimed frames by a Lorentz transformation.

In three-vector notation, as will be shown below,

$$a_2^\mu(\tau_2) = (0; \mathbf{a}_2), \qquad a_1^\mu(\tau_1) = (0; \mathbf{a}_1);$$

therefore, when $\tau_2 = \tau_1 + d\tau$, the definition of uniform acceleration, $a_2^\mu(\tau_2) = a_1^\mu(\tau_1)$, can be stated as

$$\mathbf{b} \equiv \frac{d\mathbf{a}}{dt} = 0 \tag{5-21}$$

in the rest system of the particle. In that system, however, [see (5-27) below]

$$\dot{a}^\mu = \left(\frac{1}{c}\, \mathbf{a}^2; \mathbf{b}\right).$$

A covariant form of (5-21) emerges from the observation that in the rest frame $v^\mu = (c; 0)$, so that

$$(0; \mathbf{b}) = \dot{a}_\perp^\mu,$$

where $\dot{a}_\perp^\mu$ is that component of $\dot{a}^\mu$ which satisfies

$$\dot{a}_\perp^\mu v_\mu = 0. \tag{5-22}$$

Since this is an invariant relation and $\dot{a}_\perp^\mu$ is a four-vector, the covariant form of (5-21) is, for any Lorentz frame,

$$\dot{a}_\perp^\mu = 0. \tag{5-23}$$

---

* A freely falling object viewed by an observer supported in a homogeneous gravitational field is not seen to undergo uniform acceleration in general. [F. Rohrlich, *Ann. Phys.* (N.Y.) **22**, 169 (1963)]. See also Section 8-3.

The identity $\qquad\qquad v_\mu \dot{a}^\mu \equiv -a_\mu a^\mu,$

which follows from $v_\mu a^\mu \equiv 0$ by differentiation, ensures that

$$\dot{a}_\perp^\mu = \dot{a}^\mu - a^\lambda a_\lambda v^\mu/c^2. \qquad (5\text{–}24)$$

The definition of uniform acceleration, (5–21) or (5–23), can therefore be stated alternatively as

$$\dot{a}^\mu - a^\lambda a_\lambda v^\mu/c^2 = 0. \qquad (5\text{–}23')$$

It remains to justify the above three-vector forms of a^μ and $\dot{a}^\mu$. If one uses three-vectors $\mathbf{v}(t)$, $\mathbf{a}(t) = d\mathbf{v}(t)/dt$, and $\mathbf{b}(t) = d\mathbf{a}/dt$ [see also (A1–35) through (A1–38)],

$$v^\mu = (\gamma c; \gamma\mathbf{v}), \qquad (5\text{–}25)$$

$$a^\mu = (\gamma^4\mathbf{v}\cdot\mathbf{a}/c; \gamma^2\mathbf{a} + \gamma^4\mathbf{v}\cdot\mathbf{a}\,\mathbf{v}/c^2), \qquad (5\text{–}26)$$

$$\dot{a}^\mu = (\dot{a}^0; \gamma^3\mathbf{b} + 3\gamma^5\mathbf{v}\cdot\mathbf{a}\,\mathbf{a}/c^2 + \dot{a}^0\mathbf{v}/c),$$
$$\dot{a}^0 = \gamma^5(\mathbf{v}\cdot\mathbf{b} + \mathbf{a}^2)/c + 4\gamma^7(\mathbf{v}\cdot\mathbf{a})^2/c^3. \qquad (5\text{–}27)$$

For $\mathbf{v} = 0$ these relations reduce to the form used previously. Furthermore, substitution into (5–23') yields*

$$\mathbf{b} + 3\gamma^2\mathbf{v}\cdot\mathbf{a}\,\mathbf{a}/c^2 = 0. \qquad (5\text{–}23'')$$

Equations (5–23), (5–23'), and (5–23'') are all equivalent.

PROBLEM 5–2

Prove Eqs. (5–26) and (5–27).

PROBLEM 5–3

Substitute (5–25) through (5–27) into (5–23') and prove that (5–23') and (5–23'') are equivalent.

PROBLEM 5–4

Prove (5–23'') directly from the definition of uniform acceleration; using the Lorentz transformation in three-vector notation from S to S' with relative velocity $\mathbf{u}$, show that the statement "$\mathbf{b}' = 0$ in the instantaneous rest frame S' ($\mathbf{v}' = 0$, $\mathbf{u} = \mathbf{v}$)" is equivalent to (5–23'').

Equation (5–23'') can be integrated with respect to t,

$$\gamma^3\mathbf{a} = \mathbf{g}, \qquad (5\text{–}28)$$

---

* The conformal group mentioned in Section 4–12 is the largest group of transformations which leave (5–23) invariant.

where **g** is a constant vector. If the initial velocity is parallel or antiparallel to **g**, then **v**, **a**, and **g** will always be collinear. In this special case, (5–28) can be written

$$\frac{d(\gamma \mathbf{v})}{dt} = \mathbf{g}. \tag{5-29}$$

This is just the equation of motion

$$\frac{d\mathbf{p}}{dt} = \mathbf{F}, \tag{5-30}$$

with $\mathbf{p} = m\gamma\mathbf{v}$ and $\mathbf{F} = m\mathbf{g}$ related to the corresponding four-vectors by [cf. Eqs. (A1–39) and (A1–49)]

$$p^{\mu} = \left(\frac{1}{c} E; \mathbf{p}\right),$$

$$\tag{5-31}$$

$$F^{\mu} = \left(\frac{1}{c} \gamma\mathbf{v}\cdot\mathbf{F}; \gamma\mathbf{F}\right).$$

F is a constant force.* In this collinear case, therefore, uniform acceleration is produced by a constant force also in relativistic dynamics.

If **g** is in the z-direction, Eq. (5–29) can be integrated to yield

$$x = y = 0, \qquad z = \sqrt{\alpha^2 + c^2 t^2}, \qquad \alpha = c^2/|\mathbf{g}|, \tag{5-32}$$

provided we choose $z(0) = \alpha$ and $v(0) = 0$. This equation is one branch of a hyperbola in the $(z\text{-}ct)$-plane. Hence, uniformly accelerated motion is often referred to as *hyperbolic motion* (see Fig. 5–4).

Assume now that the point mass which performs hyperbolic motion carries a charge e. The corresponding current is given by

$$j^{\mu}(x) = ec \int_{-\infty}^{\infty} \delta(x - z(\tau)) v^{\mu}(\tau)\, d\tau, \tag{4-82}$$

with

$$z^{\mu} = (ct; 0, 0, z), \qquad v^{\mu} = (\gamma c, \gamma\mathbf{v}), \qquad \mathbf{v} = \frac{c^2 t \hat{\mathbf{k}}}{\sqrt{\alpha^2 + c^2 t^2}} \tag{5-33}$$

and z given by (5–32).

The electromagnetic fields produced by this current follow from our general expression (4–98). The somewhat tedious calculations need not be presented

---

* Since g could be a function of the rest mass $\dot{m}$, **F** is in general not proportional to the mass. For example, if **F** is due to an electrostatic field, it is independent of m, while **g** is proportional to $1/m$.

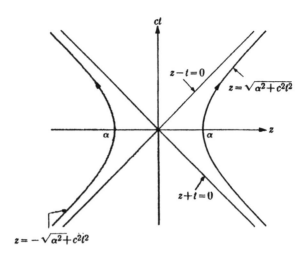

FIG. 5–4. Hyperbolic motion.

here. The result* is in cylindrical coordinates† ρ, φ, z,

$$E_\rho = 8e\alpha^2\rho z/\xi^3, \qquad E_\varphi = 0,$$
$$E_z = -4e\alpha^2[\alpha^2 + (ct)^2 + \rho^2 - z^2]/\xi^3; \tag{5–34}$$

$$B_\rho = B_z = 0, \qquad B_\varphi = 8e\alpha^2\rho ct/\xi^3. \tag{5–35}$$

In these equations ξ stands for

$$\xi \equiv \{4\alpha^2\rho^2 + [\alpha^2 + (ct)^2 - \rho^2 - z^2]^2\}^{1/2}. \tag{5–36}$$

The above fields were first obtained in 1909 by Max Born. They represent the *retarded* fields produced by the point charge whose motion is uniformly accelerated, (5–32). However, such a charge can produce retarded fields only in the space-time region $z + t > 0$, while the above fields have nonvanishing values also for $z + t < 0$ (Fig. 5–4), as can easily be verified. This fact has caused some concern until it was recognized that these same equations, (5–34) and (5–35), also describe the *advanced* fields of a charge $-e$ moving according to

$$z = -\sqrt{\alpha^2 + c^2t^2}. \tag{5–37}$$

This represents the other branch of the hyperbola. The origin of the difficulty lies in the fact that the field strengths (5–34) and (5–35) do not contain the

---

* T. Fulton and F. Rohrlich, *Ann Phys.* (N.Y.) **9**, 499 (1960).

† The coordinate $\rho = (x^2 + y^2)^{1/2}$ is not to be confused with the invariant $\rho = u_\lambda R^\lambda$ introduced in Eq. (4–90). The invariant ρ is used nowhere in this section. Note also that z and t are independent variables referring to a field point; they are not restricted by (5–32), which characterizes the location of the point charge. The coordinates of the latter should be appropriately labeled, such as t_Q (retarded time), when they occur in the same equation as the field point: $z^\mu = (ct_Q; \rho_Q, \varphi_Q, z_Q)$, $x^\mu = (ct; \rho, \varphi, z)$.

restriction $z + ct > 0$. This restriction is *necessary*, however, in order to ensure that the field strengths are retarded. It must therefore be added to these equations to provide the complete retarded solution.

PROBLEM 5-5

Show that for the motion (5-32) the quantity $R^0 \equiv x^0 - z^0$ of (4-86) is given by

$$R^0 = \frac{1}{2} \frac{[z\xi - ct(\alpha^2 + c^2t^2 + \rho^2 - z^2)]}{z^2 - c^2t^2} .$$

Show that $R^0 > 0$ ($R^0 < 0$) is equivalent to $z + ct > 0$ ($z + ct < 0$) and that therefore, according to (4-88), $z + ct > 0$ is a necessary condition for retardation.

It is clear that the motion (5-32) cannot take place over an infinite time interval, because $|t| \to \infty$ implies $v \to c$, as is obvious from (5-33). A mass point can therefore execute hyperbolic motion only for a finite time interval. Both before and after that interval the motion is necessarily *not* uniformly accelerated. This requires care in computing the retarded fields produced near the end points of that interval, but does not affect the fields produced inside the finite open time interval of uniform acceleration. Idealization of hyperbolic motion as motion over an *infinite* world line leads to problems in the associated limiting procedures, and has been the source of much confusion.

We now turn to the question of radiation emission. At first, one would think that there is no radiation at all: in the instantaneous rest system, i.e. at $t = 0$, we have $\mathbf{B} = 0$, so that no radiation can occur at that instant ($\mathbf{E} \times \mathbf{B} = 0$). Since this is true for all instantaneous rest systems, there can be no radiation at all during hyperbolic motion.

This argument is fallacious. That it *must* be so follows from the fact that the motion (5-32) involves a nonvanishing acceleration, $a_\lambda a^\lambda = \mathbf{g}^2 \neq 0$ and therefore leads to a nonvanishing rate of energy emission, $\mathcal{R}$, according to (5-14). The fallacy of the above argument is most easily explained by means of Fig. 4-2. When $\mathbf{B} = 0$ on the hyperplane $t = 0$, it means that the fields emitted at Q' happen to have a vanishing induction $\mathbf{B}$ at P', those emitted at Q'' have vanishing $\mathbf{B}$ at P'', and so on. Thus, the fact that $\mathbf{B} = 0$ on $t = 0$ is perhaps a curiosity but has nothing to do with a characterization of radiation emitted from one typical point of the world line; it refers to the complete past history. If $\mathbf{B} = 0$ were valid along the future light cone with vertex at Q', say, one could indeed conclude that there cannot be radiation. But this is not the case.

Another fallacious argument states that there cannot be radiation because at $t = 0$, $\mathbf{B} \to 0$ for $\rho \to \infty$; if $\mathbf{B}$ vanishes asymptotically, there cannot be a wave zone. The fallacy here lies in the fact that the limit to the wave zone must be taken as $\rho \to \infty$ together with $t \to \infty$, such that one proceeds *along the future light cone*. The limit $\rho \to \infty$ at $t = 0$ could establish at best (cf. Fig. 4–2) that there was no radiation emitted at the (retarded) instant $\tau = -\infty$! But this limit is meaningless because, as was explained earlier, hyperbolic motion cannot last over an infinite time interval.

There are three radiation criteria, all discussed previously and all leading to the same conclusion that a uniformly accelerated charge radiates at every instant during its hyperbolic motion. The first is the fact that $\Re \neq 0$ because $a_\lambda a^\lambda = \mathbf{g}^2 \neq 0$. It gives

$$\Re = \frac{2}{3} \frac{e^2}{c^3} \mathbf{g}^2 = \text{const}, \tag{5-38}$$

according to (5–14). The second is the presence of radiation fields ($\sim 1/\rho$ for fixed z) in the limit $\rho \to \infty$, $t \to \infty$ along any future light cone $R_\lambda R^\lambda = 0$, $R^0 > 0$. For example, the fields emitted at $t = 0$ ($z = \alpha$) become, asymptotically along $(z - \alpha)^2 + \rho^2 = c^2 t^2$, $t > 0$,

$$E_z \to -\frac{e}{\alpha} \frac{\rho^2 + \alpha(\alpha - z)}{(\rho^2 + z^2)^{3/2}} \to -\frac{e}{\alpha\rho},$$

$$E_\varphi \to \frac{e}{\alpha} \frac{\rho z}{(\rho^2 + z^2)^{3/2}} \to \frac{ez}{\alpha\rho^2}, \tag{5-39}$$

$$B_\varphi \to \frac{e}{\alpha} \frac{\rho}{\rho^2 + z^2} \to \frac{e}{\alpha\rho},$$

according to (5–34) and (5–35). Thus, for fixed z the radiation zone involves E_z and B_φ; $E_\varphi \sim 1/\rho^2$.

The third criterion is the local one discussed in Section 5–2. We compute $\Re$ according to (5–20) and find exactly (5–38).

PROBLEM 5–6

Substitute the fields (5–34) and (5–35) into (5–20) and verify (5–38). [*Hint:* Shift the origin of your cylindrical coordinate system to $z = \alpha$.]

PROBLEM 5–7

Find the angular distribution of the radiation, as seen by an inertial observer at an arbitrary instant t, when the charges have velocity $\mathbf{v} = v\hat{\mathbf{k}}$: show that the energy flux into a solid angle $R^2 \, d\Omega$ per unit time t is

$$\mathbf{S} \cdot \hat{\mathbf{n}} \, R^2 \, d\Omega = \frac{e^2}{c^3} \frac{\mathbf{a}^2 \sin^2 \theta}{\left(1 - \frac{v}{c} \cos \theta\right)^6} \frac{d\Omega}{4\pi}. \tag{5-40}$$

Multiply this expression by $dt/d\tau$ to obtain the flux per unit time τ and integrate. Show that the result is again (5–38).

5–4 SYNCHROTRON RADIATION

The emission of radiation from charged particles in circular orbits is treated extensively in the literature. Nevertheless, we shall briefly discuss the total rate of this radiation for two reasons. We want to demonstrate the ease with which rigorous results can be obtained from the covariant radiation theory of the previous Sections 5–1 and 5–2; and we want to provide results which will be needed later in other contexts (Sections 6–7 and 6–15).

The total Lorentz-invariant radiation rate $\mathcal{R}$ is, according to Eqs. (5–15) and (5–14),

$$\mathcal{R} = \frac{dW_{\text{rad}}}{dt} = \frac{2}{3}\frac{e^2}{c^3}\, a_\lambda a^\lambda. \tag{5–41}$$

It can be expressed in terms of the three-vector $\mathbf{a} = d\mathbf{v}/dt$ by means of Eq. (5–26),

$$a_\lambda a^\lambda = \gamma^4\left[\mathbf{a}^2 + \frac{1}{c^2}\gamma^2(\mathbf{v}\cdot\mathbf{a})^2\right]. \tag{5–42}$$

In circular motion $\mathbf{a}$ is directed radially toward the center while $\mathbf{v}$ is tangential, so that $\mathbf{v}\cdot\mathbf{a} = 0$. Equation (5–41) is therefore

$$\mathcal{R} = \frac{2}{3}\frac{e^2}{c^3}\gamma^4\mathbf{a}^2 = \frac{2}{3}\frac{e^2}{c^3}\left(\frac{E}{mc^2}\right)^4(\omega^2 r)^2. \tag{5–43}$$

The total kinetic energy $E = \gamma mc^2$ and the circular frequency $\omega = v/r = a/v$ are introduced here. The radius of the circle, r, and E and ω are related. One of these three quantities can therefore be eliminated by means of

$$(\omega r)^2 = v^2 = 1 - \frac{1}{\gamma^2} = 1 - \left(\frac{mc^2}{E}\right)^2. \tag{5–44}$$

The result (5–43) can thus be expressed in the following alternative ways,

$$\mathcal{R} = \frac{2}{3}\frac{e^2\omega^2}{c}\left(\frac{E}{mc^2}\right)^2\left[\left(\frac{E}{mc^2}\right)^2 - 1\right] = \frac{2}{3}c\frac{e^2}{r^2}\left[\left(\frac{E}{mc^2}\right)^2 - 1\right]^2. \tag{5–45}$$

For high energies, $E \gg mc^2$,

$$\mathcal{R} = \frac{2}{3}\frac{e^2\omega^2}{c}\left(\frac{E}{mc^2}\right)^4 = \frac{2}{3}c\frac{e^2}{r^2}\left(\frac{E}{mc^2}\right)^4. \tag{5–46}$$

In first approximation the effect of radiation emission on the shape of the orbit can be neglected. Approximately, a constant induction field $\mathbf{B}$ will thus give rise to a circular orbit in a plane perpendicular to $\mathbf{B}$. According to the Lorentz force equation with a relativistic momentum,*

$$\frac{d\mathbf{p}}{dt} = \frac{e}{c}\mathbf{v}\times\mathbf{B}, \qquad \mathbf{p} = m\gamma\mathbf{v}, \tag{5–47}$$

---

* This is exactly the Lorentz-Dirac equation (2–31) or (6–57) without the Abraham four-vector. The force necessary to produce a circular orbit when radiation emission and its effect on the orbit is fully taken into account, is treated in Section 6–15.

we can easily express ω in terms of v and B. Since $\mathbf{v} \perp \mathbf{B}$ and the acceleration $\mathbf{v} \times \mathbf{B}$ is orthogonal to both, v is constant and

$$m\gamma \left| \frac{d\mathbf{v}}{dt} \right| = m\gamma \frac{v^2}{r} = \frac{e}{c} vB.$$

Thus,

$$\omega = \frac{v}{r} = \frac{eB}{\gamma mc}. \tag{5-48}$$

The radiation rate (5–45) can therefore be expressed in terms of B and the total energy E as

$$\mathcal{R} = \frac{2}{3} \frac{e^2}{c} \left(\frac{eB}{mc} \right)^2 \left[\left(\frac{E}{mc^2} \right)^2 - 1 \right]. \tag{5-49}$$

CHAPTER 6

The Charged Particle

Electrodynamics can be defined as the physical theory which accounts for the interaction of charged particles with one another and with radiation. The autonomous nature of radiation has become apparent in the previous chapter. The question now arises as to what extent classical physics can describe charged particles and their interactions so that a consistent classical electrodynamics emerges. This problem consists of two parts, viz. the *existence* of classical fundamental charged particles and the *dynamics* of these particles. The existence problem refers to the classical model that can be constructed in view of the realization that classical physics is only a limiting case of quantum physics. The awareness of this asymptotic nature of classical physics is the key to the construction of a physically meaningful classical electrodynamics. It is also basic to the second problem, that of the dynamics of a charged particle. The present chapter is devoted to these questions.

A. Structure and Kinematics

6–1 THE ABRAHAM-LORENTZ-POINCARE MODEL

Macroscopic Maxwell electrodynamics knows only charge *distributions*. In this theory electrostatic charge is derived from a charge *density* (linear, surface, or volume density). The concept of charge as an aggregate of elementary charged particles is foreign to it. How, then, can a microscopic electrodynamics succeed on the basis of a particle picture?

This was the problem faced by Abraham, Lorentz, and Poincaré in the first few years of this century. The most obvious model of a charged particle is a sphere carrying a spherically symmetrical charge distribution. While such a model is meant (and was indeed proposed) as a picture of a charged *elementary* particle (an electron, for example), it is obvious that it is basically a macroscopic charged body, only much smaller. There is nothing "elementary" about it. To see this clearly, let us consider a macroscopic charged sphere in more detail.

Consider a sphere of radius R and spherically symmetric charge density $\rho(\mathbf{r})$. Its total charge is

$$e = \int_V \rho(\mathbf{r})\, d^3x, \qquad (6-1)$$

the integral extending over the volume $V = 4\pi R^3/3$. Since $e \neq 0$ by assumption, there must exist repulsive Coulomb forces between "parts" of the total charge. Symmetry would therefore require that this charged sphere expand like a balloon and continue to do so to arbitrarily large R. The fact that macroscopic charged spheres are actually stable and do not "explode" is entirely due to interatomic and intermolecular forces which hold the sphere rigidly together and *bind* the charge distribution ρ into a rigid sphere. The structure of macroscopic matter is essential for the stability of the charged sphere.

Lorentz and others envisioned an elementary charged particle as a purely electromagnetic object. Since the binding effect of macroscopic matter is then absent, this object is necessarily unstable. But it is just the purely electromagnetic nature which makes this particle "elementary" and distinguishes it from a miniature charged macroscopic sphere of matter. Thus, Lorentz's model of an elementary charge fails because it is not stable.

These considerations can be made quantitative by means of the electromagnetic energy tensor $\Theta_{\text{elm}}^{\mu\nu}$, Eq. (4–114). If the particle is purely electromagnetic, its stability will be determined by this tensor. Since the trace of $\Theta_{\text{elm}}^{\mu\nu}$ vanishes according to (4–116), it follows that in the rest system of the spherically symmetric charge

$$\int \Theta_{\mu\nu}^{(0)} \, d^3x = 0 \qquad (\mu \neq \nu) \tag{6–2a}$$

and

$$\int \Theta_{11}^{(0)} \, d^3x = \int \Theta_{22}^{(0)} \, d^3x = \int \Theta_{33}^{(0)} \, d^3x = \tfrac{1}{3} \int \Theta_{00}^{(0)} \, d^3x. \tag{6–2b}$$

Furthermore, since $\Theta_{00} = -U < 0$, (4–118), the *self-stress* $\int \Theta_{11}^{(0)} \, d^3x < 0$ and does not vanish.

The effect of the nonvanishing self-stress is most easily seen from Eq. (4–138). For our model in the rest system $\mathbf{B} = 0$ and

$$\nabla \cdot \mathbf{T} = \rho \mathbf{E}. \tag{6–3}$$

If we assume a sphere of radius R and a uniform surface charge distribution, then

$$\mathbf{E} = \frac{e}{r^2}\,\hat{\mathbf{r}} \quad (r \geq R); \qquad \mathbf{E} = 0 \quad (r < R). \tag{6–4}$$

The self-force can now be computed from (6–3),

$$2d\mathbf{F}_{\text{self}} \equiv d\Omega \int \rho \mathbf{E} r^2 \, dr = d\Omega \int \nabla \cdot \mathbf{T} \, r^2 \, dr = d\Omega \int_R^\infty \frac{d}{dr}\left(\frac{E^2 \hat{\mathbf{r}}}{8\pi}\right) r^2 \, dr$$

$$= \frac{e^2}{4\pi R^2}\,\hat{\mathbf{r}}\, d\Omega.$$

The internal pressure, i.e. the self-force per unit surface element $R^2 \, d\Omega$ is

$$\mathfrak{p} \equiv \frac{d\mathbf{F}_{\text{self}}}{R^2 \, d\Omega} = \frac{e^2}{8\pi R^4}\,\hat{\mathbf{r}} = 2\pi\sigma^2\hat{\mathbf{r}}, \tag{6–5}$$

where σ is the surface charge density. The last equality employs (4–117) and (6–4). Thus, the nonvanishing self-stress leads to a self-force which is radially outward and tries to expand the sphere.

The electrostatic self-energy follows from (4–125)[(0)],

$$W_{\text{self}}^{(0)} = \frac{1}{8\pi} \int \mathbf{E}^2 \, d^3x = \frac{e^2}{2R},$$ (6–6)

so that the pressure becomes

$$\mathfrak{p} = \frac{1}{3} \frac{W_{\text{self}}^{(0)}}{V} \hat{\mathfrak{r}}.$$ (6–7)

The nonvanishing self-stress is therefore equivalent to a nonvanishing self-energy as well as to a nonvanishing self-force.

If the charged particle is purely electromagnetic, its total energy must be given by (6–6),

$$mc^2 = W_{\text{self}}^{(0)}.$$ (6–8)

Its entire rest mass m is therefore of purely electromagnetic origin (*electromagnetic mass*).

Equations (6–6) and (6–8) determine the particle radius,

$$R = \frac{1}{2} \frac{e^2}{mc^2}.$$ (6–9)

Obviously, a different spherical charge distribution will not modify these conclusions. The corresponding W_{self} will differ from the right-hand side of (6–6) only by a numerical factor of order one. For example, for a uniform volume distribution of charge it would differ by a factor of $\frac{6}{5}$. Correspondingly, the radius R in (6–9) will contain the same factor: $R = (\frac{3}{5})(e^2/mc^2)$ for a volume distribution. For this reason, the quantity

$$r_0 = \frac{e^2}{mc^2}$$ (6–10)

has been designated as *the* classical electron radius when m is the electron rest mass. It gives the correct order of magnitude of the radius resulting from any charge distribution.

The structure of a classical charged particle is thus specified by its charge distribution. But the characteristic theoretical difficulty connected with this particle, viz. its nonvanishing electromagnetic self-stress and corresponding instability (tendency to explode), is independent of this structure.

In the first decade of this century, when this theory was at its peak of interest and development, there was no experimental evidence to confirm or contradict any particular particle size or structure of the order of r_0. Therefore, the only obvious and serious difficulty was the qualitative feature of *instability*.

The solution to this difficulty appeared to be a structure that was not purely electromagnetic. In a way, this was a return to the minute macroscopic particle. The internal forces responsible for the rigidity of a solid, and consequently for the stability of a macroscopic charge, are reintroduced. Except that now they appear to be of mysterious origin, having no other manifestation but the ad hoc function of furnishing the "glue" which prevents the elementary charge from exploding.   Mathematically, this idea was first attempted by Poincaré.   A nonelectromagnetic symmetric tensor $\Pi^{\mu\nu}$ is postulated* which, when combined with $\Theta^{\mu\nu}_{\text{elm}}$, has the property

$$\partial_\mu T^{\mu\nu} = 0, \qquad T^{\mu\nu} = \Theta^{\mu\nu}_{\text{elm}} + \Pi^{\mu\nu}. \qquad (6\text{–}11)$$

The tensor $T^{\mu\nu}$ can be regarded as the energy tensor of the total system "charged particle"; this equation can then be interpreted as the differential conservation law of energy and momentum:

$$P^\mu \equiv \int_{\substack{\text{spacelike} \\ \text{plane } \sigma}} T^{\mu\nu}\, d^3\sigma_\nu. \qquad (6\text{–}12)$$

It yields

$$\frac{dP^\mu}{d\tau} = 0 \qquad (6\text{–}13)$$

in view of (6–11) and (A1–62).   In the rest system S_0 of the particle, $T^{\mu\nu}$ is *by definition* supposed to have the form

$$T^{(0)}_{\mu\nu} = \begin{cases} -U^{(0)}_{\text{total}} & (\mu = \nu = 0) \\ 0 & \text{otherwise.} \end{cases} \qquad (6\text{–}14)$$

This means that the stress components Θ^{kk}_{elm} are exactly compensated by Π^{kk} in the system S_0.   The total energy of the particle at rest is then

$$mc^2 = E^{(0)} = \int U^{(0)}_{\text{total}}\, d^3x = -\int (\Theta^{(0)}_{00} + \Pi^{(0)}_{00})\, d^3x. \qquad (6\text{–}15)$$

The mass now consists of an electromagnetic and a "cohesion" part.

The postulate of the existence of $\Pi^{\mu\nu}$ with the prescribed properties solves the stability problem, because $T^{(0)}_{kk} = 0$ by definition of $\Pi^{\mu\nu}$.   But this tensor is obviously obtained by ad hoc assumption; furthermore, it destroys the attractive idea of a fundamental charged particle of purely electromagnetic energy content.

What is the reason for this failure to construct a stable classical charged particle?   We can answer this question today with the aid of over fifty years of hindsight: what was attempted then was impossible to achieve because the structure of elementary particles lies outside the domain of classical physics. We know today that classical physics is an asymptotic limit of a more fundamental level of physics which we might call quantum physics, embracing

---

* In his fundamental paper, [*Rend. Circ. Mat. Palermo* **XII**, 129 (1906)] Poincaré derives the cohesion forces from a potential, but such that they transform under Lorentz transformations exactly like the electromagnetic forces which are responsible for the instability.

quantum mechanics (nonrelativistically) and quantum field theory (relativistically). While quantum physics itself is deducible from a still more fundamental level of physics which is yet to be understood, we now realize that we cannot expect a satisfactory description of elementary charged particles on the classical level unless we are able to avoid trespassing the limits of classical physics. This means, in particular, that we must formulate such a theory *without reference to the structure* of the particles involved.

PROBLEM 6-1

A charge e is uniformly distributed in the shape of a spherical shell of radius R. Inside and outside the shell there is a vacuum. The corresponding electromagnetic mass $m = e^2/(2Rc^2)$ [cf. Eq. (6-9)]. Assume that the shell rotates with angular velocity ω.

(a) Show that the field at large distance is that due to a magnetic moment

$$\mu = \frac{eR^2}{3c} \omega.$$

[*Hint:* Consider the shell as a sum of circular current loops.]

(b) Find **L** and **Σ**, the electromagnetic orbital and spin angular momenta, (4–165) and (4–166), and verify that the gyromagnetic equation

$$\mu = \frac{e}{2mc} \mathbf{J}$$

holds also in this case.

(c) Define an electromagnetic moment of inertia by

$$\Sigma = I\omega$$

and show that this is exactly the same I that would be obtained if the rotating spherical shell of radius R were made of a uniform mass distribution $m/(4\pi R^2)$, where m is the electromagnetic mass.

(d) Assume that this shell is used as a classical model of an electron with mass m, charge e, and magnetic moment $e\hbar/2mc$ ($\hbar = 1/2\pi$ times Planck's constant). Find R and ω numerically.

6-2   CHARGED PARTICLES WITHOUT STRUCTURE

A spherically symmetric charge distribution which is confined to a sphere of radius R gives rise, in its rest system, to a Coulomb field

$$\mathbf{E} = \frac{e}{r^2} \hat{\mathbf{r}} \qquad\qquad (6\text{--}16)$$

at all points $r > R$. This field reveals nothing whatsoever about the size of R (except that $R < r$) or the charge distribution ρ (except that it is spherically symmetric). In probing the structure of such a system it is obviously necessary to penetrate to distances $r < R$. But if we are now told that there exist no experimental means *within the realm of classical physics* to penetrate so deeply, we must reluctantly conclude that classically this structure is a physically meaningless concept. It has no observational significance.

Does this situation prevent us from describing other (nonstructural) features associated with this system? By no means. The dynamics of this system, its behavior under the influence of various kinds of (classical) forces is clearly meaningful. A theory must therefore exist which accounts for the dynamics of this charge but which at the same time tells nothing about its structure other than that it is out of reach. This is exactly the situation in the classical dynamics of charged atomic particles.

When in the following we study *charged particles* we have in mind specifically those particles, usually called "atomic" or "subatomic," whose size confines their structure to quantum physics (as in the case of atomic or molecular ions) or to subquantum physics (as in the case of electrons and charged mesons or protons). Many physical phenomena of great importance are due to the *classical* behavior of these particles, such as the dynamics of beams in electron optics, the orbit theory of charges in particle accelerators, most phenomena associated with plasmas, etc.

The practicing physicist working in these fields very often thinks of electrons, protons, ions, etc., as *point particles*. This is satisfactory and correct in many respects, but it is a very dangerous concept. When used indiscriminately it implies an infinite field strength (Eq. 6–16) at the position of the particle, infinite self-energy [$R \to 0$ in (6–6)], and, correspondingly, an infinite self-stress and self-force. All this is clearly meaningless. The guiding principle in the construction of a theory of structureless particles must be that it be independent of the radius R of the sphere within which the (quantum-mechanical) structure would appear. It is a customary procedure to formulate the theory first for a particle of finite radius R and then pass to the point-particle limit $R \to 0$. This procedure is unsatisfactory. Either the theory contains quantities which depend on R, or it does not. If it does, the limit $R \to 0$ may lead to meaningless infinite results; if it does not, there is no point in attempting a finite R formulation in the first place. Thus, *the problem is to find a formulation of classical charged particle theory which does not require any reference to, or assumptions about, the particle structure, its charge distribution, and its size.*

Another (also unsatisfactory) approach to this problem is the following. One assumes some radius and structure; one formulates the theory with the aid of these concepts; one eliminates the structure-dependent terms, and one points at the structure-independent features as the ones of "true" physical meaning. The structure elimination in this approach has been carried out over the years on many different levels of sophistication. The most naïve method consists of simply dropping the undesirable terms. Other methods go under the names of "cutoff procedure" and "renormalization technique." Some of these have been developed to a much greater extent in quantum field theory where the same problems occur as long as one insists on a formulation in close analogy to classical mechanics. These methods provide no solution to the problem. Their accomplishment lies solely in arriving at those physically meaningful structure-independent results which would have been obtained had the theory been formulated in a structure-independent way from the very beginning. Consequently, they are very valuable for the experimental physicist who is interested only in the results and predictions of the theory. But they are entirely unsatisfactory from the point of view of physical theory.

6-3 THE RELATIVISTIC LORENTZ ELECTRON

It is very unfortunate that relativity theory was not developed *before* Lorentz conceived of his ambitious theory of electrons.* The fact is that Lorentz and others took great pains to construct an electron model consistent with the Lorentz-Fitzgerald contraction hypothesis. But the Lorentz transformation was poorly understood at the time, the transformation properties of various physical quantities were not sufficiently well known, and physicists did not have today's facility in handling the corresponding mathematics. The latter we owe to a large extent to Minkowski, whose first paper on it did not appear until 1908. As a consequence, much work which was meant to be relativistic was actually incorrect, and wrong conclusions were drawn from some of these results.

The most important flaw in the early literature on this subject concerns the confusion which existed between the transformation properties of the electromagnetic energy and momentum, and the need for a vanishing self-stress of the purely electromagnetic charged particle. Unfortunately, this confusion has survived for many years and can still be found in some of the current literature. The present section is therefore devoted to the relativistic kinematics of a purely electromagnetic charged particle in the Lorentz sense. The stability problem will then emerge as completely independent of this kinematics.

As was remarked previously, a clear distinction must be made between the radiation fields and the velocity fields of a given charged particle. The radiation fields are *emitted* in the sense that they assume an independent existence as soon as they are produced, while the velocity fields are "tied" to their source permanently and move along with it. In fact, it would be physically meaningless to separate a charge from its velocity field. Neither can be observed without the other: a charge at rest is *always* surrounded by a Coulomb field and, conversely, every Coulomb field has a source.

This point must be stressed repeatedly, because the customary structure of our electrodynamic theories is such that this separation is introduced from the very beginning. In the usual theories it is not difficult to write down equations that describe a charged particle which does not interact with others, i.e. which has no field.

---

* It should be noted that many authors, including Abraham and Lorentz, used the term "electrons" to designate both positively and negatively charged particles.

Whether or not a charged fundamental particle (an electron, say) consists of purely electromagnetic mass, the associated electromagnetic energy and momentum must transform under Lorentz transformation exactly as for any other particle in relativistic mechanics.   If the particle consists of "matter" and "electromagnetic fields," this transformation property is necessary in order that these two parts stay together throughout the motion and irrespective of the velocity of the observer.

Let us consider a charged particle in uniform motion.   The only electromagnetic fields associated with it are velocity fields.   From these one can construct an energy tensor $\Theta_{\text{elm}}^{\mu\nu}$ which has vanishing divergence, Eq. (4–115), in every space-time volume not containing the particle.   Outside the spherically symmetrical particle the field strengths are given by Eq. (4–98) and are indistinguishable from those of a point particle.   The energy tensor is therefore, according to Eq. (5–8),

$$\Theta_{\text{elm}}^{\mu\nu} = \frac{e^2}{4\pi\rho^4}\left(u^\mu u^\nu - \frac{v^\mu v^\nu}{c^2} - \tfrac{1}{2}\eta^{\mu\nu} \right). \tag{6–17}$$

In analogy to Eqs. (4–123) and (5–3) we now define

$$P_{\text{elm}}^{\mu} \equiv \frac{1}{c}\int_{(\sigma)} \Theta_{\text{elm}}^{\mu\nu}\, d\sigma_\nu, \tag{6–18}$$

where the integration is extended over the spacelike plane σ *with a hole cut out* at the particle position (sphere of radius R in the rest system).   This is indicated by the notation (σ) instead of σ under the integral sign.   Thus, the integral extends over the whole comoving velocity field "outside" the particle.   In order that P_{elm}^{μ} be a vector, $d\sigma^\mu$ must transform like a timelike four-vector.   On the other hand, $d\sigma^\mu = (d^3x; 0, 0, 0)$ whenever the observer is in the rest system of the particle.   Therefore, we must have

$$d\sigma^\mu = \frac{v^\mu}{c}\, d\sigma, \tag{6–19}$$

where $d\sigma$ is the invariant surface element.

The requirement (6–19) makes P_{elm}^{μ} in (6–18) *dependent* on the surface, in apparent contradiction to the surface independence usually required as a condition for the vector character of P_{elm}^{μ}.   However, in (6–18) the integration does not extend over the *whole* surface, since a piece has been cut out.   This piece is a sphere in the rest system and must be the Lorentz transform of a sphere in any other Lorentz system.   For an infinitesimally small sphere this is ensured by (6–19).   Since we take Lorentz transformations in the strict sense, i.e. between rectilinear orthogonal Cartesian coordinates, and do not admit general curvilinear coordinate systems in our pseudo-Euclidean space, all spacelike surfaces are planes.   They are therefore fixed by choosing the normal at one point, viz. at the particle position, according to (6–19).   Geometrically, the choice (6–19) requires that σ be orthogonal to the world line of the particle. Physically, (6–19) expresses the *rigidity* of the charged particle in the relativistic sense: the fields on two different spacelike planes through the particle are not the same. The integration is to be carried out exactly for the static case; i.e. the plane orthogonal to the world line is singled out. Only this plane is orthogonal to the retarded light cones whose vertices are the past position of the uniformly moving charge.

Substitution of (6–17) and (6–19) into (6–18) yields the electromagnetic momentum four-vector of the velocity fields,

$$P^\mu_{\text{elm}} = \frac{e^2}{4\pi c} \int_{(\sigma)} \frac{1}{\rho^4} \left(\frac{v^\mu}{c} - \frac{1}{2} \frac{v^\mu}{c} \right) d\sigma = \frac{e^2}{8\pi c^2} v^\mu \int_{(\sigma)} \frac{d\sigma}{\rho^4}, \qquad (6\text{–}20)$$

where we used (4–89). This result can be written in the form

$$P^\mu_{\text{elm}} = m_{\text{elm}} v^\mu, \qquad (6\text{–}21)$$

in which

$$m_{\text{elm}} \equiv \frac{e^2}{8\pi c^2} \int_{(\sigma)} \frac{d\sigma}{\rho^4}$$

is an invariant and can be regarded as the mass equivalent of the electromagnetic energy of the velocity field outside the charged particle. It can be evaluated explicitly. Being an invariant,

$$m_{\text{elm}} = \frac{e^2}{8\pi c^2} \int_R^\infty \frac{1}{r^4} r^2 \, dr \, d\Omega = \frac{e^2}{2Rc^2}. \qquad (6\text{–}22)$$

The total energy-momentum four-vector of the particle consists of the above P^μ_{elm} and P^μ_{inside} due to the internal energy content of the particle. The latter requires specific assumptions about the particle structure. For example, if it is a sphere with surface charge, the electromagnetic part of P^μ_{inside} vanishes because $F^{\mu\nu} = 0$ inside and (6–22) is the total mass of the particle. Comparison with (6–6) then shows that

$$m_{\text{elm}} = W^{(0)}_{\text{self}}/c^2. \qquad (6\text{–}23)$$

Moreover, Eqs. (6–18) and (6–19) specify

$$P^\mu_{\text{elm}} = \frac{1}{c^2} \int \Theta^{\mu\nu}_{\text{elm}} v_\nu \, d\sigma, \qquad (6\text{–}24)$$

so that in the rest system, where $v^\mu = (1;0,0,0)$,

$$P^{\mu(0)}_{\text{elm}} = -\frac{1}{c} \int \Theta^{\mu 0}_{(0)} \, d^3x = \left(\frac{1}{c} \int U^{(0)} \, d^3x; 0, 0, 0 \right) \qquad (6\text{–}25)$$

[cf. (4–124)$^{(0)}$ through (4–126)$^{(0)}$]. The space-part vanishes because $\mathbf{B} = 0$ in the rest system, so that $\mathbf{S}^{(0)} = 0$.

In general, P^μ_{inside} must satisfy a relation analogous to (6–21),

$$P^\mu_{\text{inside}} = m_{\text{inside}}\, v^\mu, \tag{6-26}$$

in order that

$$P^\mu = P^\mu_{\text{elm}} + P^\mu_{\text{inside}} = mv^\mu, \tag{6-27}$$

$$m = m_{\text{elm}} + m_{\text{inside}}. \tag{6-28}$$

Thus, the covariant specification of the shape of the particle (sphere in the rest system) requires that the energy contents of both outside and inside each yield momentum four-vectors.

The observed mass of the particle is m. There is no experiment by which m_{elm} and m_{inside} can be observed separately.* This is an example of the occurrence of structure-dependent aspects of the theory which are physically meaningless.

The difficulty encountered by the founders of the classical electron theory can best be exhibited in noncovariant notation. Let us again assume a purely electromagnetic particle with surface charge, so that

$$P^\mu = P^\mu_{\text{elm}} = \frac{W^{(0)}_{\text{self}}}{c^2} v^\mu \tag{6-29}$$

according to the above results. This result is based on the fact that P^μ_{elm} as defined by (6–24) is a four-vector. In three-vector notation this definition becomes [cf. (4–124) through (4–126)]

$$W = \gamma \int_{(\sigma)} U\, d\sigma - \frac{\gamma}{c^2} \int_{(\sigma)} \mathbf{S} \cdot \mathbf{v}\, d\sigma, \tag{6-30}$$

$$\mathbf{P} = \frac{\gamma}{c^2} \int_{(\sigma)} \mathbf{S}\, d\sigma + \frac{\gamma}{c^2} \int_{(\sigma)} \mathsf{T} \cdot \mathbf{v}\, d\sigma, \tag{6-31}$$

$$P^\mu = \left(\frac{1}{c}\, W;\, \mathbf{P}\right). \tag{6-32}$$

These relations must be compared with the definitions used by Abraham, Lorentz, and others:†

$$W = \int_{(\sigma)} U\, d^3x, \tag{6-33}$$

$$\mathbf{P} = \frac{1}{c^2} \int_{(\sigma)} \mathbf{S}\, d^3x. \tag{6-34}$$

---

* Of course, m_{inside} may be partially or wholly electromagnetic. This, however, is irrelevant, and the notation chosen here will be convenient later on.

† See, for example, H. A. Lorentz, *The Theory of Electrons*, Second Edition, Dover, 1952, §24.

As is obvous from (4–118) through (4–120), $(1/c)W$ and $\mathbf{P}$ defined in this way cannot form a four-vector of the type (6–32). Since the importance of the transformation properties of such quantities was not realized at the time, one was guided by physical arguments. Thus, Abraham defined $\mathbf{P}$ in (6–34) as the "electromagnetic momentum," because this definition proved satisfactory for the study of *radiation* phenomena. Without further questioning, this same definition was then applied to the velocity fields. The result was that, instead of (6–29), one obtained results of the form*

$$P^\mu = \frac{4}{3} \frac{W_{\text{self}}^{(0)}}{c^2} v^\mu f\left(\frac{v}{c}\right).$$

The function $f(v/c)$ is not an invariant and, consequently, P^μ is not a four-vector. In the nonrelativistic limit this expression reduces to

$$W = W_{\text{self}}^{(0)}, \qquad \mathbf{P} = \frac{4}{3}\left(\frac{W_{\text{self}}^{(0)}}{c^2}\right)\mathbf{v}, \tag{6–35}$$

in contradiction to the momentum-velocity relation of Newtonian particle mechanics. A simple calculation shows† that the additional terms in (6–31) reduce the incorrect factor $\tfrac{4}{3}$ to 1.

These unsatisfactory results were not the only consequence of the incorrect definitions (6–33) and (6–34). If these equations are regarded as valid for an arbitrary Lorentz system, a transformation to the rest system reveals that the expected result (6–25) can be obtained only if $\Theta_{kk}^{(0)}$ ($k = 1, 2, 3$) vanishes. Thus, one is incorrectly led to believe that the stability of the charged particle is a necessary condition for the correct transformation properties of $\mathbf{P}$ and W.

These erroneous conclusions of long standing should be contrasted with the present satisfactory equations, (6–21) and (6–23), which emerge in a natural manner once the requirements of Lorentz covariance are heeded. The latter follow without any assumptions whatever concerning the self-stress. *It is possible to satisfy the requirements of relativity and still have an unstable particle.*

It is instructive to see the above difficulties from a somewhat different point of view. Assume that we adopt the Poincaré tensor so that we have a charged particle model of finite extent with energy tensor $T^{\mu\nu}$ satisfying (6–11) and (6–14). The corresponding momentum four-vector P^μ, Eq. (6–12), is independent of σ. However, when it is written as the sum of two integrals,

$$P^\mu = P^\mu_{\text{elm}} + P^\mu_{\text{inside}} = \int \Theta^{\mu\nu}_{\text{elm}} \, d\sigma_\nu + \int \Pi^{\mu\nu} \, d\sigma_\nu, \tag{6–36}$$

each integral by itself is *not* independent of σ. Hereby it is irrelevant whether this separation is made on the basis of electromagnetic and nonelectromagnetic

---

* H. A. Lorentz, loc. cit, §§26 through 28.

† For details the reader is referred to F. Rohrlich, *Am. J. Phys.* **28**, 639 (1960).

fields or on a geometrical basis, by restricting the first integral in the rest system to $r \geq R$ (purely electromagnetic fields) and the second one to $r < R$, as was assumed in (6–18). The σ-independence of the *sum* in (6–36) permits one to choose $d\sigma^{(0)} = d^3x$ for the plane, even when $\mathbf{v} \neq 0$ to obtain the correct P^μ. But P^μ_{elm} alone, computed in this way, will yield the noncovariant result (6–35). This means that the separation (6–36) of P^μ into two integrals is not covariant for an arbitrary choice of σ. As we have seen, it *is* covariant for the particular choice (6–19) of σ. The fact that P^μ_{elm} depends on σ is related to the fact that $\Theta^{\mu\nu}_{\text{elm}}$ is not divergence-free *everywhere* on σ and requires $\Pi^{\mu\nu}$ for stability; but this does not prevent one from defining P^μ_{elm} covariantly by letting σ be determined by the motion of the particle, Eq. (6–19).

PROBLEM 6–2

Consider a nonrelativistic charged particle in uniform motion. Find $\mathbf{E}$ and $\mathbf{B}$ and compute $\mathbf{P}$ and W according to the definitions (6–33) and (6–34); repeat the calculations with the correct definitions, (6–30) and (6–31). Compare the results and exhibit the source of the factor $\frac{4}{3}$ in (6–35).

PROBLEM 6–3

With the aid of the transformation for the tensor $\Theta^{\mu\nu}_{\text{elm}}$, find W and $\mathbf{P}$ in terms of their values in the rest system, using the definitions (6–33) and (6–34). Show that the desired energy-momentum relations result if and only if the self-stress vanishes in the rest system.

B. Dynamics of a Single Charged Particle

6–4   THE ASYMPTOTIC CONDITIONS

In Part A of this chapter we have exhibited the severe difficulties arising from the dichotomy of particle and field. It seems impossible to bridge the gap between structureless point particles and finite field energies; there does not seem to exist a consistent theory of particles which carry a Coulomb field and which are at the same time purely electromagnetic.

As in other dichotomies in physics, the bridge between particle and field is spanned by a higher-level theory in which these two concepts are synthesized into a "particle-field." This is accomplished in quantum field theory. A somewhat analogous dichotomy is that of particle and wave in classical physics which is bridged in quantum mechanics. In both cases the essential step lies in the realization of the *inseparability* of the two apparently opposing concepts. Although one of the two concepts usually dominates the description, depending on the particular measurement at hand, the theory provides a complete synthesis which stands entirely beyond this dichotomy.

We note, however, that in addition to the quantum field-theoretic description of the *electromagnetic field* which synthesizes the particle (photon) properties and the wave properties of that field, there also exists the classical electromagnetic description of light synthesizing ray optics and wave optics. As long as each description is limited to its domain of applicability, depending on the system and the experiment, no difficulty arises from the apparently contradictory aspects of the same entity *light*.

The particle-field dichotomy for *charged particles* can also be resolved in the classical domain, so as to permit the construction of a consistent classical theory of charged particles. That this must be possible can be deduced from two points: the existence of a consistent quantum field-theoretic formulation* and the logical necessity of being able to derive the lower-level classical theory from the higher-level quantum theory in a suitable limit.

The resolution lies in the observation that the theory at hand is *phenomenological* as far as the structure and properties of the charged particles are concerned. These are not explained by the theory, but are to be given explicitly. While the structure is, in fact, outside the competence of classical theory, as was discussed in the previous section, the mass and charge must be determined experimentally and must be *inserted* into the theoretical framework at an appropriate place.

Consider a typical scattering event. A charged particle is sent into a region in which there exists a known ("external") force field; it is given a certain momentum and energy transfer, after which it leaves the force field region again. The identity of the particle is clearly established before or after its interaction (or at both times) but certainly not during the scattering process. If the particle is asymptotically free, i.e. the action of the force is reasonably limited in space-time, we can make a measurement of the free particle in order to identify it. As far as the scattering process is concerned, however, there is no doubt that this identification takes place in the *asymptotic* region of the process.

It is therefore our task to construct the theory in such a way that it will permit us to feed into it the identifying constants as asymptotic conditions.

The asymptotic conditions therefore consist of two parts, one dynamic, the other kinematic. The dynamic part is the statement that the interaction ceases asymptotically, so that the particle is asymptotically free and moves in uniform motion (law of inertia). This is expressed in terms of the acceleration four-vector as

$$\lim_{|\tau|\to\infty} a^\mu(\tau) = 0. \tag{6–37}$$

It clearly implies certain conditions on the imposed forces.

---

* The divergence difficulties in self-energies, etc. can be completely overcome by reformulation as an "already renormalized" field theory. [See for instance, R. E. Pugh, *Ann. Phys.* (N.Y.) **23**, 335 (1963) whose work develops the asymptotic quantum field theory first formulated by Lehmann, Symanzik, and Zimmermann.]

An objection could be raised here. Is this condition not too strong? Does it not rule out many important physical systems such as a charged particle in a periodic field or any stable bound system? This objection is not valid. The charged particle in a periodic field belongs to a class of physical systems which are idealized for convenience and which cannot exist in this form in nature. It involves systems with infinite energy supply, since energy (radiation) is being emitted in each period: no imposed force can do work over an infinite time interval without vanishing asymptotically unless its energy supply is unlimited; a periodic external field must damp out eventually. The point that is made here, and whose necessity will become evident in the following, is simply that such idealization is not admissible when the physical system involves charged particles. A special case of this situation was already encountered in the radiation from uniformly accelerated charges (see the comments following Problem 5–5).

The second class of systems which seem to be excluded are bound systems. Again, this is not a limitation, because there are no stable bound systems for classical charges. Any such system necessarily involves an accelerated charge; according to the Maxwell-Lorentz equations such a charge must lose energy due to radiation and therefore cannot be part of a stable bound system. A typical example is the "classical hydrogen atom." A charged particle in a Kepler orbit around a Coulomb center of attraction necessarily spirals toward this center because of the potential-energy loss by radiation. Clearly, the same would be true if the force of attraction were a nonelectromagnetic one (Kepler motion of a charged satellite). These matters are studied in detail in Section 6–15.

The kinematic asymptotic condition states our observation of the free particle. It tells us, for example, that it has a rest mass m which can be measured. Consequently, one observes the asymptotic particle momenta

$$p_{in}^{\mu} = mv_{in}^{\mu} = \lim_{\tau \to -\infty} mv^{\mu}(\tau), \tag{6–38}_{in}$$

$$p_{out}^{\mu} = mv_{out}^{\mu} = \lim_{\tau \to +\infty} mv^{\mu}(\tau). \tag{6–38}_{out}$$

Moreover, the charged particle is at all times surrounded by an electromagnetic field. Asymptotically, this field is a pure Coulomb field

$$\lim_{\tau \to -\infty} F_{particle}^{\mu\nu} = F_{Coul, in}^{\mu\nu}, \tag{6–39}_{in}$$

$$\lim_{\tau \to +\infty} F_{particle}^{\mu\nu} = F_{Coul, out}^{\mu\nu}. \tag{6–39}_{out}$$

Associated with these fields are of course certain four-momenta $P_{Coul, in}^{\mu}$ and $P_{Coul, out}^{\mu}$. However, *it will not be necessary to establish or assume any relationship* between P_{Coul}^{μ} and p^{μ} for either the *in* or the *out* states. This point is of crucial importance, because any such relationship is a statement on the structure of the charged particle. For example, it is often assumed that the particle consists of a "matter" center, called *bare particle* of mass m_0 and a Coulomb field around it. In such a case

$$p^\mu = p^\mu_{\text{bare}} + P^\mu_{\text{Coul}} \qquad (6\text{-}40)$$

for both *in* and *out* states. Since the bare particle and its Coulomb field always move together, this means [cf. (6-28)]

$$m = m_{\text{bare}} + m_{\text{Coul}}. \qquad (6\text{-}40')$$

The quantity m_{Coul} is the mass equivalent of the electrostatic self-energy. Since this quantity diverges for a point charge, the corresponding m_{bare} would have to be $-\infty$ so that m has the observed value. This difficulty does not exist when the relation (6-40) or (6-40') is not needed. Nor do we thereby dodge a divergent term because, according to our discussion in Section 6-2, a point particle is simply one whose "radius" is too small to be observed. This does not imply that its radius vanishes, i.e., it does not imply that m_{Coul} is infinite. However, it *is* undetermined. For this reason it is highly satisfactory that no relation of the type (6-40) needs to be specified; i.e. no *mass renormalization* is needed.

The kinematic asymptotic conditions therefore consist of a statement about the particle, (6-38), and a statement about the field strengths, (6-39).

Finally, it should be emphasized that the term *asymptotic condition* is misleading. What is meant here is not a condition which is arbitrarily imposed. Rather, these asymptotic conditions are an *essential part of the description of charged particles*. For example, (6-38) is necessary for the identification of the particle. It would not be meaningful to look for equations of motion and their solutions *apart* from these conditions. No physical meaning can be attached to such solutions. In fact, we shall see later that the most satisfactory formulation of the theory combines these asymptotic conditions with certain differential equations so that only the *combined* system (a set of integrodifferential equations) can be meaningfully identified as the equations of motion.

6-5 DEDUCTIONS FROM THE MAXWELL-LORENTZ EQUATIONS AND THE CONSERVATION LAWS

Consider a point charge e of mass m in an external (i.e. given) force field F^μ_{ext}. Let its world line be $z^\mu(\tau)$ with corresponding velocity $v^\mu = \dot{z}^\mu$ and acceleration $a^\mu = \ddot{z}^\mu$, the dots indicating derivatives with respect to the proper time τ. What can be deduced from the Maxwell-Lorentz equations and the conservation laws concerning the motion of this particle?

This problem was solved by Dirac[*] in 1938. An important contribution to it was made by Haag[†] in 1955. I shall first present the line of thought which leads to the result (6-57). The proofs will be postponed until the end of this section.

[*] P. A. M. Dirac, *Proc. Roy. Soc.* (*London*) **A 167**, 148 (1938).

[†] R. Haag, *Z. Naturforsch.* **10a**, 752 (1955).

In general, the charged particle may interact with an incident radiation field $F_{in}^{\mu\nu}$ as well as with external forces F_{ext}^{μ}. The latter may be partially or wholly electromagnetic. The general system is therefore an *open* system; F_{ext}^{μ} is controlled from the outside and, in the most general case, is able to give or take energy, momentum, and angular momentum. It will be convenient to consider first only the presence of $F_{in}^{\mu\nu}$; it forms a *closed* system together with the charged particle. The necessary generalizations to include F_{ext}^{μ} can be supplied easily.

Asymptotically, for $\tau \to -\infty$ only $F_{in}^{\mu\nu}$ is present in addition to the particle. During the motion the particle will emit radiation whenever it is accelerated (Chapter 5). Asymptotically, as $\tau \to +\infty$, the total free electromagnetic field will therefore be

$$F_{out}^{\mu\nu} = F_{in}^{\mu\nu} + F_{rad}^{\mu\nu}. \tag{6-41}$$

The only other fields present asymptotically are the $F_{Coul}^{\mu\nu}$ of the particle, but these do not satisfy the free field equations.

The Maxwell-Lorentz equations are

$$\partial_\mu F^{\mu\nu}(x) = -\frac{4\pi}{c} j^\nu(x), \tag{6-42}$$

where $j^\mu(x)$ is the current density four-vector of the point charge,

$$j^\mu(x) = ec \int_{-\infty}^{\infty} \delta(x - z) v^\mu(\tau) \, d\tau, \tag{4-82}$$

as was discussed in Section 4–7. The solution of this equation which satisfies the condition that at $\tau = -\infty$ the only free field (i.e. field which satisfies the homogeneous equation) was $F_{in}^{\mu\nu}$ is

$$F^{\mu\nu} = F_{ret}^{\mu\nu} + F_{in}^{\mu\nu}. \tag{6-43}$$

Similarly, knowing that $F_{out}^{\mu\nu}$ are the only free fields at $\tau = +\infty$, we have

$$F^{\mu\nu} = F_{adv}^{\mu\nu} + F_{out}^{\mu\nu}. \tag{6-44}$$

From this and (6–41) it follows that

$$F_{out}^{\mu\nu} - F_{in}^{\mu\nu} = F_{ret}^{\mu\nu} - F_{adv}^{\mu\nu}. \tag{6-45}$$

For the following it is convenient to define

$$F_+^{\mu\nu} \equiv \tfrac{1}{2}(F_{ret}^{\mu\nu} + F_{adv}^{\mu\nu}), \tag{6-46}_+$$

$$F_-^{\mu\nu} \equiv \tfrac{1}{2}(F_{ret}^{\mu\nu} - F_{adv}^{\mu\nu}), \tag{6-46}_-$$

$$\overline{F}^{\mu\nu} \equiv \tfrac{1}{2}(F_{in}^{\mu\nu} + F_{out}^{\mu\nu}). \tag{$\overline{6-46}$}$$

We can now carry out a study of the electromagnetic momentum four-vector which escapes from the particle during a given proper time interval $\tau_2 - \tau_1$.

To this end we surround the world line of the particle by a narrow tube of radius ϵ (invariantly defined) and compute $\int_1^2 \Theta_{elm}^{\mu\nu} \, d\sigma_\nu$, where $\Theta_{elm}^{\mu\nu}$ is the energy tensor (4–114) and $d\sigma_\nu$ is the (three-dimensional) timelike surface of the narrow tube. The result of this calculation (see end of section) is

$$-\frac{1}{c}\int_1^2 \Theta_{elm}^{\mu\nu} \, d\sigma_\nu = \int_1^2 \left(\frac{dP_{Coul}^\mu}{d\tau} - \frac{e}{c} F^{\mu\nu} v_\nu\right) d\tau, \qquad (6\text{–}47)$$

where P_{Coul}^μ is the four-momentum of the Coulomb field associated with the moving charge. This formula holds irrespective of the choice of the two points 1 and 2 on the world line. It is therefore a function of the end points only, and both terms of the integrand must be total differentials. If we extend the integral over the whole world line,

$$\int_{-\infty}^{\infty} \left(\frac{dP_{Coul}^\mu}{d\tau} - \frac{e}{c} F^{\mu\nu} v_\nu\right) d\tau = (P_{out}^\mu + P_{Coul,\,out}^\mu)_\infty - (P_{in}^\mu + P_{Coul,\,in}^\mu)_{-\infty},$$

since the fields $F_{out}^{\mu\nu}$, $F_{in}^{\mu\nu}$, $F_{Coul}^{\mu\nu}$, give rise to corresponding momenta P_{out}^μ, P_{in}^μ, P_{Coul}^μ. Here, the asymptotic condition (6–39) is used.

It is now clear that the P_{Coul}^μ terms cancel and one has

$$\frac{e}{c}\int_{-\infty}^{\infty} F^{\mu\nu} v_\nu \, d\tau = P_{in}^\mu - P_{out}^\mu. \qquad (6\text{–}48)$$

This expression is independent of the choice of the tube radius ϵ, since ϵ occurs only in P_{Coul}^μ. The cancellation of P_{Coul}^μ is therefore crucial in obtaining a *structure-free* equation (6–48).

As mentioned, the integrand (6–47) is a total derivative. It expresses the change in the electromagnetic fields over a proper time interval $d\tau$ as a sum of contributions of the bound (Coulomb) field and the free field,

$$\left(\frac{dP_{Coul}^\mu}{d\tau} - \frac{e}{c} F^{\mu\nu} v_\nu\right) d\tau = dP_{Coul}^\mu + dP^\mu.$$

Here, dP^μ is the change in the momentum of the free electromagnetic field (radiation field); we learn that

$$dP^\mu = -\frac{e}{c} F^{\mu\nu} v_\nu \, d\tau. \qquad (6\text{–}49)$$

This is the differential equivalent to (6–48), since

$$\int_{-\infty}^{\infty} dP^\mu = P_{out}^\mu - P_{in}^\mu.$$

Again, the Coulomb field momentum does not enter in (6–49). The point character of the particle causes no difficulties here. Equation (6–49) is independent of the structure of the particle.

So far, the consideration was limited to electromagnetic fields. The particle momenta now enter the picture through the conservation laws as follows.

Since we are considering a closed system, we can make use of momentum conservation: the total momentum of the particle, $p^\mu = mv^\mu$, and of the radiation fields, P^μ, must be conserved,

$$dp^\mu + dP^\mu = 0. \tag{6–50}$$

Combining (6–49) and (6–50) yields

$$ma^\mu = \bar{F}^\mu \equiv \frac{e}{c}\,\bar{F}^{\mu\nu}v_\nu. \tag{6–51}$$

This equation is written in an obvious, suggestive form, but it is *not* the equation of motion of the charged particle, because it does not ensure the asymptotic conditions. Specifically, Eq. (6–37) would have to be combined with it: in the remote past and future the incident radiation field and the particle cannot interact. The equation of motion will be derived from (6–51) and (6–37) in the next section.

If we do not have a closed system, but an external force F^μ_{ext}, the conservation law (6–50) is replaced by

$$dp^\mu + dP^\mu = F^\mu_{\text{ext}}\,d\tau \tag{6–52}$$

in the spirit of classical mechanics [cf. (3–12) and (4–142)]. Instead of (6–51) there results from (6–52) and (6–49)

$$ma^\mu = \bar{F}^\mu + F^\mu_{\text{ext}}. \tag{6–53}$$

Again, this in itself is *not* an equation of motion. The same comments apply as for (6–51).

The difference between Eqs. (6–51) and (6–53) is not only the term F^μ_{ext}. An important difference lies in the fact that in the derivation of (6–53) it was necessary to appeal to Newton's second law (or, rather, to its relativistic generalization), while this was not necessary for the derivation of (6–51). The latter is deduced *entirely* on the basis of the Maxwell-Lorentz equations, the conservation laws, and the asymptotic conditions (6–38), (6–39). We shall return to this observation in Section 6–6.

One more deduction can be made on the basis of the Maxwell-Lorentz equations. It will be shown that $\bar{F}^\mu$ can be written as

$$\bar{F}^\mu \equiv F^\mu_{\text{in}} + \Gamma^\mu, \tag{6–54}$$

where the *Abraham four-vector*, Γ^μ, is given by the important equation

$$\Gamma^\mu = \frac{2}{3}\frac{e^2}{c^3}\left(\dot{a}^\mu - \frac{1}{c^2}a^\lambda a_\lambda v^\mu\right), \tag{6–55}$$

and F^μ_{in} is defined by

$$F^\mu_{\text{in}} \equiv \frac{e}{c}\,F^{\mu\nu}_{\text{in}}v_\nu. \tag{6–56}$$

Equation (6–53) can therefore be expressed in the form

$$ma^\mu = \frac{e}{c} F^{\mu\nu}_{\text{in}} v_\nu + F^\mu_{\text{ext}} + \frac{2}{3}\frac{e^2}{c^3}\left(\dot{a}^\mu - \frac{1}{c^2} a^\lambda a_\lambda v^\mu\right). \qquad (6\text{–}57)$$

This equation is identical with an equation first derived by Lorentz for an extended electron, provided one shrinks the radius of the particle to zero (cf. Section 2–3). It was derived by Dirac on the basis of the Maxwell-Lorentz equations, the conservation law of electromagnetic fields as well as additional assumptions of simplicity [cf. Eq. (2–31)]. The latter were not used in our proof. Instead, we used the conservation laws (6–50) and (6–52) for open and closed systems, respectively.

Equation (6–57) is known as the *Lorentz-Dirac equation*. It will be used to derive the equations of motion by combining it with the asymptotic conditions in the following section.

In the above derivation of the Lorentz-Dirac equation the proofs of Eqs. (6–47) and (6–55) were omitted. They will be given now, though lengthy steps of elementary algebra will be left to the reader.

The proof will consist of two parts. In the first part the fields produced by the moving charge, $F^{\mu\nu}_{\text{ret}}$ and $F^{\mu\nu}_{\text{adv}}$, will be computed in the neighborhood of the world line. This will give us (6–55). The second part will consist of the evaluation of the integral (6–47) of the electromagnetic energy tensor, using the results obtained in the first part. For convenience the units will be so chosen that the velocity of light $c = 1$. The necessary factors of c can then easily (and uniquely) be inserted afterward on the basis of dimensional considerations.

The starting point is Eq. (4–101) for the field strengths. In the notation (5–2) this equation reads

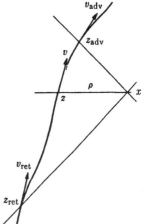

$$F^{\mu\nu}_{\substack{\text{ret}\\\text{adv}}}(x) = \pm \frac{2e}{\rho}\frac{d}{d\tau}\left(\frac{v^{[\mu}R^{\nu]}}{\rho}\right). \qquad (4\text{–}101)$$

Equation (4–89) and (4–90) for u^μ, ρ, and R^μ will be used extensively in the following.

Figure 6–1 illustrates the geometry. The field point x is simultaneous with a point τ_0 on the world line, as seen in the rest frame of the particle. If, for convenience, we take $\tau_0 = 0$ and omit the argument (0) in all quantities referring to $\tau = 0$, then

$$v^\mu(z_\mu - x_\mu) = 0.$$

The vector from z to x is therefore

$$x^\mu - z^\mu = \rho u^\mu \qquad (6\text{–}58)$$

in the notation of (4–89).

FIG. 6–1. Notation for the evaluation of the fields in the neighborhood of the world line.

We are interested in the limit $\rho \to 0$ and shall therefore expand all quantities in the vicinity of $\tau = 0$. The retarded position associated with x is $z(-\tau)$,

$$z^\mu(-\tau) = z^\mu - \tau v^\mu + \frac{\tau^2}{2} a^\mu - \frac{\tau^3}{6} \dot{a}^\mu + \cdots$$

The small quantity τ is positive throughout. In (4–101) we must now differentiate with respect to $-\tau$. Similar statements hold for the advanced case with position $z(\tau)$ $(\tau > 0)$. Thus, (4–101) is conveniently rewritten

$$F^{\mu\nu}_{\substack{\text{ret} \\ \text{adv}}}(x) = \pm \frac{2e}{\rho(\mp\tau)} \frac{d}{d(\mp\tau)} \left(\frac{v^{[\mu}(\mp\tau) R^{\nu]}(\mp\tau)}{\rho(\mp\tau)} \right)$$

$$= \frac{-2e}{\rho(\mp\tau)} \frac{d}{d\tau} \left(\frac{v^{[\mu}(\mp\tau) R^{\nu]}(\mp\tau)}{\rho(\mp\tau)} \right). \tag{6-59}$$

The expansions needed are (to the required order of $\tau \sim \rho$)

$$v^\mu(\mp\tau) = v^\mu \mp \tau a^\mu + \frac{\tau^2}{2} \dot{a}^\mu + \cdots$$

and

$$R^\mu(\mp\tau) = x^\mu - z^\mu(\mp\tau) = \rho u^\mu \pm \tau v^\mu - \frac{\tau^2}{2} a^\mu \pm \frac{\tau^3}{6} \dot{a}^\mu + \cdots$$

From these follows $\rho(\mp\tau)$ to third order,

$$\rho(\mp\tau) = \mp v_\mu(\mp\tau) R^\mu(\mp\tau) = \tau(1 + \rho a_u) \mp \frac{\rho\tau^2}{2} \dot{a}_u + \frac{\tau^3}{6} a^2,$$

where

$$a_u \equiv a_\lambda u^\lambda, \qquad \dot{a}_u \equiv \dot{a}_\lambda u^\lambda, \qquad a^2 \equiv a_\lambda a^\lambda.$$

The relation between τ and ρ is given by the requirement that $R(\mp\tau)$ is a null vector,

$$R_\mu(\mp\tau) R^\mu(\mp\tau) = 0.$$

This yields

$$\rho^2 = \tau^2 \left(1 + \rho a_u \mp \frac{\rho\tau}{3} \dot{a}_u + \frac{\tau^2}{12} a^2 \right).$$

To first order, $\rho = \tau$, so that

$$\rho^2 = \tau^2 \left[1 + \rho a_u + \frac{\rho^2}{3} \left(\frac{a^2}{4} \mp \dot{a}_u \right) \right]$$

and*

$$\left(\frac{\tau}{\rho} \right)^2 = 1 - \rho a_u + \rho^2 a_u^2 - \frac{\rho^2}{3} \left(\frac{a^2}{4} \mp \dot{a}_u \right) = \frac{1}{1 + \rho a_u} \left[1 - \frac{\rho^2}{3} \left(\frac{a^2}{4} \mp \dot{a}_u \right) \right].$$

Thus,

$$\frac{\tau}{\rho} = \frac{1 - [(\rho^2/6)(a^2/4 \mp \dot{a}_u)]}{\sqrt{1 + \rho a_u}}. \tag{6-60}$$

---

* One thereby achieves the elimination of the first-order terms from the parentheses.

The rest of the evaluation of $F^{\mu\nu}$ in (6–59) is a matter of substitution. Conveniently writing*

$$\frac{1}{\rho(\mp\tau)} = \frac{1}{\tau}\frac{1}{1+\rho a_u}(1 \pm \tfrac{1}{2}\rho\tau\dot{a}_u - \tfrac{1}{6}\tau^2 a^2),\qquad(6\text{–}61)$$

which is easily seen to agree with the above expansion for $\rho(\mp\tau)$, we have

$$\frac{\nu^{[\mu}(\mp\tau)R^{\nu]}(\mp\tau)}{\rho(\mp\tau)} = \frac{1}{1+\rho a_u}\left(\frac{\rho}{\tau}\,v^{[\mu}u^{\nu]} \mp \rho a^{[\mu}u^{\nu]} + \frac{\tau}{2}\,v^{[\mu}a^{\nu]} \pm \tfrac{1}{2}\rho^2\dot{a}_u v^{[\mu}u^{\nu]}\right.$$
$$\left. - \tfrac{1}{6}\rho\tau a^2 v^{[\mu}u^{\nu]} \pm \frac{\tau^2}{3}\,\dot{a}^{[\mu}v^{\nu]} + \frac{\rho\tau}{2}\,\dot{a}^{[\mu}u^{\nu]}\right).$$

Differentiation with respect to τ and multiplication by (6–61) yields

$$F_{\substack{\text{ret}\\\text{adv}}}^{\mu\nu}(x) = \frac{2e}{(1+\rho a_u)^2}\left(-\frac{\rho}{\tau^3}\,v^{[\mu}u^{\nu]} + \frac{1}{2\tau}\,v^{[\mu}a^{\nu]}\right.$$
$$\left. \mp \frac{\rho^2}{2\tau^2}\,\dot{a}_u\,v^{[\mu}u^{\nu]} \pm \tfrac{2}{3}\dot{a}^{[\mu}v^{\nu]} + \frac{\rho}{2\tau}\,\dot{a}^{[\mu}u^{\nu]}\right).$$

We can now make use of the relation (6–60) to eliminate τ. Since $\tau\sqrt{1+\rho a_u}$ occurs naturally, it is best to take a factor $(1+\rho a_u)^{3/2}$ in the denominator inside the parentheses. The result is

$$F_{\substack{\text{ret}\\\text{adv}}}^{\mu\nu}(x) = \frac{2e}{\sqrt{1+\rho a_u}}\left(\frac{1}{\rho^2}\,v^{[\mu}u^{\nu]} - \frac{1}{2\rho}\,v^{[\mu}a^{\nu]} + \tfrac{1}{2}a_u v^{[\mu}a^{\nu]} + \frac{a^2}{8}\,v^{[\mu}u^{\nu]}\right.$$
$$\left. - \tfrac{1}{2}\dot{a}^{[\mu}u^{\nu]} \mp \tfrac{2}{3}\dot{a}^{[\mu}v^{\nu]}\right).\qquad(6\text{–}62)$$

The remarkable observation first made by Dirac is that $F_{\text{ret}}^{\mu\nu}$ and $F_{\text{adv}}^{\mu\nu}$ differ from each other for small ρ only in the last term of (6–62); in the limit $\rho \to 0$,

$$F_{-}^{\mu\nu}(z) \equiv \tfrac{1}{2}(F_{\text{ret}}^{\mu\nu}(z) - F_{\text{adv}}^{\mu\nu}(z)) = -\frac{4e}{3c^4}\,\dot{a}^{[\mu}v^{\nu]}.\qquad(6\text{–}63)$$

The factors of c have here been supplied again.

Now Γ^{μ} is defined by (6–54), (6–51), and (6–56), so that

$$\Gamma^{\mu} \equiv \bar{F}^{\mu} - F^{\mu}{}_{\text{in}} \equiv \frac{e}{c}\,(\bar{F}^{\mu\nu} - F^{\mu\nu}_{\text{in}})v_{\nu}.$$

Equations (6–45) and (6–46) permit us to write this as

$$\Gamma^{\mu} = \frac{e}{c}\,\tfrac{1}{2}(F_{\text{out}}^{\mu\nu} - F_{\text{in}}^{\mu\nu})v_{\nu} = \frac{e}{c}\,F_{-}^{\mu\nu}v_{\nu}.\qquad(6\text{–}64)$$

The result (6–63) can be used in this expression to yield

$$\Gamma^{\mu} = -\frac{2}{3}\frac{e^2}{c^5}\,(-\dot{a}^{\mu}c^2 - v^{\mu}\dot{a}^{\lambda}v_{\lambda}).$$

---

* One thereby achieves the elimination of the first-order terms from the parentheses.

This agrees with (6-55) since $v_\lambda v^\lambda = -c^2$ implies $a_\lambda v^\lambda = 0$ and $\dot{a}^\lambda v_\lambda = -a^\lambda a_\lambda$ for any point τ on the world line. Equation (6-55) is thus proven and we can turn to the second part, the proof of Eq. (6-47).

Consider a tube of radius ρ surrounding the world line of the charged particle between times τ_1 and τ_2. Thus, the point x in Fig. 6-1 is on the surface of this tube. The surface is a three-dimensional timelike surface with normal vector u^μ which is spacelike,

$$d^3\sigma^\mu = u^\mu \, d^3\sigma. \tag{6-65}$$

The three-dimensional surface element $d^3\sigma$ consists of the two-dimensional surface element $d^2\sigma = \rho^2 \, d\Omega$ ($d\Omega$ is the element of solid angle), which belongs to the surface of a sphere in ordinary three-dimensional space, and the magnitude $\alpha \, d\tau$ of a timelike vector parallel to v^μ. [Compare the similar but simplified problem in connection with Eq. (5-11).]

To find α we note that if the vector $x^\mu - z^\mu = \rho u^\mu$ in Fig. 6-1 is slightly displaced along the world line, it must remain orthogonal to v^μ:

$$\frac{d}{d\tau}[(x^\mu - z^\mu)v_\mu] = 0.$$

This gives

$$(dx^\mu - v^\mu \, d\tau)v_\mu + \rho u^\mu a_\mu \, d\tau = 0$$

or

$$v_\mu \, dx^\mu = -(1 + \rho a_u) \, d\tau.$$

Thus, the displacement dx^μ of x^μ which keeps $x^\mu - z^\mu$ orthogonal to v^μ has a component along v^μ which is $(1 + \rho a_u)v^\mu \, d\tau$. Hence $\alpha = 1 + \rho a_u$, and

$$d^3\sigma = (1 + \rho a_u) \, d\tau \rho^2 \, d\Omega. \tag{6-66}$$

Consider now the momentum of the electromagnetic field crossing the tube between τ_1 and τ_2,

$$-\int_1^2 \Theta^{\mu\nu} \, d^3\sigma_\nu = -\int_{\tau_1}^{\tau_2} d\tau \int d\Omega \, \Theta^{\mu\nu} u_\nu \rho^2 (1 + \rho a_u). \tag{6-67}$$

The energy tensor $\Theta^{\mu\nu}$ is to be constructed according to (4-114) by means of the solution (6-43) of the Maxwell-Lorentz equation (6-42). This solution can be written

$$F^{\mu\nu} = F_{\text{in}}^{\mu\nu} + F_+^{\mu\nu} + F_-^{\mu\nu} = F_+^{\mu\nu} + \overline{F}^{\mu\nu},$$

using (6-45) and (6-46). $\overline{F}^{\mu\nu}$ is of course given, but $F_+^{\mu\nu}$ must be obtained from (6-62). The substitution into (6-67) is straightforward but tedious. The fact that only the component in the u-direction is needed gives rise to considerable simplification. For small ρ, keeping only terms of order $1/\rho^n$ ($n \geq 2$), one obtains

$$\Theta^{\mu\nu} u_\nu = \frac{e^2/4\pi}{1 + \rho a_u}\left[\left(\frac{1}{2\rho^4} - \frac{a^2}{2\rho^2}\right)u^\mu - \left(\frac{1}{2\rho^3} - \frac{3a_u}{4\rho^2}\right)a^\mu\right] + \frac{e/4\pi}{\rho^2}\overline{F}^{\mu\nu}v_\nu. \tag{6-68}$$

The first term is due to $\Theta^{\mu\nu}(F_+)$; the second term is due to the cross term in $\Theta^{\mu\nu}$ between $F_+^{\alpha\beta}$ and $\overline{F}^{\alpha\beta}$; there is no contribution in the desired order from $\Theta^{\mu\nu}(\overline{F})$.

The integration in (6-67) gives zero for all terms involving u linearly. The $(1 + \rho a_u)$-factor in the denominator cancels and the only surviving terms are

$$-\int_1^2 \Theta^{\mu\nu} d^3\sigma_\nu = \frac{e^2}{4\pi} \int_1^2 \frac{d\Omega}{2\rho} a^\mu d\tau - \frac{1}{4\pi} \int_1^2 \overline{F}^\mu d\Omega\, d\tau.$$

Here we used the definition (6-51). Since a spherical charge distribution contained inside a sphere of radius ρ has an electrostatic energy outside this sphere which is $e^2/(2\rho)$, one makes the identification [cf. (6-21) and (6-22)]

$$\frac{dP^\mu_{\text{Coul}}}{d\tau} = \frac{d}{d\tau}\left(\frac{e^2}{2\rho} v^\mu\right). \tag{6-69}$$

P^μ_{Coul} is exactly the electromagnetic momentum P^μ_{elm} encountered in Section 6-3; in that section it was derived from the velocity fields. Since we have also an acceleration field present, the more specific index "Coul" is preferable in the present context. In this derivation P^μ_{Coul} is entirely due to $F_+^{\mu\nu}$.

The integration over $d\Omega$ is trivial, yielding

$$-\int_1^2 \Theta^{\mu\nu} d^3\sigma_\nu = \int_1^2 \left(\frac{dP^\mu_{\text{Coul}}}{d\tau} - \overline{F}^\mu\right) d\tau, \tag{6-47}$$

which is the equation we set out to prove.

Consider two tubes surrounding the world line, so that they have the same ends at 1 and at 2 but different shapes between 1 and 2. It is easily seen that the result (6-47) is independent of this shape, for the four-dimensional toroidal volume enclosed by the two tubes involves no sources, so that $\partial_\mu\Theta^{\mu\nu} = 0$ at every point in it. The flux through one tube is therefore the same as through the other. Equation (6-47) therefore depends only on the end points 1 and 2.

6-6 THE EQUATIONS OF MOTION*

The Lorentz-Dirac equation

$$ma^\mu = F^\mu_{\text{in}} + F^\mu_{\text{ext}} + \Gamma^\mu \tag{6-57}$$

with

$$F^\mu_{\text{in}} = \frac{e}{c} F^{\mu\nu}_{\text{in}} v_\nu \tag{6-56}$$

and

$$\Gamma^\mu = \frac{2}{3}\frac{e^2}{c^3}\left(\dot{a}^\mu - \frac{1}{c^2} a^\lambda a_\lambda v^\mu\right) \tag{6-55}$$

---

* F. Rohrlich, *Ann. Phys.* (N.Y.) **13**, 93 (1961).

must be combined with the asymptotic condition

$$\lim_{\tau \to \infty} a^{\mu}(\tau) = 0 \qquad (6\text{-}37)$$

in order to yield an equation of motion. In Section 6–4 it was emphasized that this asymptotic condition must be an integral part of an equation of motion.

We define

$$\tau_0 \equiv \frac{2}{3} \frac{e^2}{mc^3} \qquad (6\text{-}70)$$

and

$$K^{\mu}(\tau) = F_{\text{in}}^{\mu} + F_{\text{ext}}^{\mu} - \frac{1}{c^2} \Re v^{\mu}, \qquad (6\text{-}71)$$

where $\Re$ is the total radiation energy rate (5–14). According to (5–13), the last term is therefore just the rate at which electromagnetic four-momentum is emitted.

The Lorentz-Dirac equation now has the form

$$m(a^{\mu} - \tau_0 \dot{a}^{\mu}) = K^{\mu}.$$

It is easy to integrate this third-order differential equation for $z^{\mu}(\tau)$ formally and to obtain a second-order equation. Multiplication by the integrating factor $e^{-\tau/\tau_0}$ yields

$$-\frac{d}{d\tau}\left(e^{-\tau/\tau_0} a^{\mu}(\tau)\right) = \frac{1}{m\tau_0} e^{-\tau/\tau_0} K^{\mu}(\tau). \qquad (6\text{-}72)$$

The asymptotic condition (6–37) implies the weaker condition

$$\lim_{\tau \to \infty} e^{-\tau/\tau_0} a^{\mu}(\tau) = 0. \qquad (6\text{-}73)$$

Equation (6–72) can therefore be integrated between τ and ∞; one finds

$$a^{\mu}(\tau) = \frac{e^{+\tau/\tau_0}}{m\tau_0} \int_{\tau}^{\infty} e^{-\tau'/\tau_0} K^{\mu}(\tau')\, d\tau'. \qquad (6\text{-}74)$$

The asymptotic condition (6–37) can be stated as the following restriction on this integral:

$$\lim_{|\tau| \to \infty} \left[e^{\tau/\tau_0} \int_{\tau}^{\infty} e^{-\tau'/\tau_0} K^{\mu}(\tau')\, d\tau' \right] = 0. \qquad (6\text{-}75)$$

Equation (6–74) is not a solution for a^{μ} because a^{μ} also occurs in the integrand,

$$a^{\mu}(\tau) = \int_{\tau}^{\infty} e^{(\tau - \tau')/\tau_0} \left[\frac{1}{m\tau_0}(F_{\text{in}}^{\mu}(\tau') + F_{\text{ext}}^{\mu}(\tau')) - \frac{1}{c^2} a^{\lambda}(\tau') a_{\lambda}(\tau') v^{\mu}(\tau') \right] d\tau'. \qquad (6\text{-}76)$$

Furthermore, this expression must again be integrable up to ∞, since $v^{\mu}(\tau)$ must exist even in the limit, according to (6–38). Thus (6–75) becomes a *necessary* condition for the existence of the momenta p^μ,

$$p^\mu(\tau) = p_{\text{in}}^\mu + \frac{1}{\tau_0} \int_{-\infty}^{\tau} e^{\tau'/\tau_0}\, d\tau' \int_{\tau'}^{\infty} e^{-\tau''/\tau_0} K^\mu(\tau'')\, d\tau'', \qquad (6\text{--}77)_{\text{in}}$$

$$p^\mu(\tau) = p_{\text{out}}^\mu - \frac{1}{\tau_0} \int_{\tau}^{\infty} e^{\tau'/\tau_0}\, d\tau' \int_{\tau'}^{\infty} e^{-\tau''/\tau_0} K^\mu(\tau'')\, d\tau''. \qquad (6\text{--}77)_{\text{out}}$$

The existence of these equations, i.e. the convergence of the integrals involved in both equations guarantees the asymptotic behavior (6–37) and (6–38). Condition (6–39) is ensured by these conditions together with the Maxwell-Lorentz equations. Thus, Eqs. (6–77) fulfill all the asymptotic conditions. They can therefore be called *the equations of motion of a charged particle.*

Equation (6–77) is not the familiar form resembling Newton's second law in relativistic form, (3–12). In fact, it corresponds more to

$$p^\mu(\tau) = p_{\text{in}}^\mu + \int_{-\infty}^{\tau} F^\mu(\tau')\, d\tau'. \qquad (6\text{--}78)$$

The more familiar form is (6–74) when multiplied by m. By introducing the variable

$$\alpha = \frac{\tau' - \tau}{\tau_0}, \qquad (6\text{--}79)$$

(6–74) becomes

$$ma^\mu(\tau) = \int_{0}^{\infty} K^\mu(\tau + \alpha\tau_0)\, e^{-\alpha}\, d\alpha. \qquad (6\text{--}80)$$

This equation, together with the asymptotic conditions (6–37) and (6–38), is equivalent to the equations of motion (6–77).

We can compare (6–80) with (3–12),

$$ma^\mu(\tau) = F^\mu(\tau). \qquad (3\text{--}12)$$

The difference is twofold. Mathematically, since $F^\mu(\tau)$ is a known function of $z^\mu(\tau)$ and possibly v^μ, (3–12) is a second-order ordinary differential equation for the vector z^μ. Equation (6–80), on the other hand, is an ordinary second-order *integrodifferential equation,* since z^μ, as well as v^μ and a^μ, occur in the integrand. Physically, (3–12) is a local equation (in time); the acceleration at the time τ depends only on the force F^μ at that time τ. Equation (6–80) is *nonlocal* in time; $a^\mu(\tau)$ depends on the force at time τ and also at all times later than τ. The implications of this new physical situation will be studied in Section 6–7, while the mathematical problems raised by (6–80) will be deferred to Section 6–8.

But there is a third difference between (6–80) and (3–12). The former *must* be accompanied explicitly by the asymptotic conditions (6–37) and (6–38) which are not ensured by (6–80). Note that (6–80) implies only the weaker condition (6–72).

We now turn to the comparison of (6–77) with (6–78). The same comments
on the mathematical structure apply as before. But now, in order to ensure
the existence of asymptotic momenta, both equations (6–77) are needed, while
the form (6–78) need not hold for the dynamics of other types of interaction
(p_{in}^μ or p_{out}^μ need not exist, as in the case of periodic motion).

One concludes that the equations of motion for charged particles (6–77) contain all the asymptotic conditions implicitly. They can be combined to yield

$$p_{out}^\mu - p_{in}^\mu = \frac{1}{\tau_0} \int_{-\infty}^{\infty} e^{\tau'/\tau_0}\, d\tau' \int_{\tau'}^{\infty} e^{-\tau''/\tau_0} K^\mu(\tau'')\, d\tau''. \tag{6-81}$$

This is an integral equation for $z^\mu(\tau)$ in terms of the given forces and the given
"initial" values, $p^\mu(-\infty) = p_{in}^\mu$ and $p^\mu(+\infty) = p_{out}^\mu$.

The usual problems in classical dynamics do not specify p_{in}^μ and p_{out}^μ, but
rather specify $z^\mu(\tau_i)$ and $v^\mu(\tau_i)$ at some finite time τ_i. The asymptotic momenta
are thereby not specified, but their existence must be guaranteed, according to
(6–38). In this case, it is more convenient to use as equations of motion (6–80)
together with the conditions (6–38). Jointly, they are equivalent to the two
equations (6–77). A formal integration of (6–80) incorporating these initial
values stated together with (6–38) gives

$$z^\mu(\tau) = z^\mu(\tau_i) + v^\mu(\tau_i)(\tau - \tau_i) + \frac{1}{m} \int_{\tau_i}^{\tau} d\tau' \int_{\tau_i}^{\tau'} d\tau'' \int_{0}^{\infty} K^\mu(\tau' + \alpha\tau_0)\, e^{-\alpha}\, d\alpha, \tag{6-82}$$

$$\lim_{\tau \to -\infty} v^\mu(\tau) = v_{in}^\mu \quad \text{exists}, \qquad \lim_{\tau \to +\infty} v^\mu(\tau) = v_{out}^\mu \quad \text{exists}.$$

This set of equations can be regarded as the *equations of motion* for the determination of the orbit $z^\mu(\tau)$ of a charged particle in a given force field and with
initial values $z^\mu(\tau_i)$ and $v^\mu(\tau_i)$.

For most problems, the set (6–82) is more convenient than the set (6–77).
However, we shall see later that a higher-level theory (quantum electrodynamics) leads to a description of a charged particle system which is much
more similar to (6–81) than to (6–82).

From these considerations emerges the fact that the equations of motion of a
charged particle, in contradistinction to other equations of motion usually encountered, can take on a variety of forms, depending on the initial conditions.
Since certain asymptotic conditions must be part of the equations which determine the dynamics, it does make a difference whether or not the given
initial values are specified asymptotically. If they are, (6–77) can be used; if
they are not, (6–82) can be used.

It is clear that (6–82) with initial values at τ_i is equivalent to (6–77) when
the "initial" values of the latter are specified as p_{in}^μ and p_{out}^μ. Furthermore, these
equations are equivalent to the Lorentz-Dirac equation together with the asymptotic conditions and either set of initial values. The equivalence follows from
the reversibility of the argument leading from (6–57) to (6–77). Differentiation
of the latter gives the Lorentz-Dirac equation.

From the derivation of the Dirac equation and this equivalence we conclude *that the equations of motion of a charged particle can be derived from the Maxwell-Lorentz equations, the momentum conservation law, and the asymptotic conditions; no other assumption needs to be made.*

PROBLEM 6–4

Contract the Lorentz-Dirac equation with a^μ and show, by an argument analogous to the one yielding (6–74), that the acceleration satisfies the integral equation

$$a^\mu(\tau)a_\mu(\tau) = \frac{1}{m}\int_0^\infty e^{-\alpha}F^\mu(\tau + \tfrac{1}{2}\tau_0\alpha)\, a_\mu(\tau + \tfrac{1}{2}\tau_0\alpha)\, d\alpha, \qquad (6\text{–}83)$$

where F is the total force.

6–7 PHYSICAL MEANING OF THE EQUATIONS OF MOTION

The integrodifferential equations of motion (6–82) contain K^μ in their integrand. This vector is given in (6–71) and has an obvious interpretation. It is the *effective force* responsible for the acceleration of the particle. K^μ consists of the impressed forces, namely the incident radiation field and the external forces, as well as the *radiation reaction force*, $-\Re v^\mu$. The latter is the negative of $\Re v^\mu$, the rate at which electromagnetic radiation momentum is emitted. This rate was derived from Maxwell's equations in Section 5–1 and is not new to us [cf. Eq. (5–13)]. However, it is to be noted that this force is a *timelike* vector, in contradistinction to the usual force four-vectors which are spacelike. There is of course no reason why all forces must be spacelike. Such a requirement can be made only if one restricts himself to certain types of equations of motion. For example, the Newtonian type equation (3–12) could not hold if F^μ were not orthogonal to the velocity v^μ, since the left-hand side has this property.

If one regards the Lorentz-Dirac equation (6–57) as the equation of motion, he is tempted to identify the Abraham four-vector Γ^μ with the radiation reaction. This, however, leads to various difficulties: it is possible that at a particular instant $a^\mu = 0$ but $\dot{a}^\mu \neq 0$; this results in no radiation emission, $\Re = 0$, but a nonvanishing "radiation reaction," Γ^μ. Conversely, it is possible that $\Gamma^\mu = 0$ but that radiation is being emitted, $\Re \neq 0$. This is the case whenever

$$\dot{a}^\mu - \frac{1}{c^2}a^\lambda a_\lambda v^\mu = 0, \qquad a^\lambda a_\lambda \neq 0.$$

We recognize this equation as the condition for uniform acceleration (cf. Problem 5–3). Thus, in uniformly accelerated motion radiation is emitted while the "radiation reaction" Γ^μ vanishes. For these reasons the interpretation of Γ^μ as radiation reaction force is to be rejected. Obviously, the radiation reaction $-\Re v^\mu$ vanishes if and only if no radiation is emitted, $\Re = 0$.

With this physical interpretation of K^μ we can now consider the equations of motion (6–82). The most interesting feature is its nonlocality in time. In order to understand it better, let us first look at (6–80). The parameter τ_0 which arises in a natural way in these equations is a very short universal-time interval. For an electron,

$$\tau_0 = \frac{2}{3}\frac{e^2}{mc^3} = 0.62 \times 10^{-23}\ \text{sec.}$$

For other elementary particles τ_0 is much smaller; for macroscopic charged particles τ_0 is smaller still. Thus, for a classical system K^μ will change very little over a time interval τ_0, so that in good approximation

$$ma^\mu(\tau) = K^\mu(\tau + \xi\tau_0) \doteq K^\mu(\tau). \qquad (6\text{–}84)$$

One thus recovers the local theory approximately. It is interesting to note that the difference between (6–84) with $\xi = 0$ and the Lorentz-Dirac equation is exactly the term

$$\frac{2}{3}\frac{e^2}{c^3}\dot{a}^\mu.$$

This term is sometimes called the *Schott term.** It is the term by which Γ^μ differs from the radiation reaction. It is also the term whose physical interpretation has been obscure. It can now be identified as responsible for the nonlocal time dependence of the equation of motion, as is evident from the derivation of the integrodifferential equation: the integrating factor eliminates the Schott term.

If we keep $\xi \neq 0$ in (6–84), the interpretation of this equation would be as follows. The acceleration at time τ is determined by the effective force at the very slightly later time $\tau + \xi\tau_0$ (ξ is of order one). Thus, the equation of motion differs from the usual Newtonian equation (3–12) by a slight time delay between acceleration and effective force.

A question often raised at this point is: does the nonlocal time behavior constitute a violation of causality? In Section 3–15 we listed three different meanings of causality. If causality means predictability, the equations of motion certainly are causal, since one can determine the future development of the system from certain initial data [use of Eq. (6–82)]. Equation (6–80) is a *second*-order integrodifferential equation, so that its solution is determined by initial position and velocity. The need to know the external forces for all future time is not an objection, since these forces must be given and we have agreed not to interfere with the motion after the initial instant; otherwise, of course, the equations cannot predict the motion.

---

* G. A. Schott's contributions to classical electrodynamics are largely contained in his book *Electromagnetic Radiation*, Cambridge University Press, 1912.

The second meaning of causality, namely that of finite signal velocities ($\leq c$), also does not pose a problem. So long as we have relativistic invariance this is ensured.

The third meaning refers to the absence of advanced effects, e.g. the absence of signals sent backward in time (with velocity $\leq c$ to be sure). Assume that there is a force acting on the particle only during the time interval $\Theta: \tau_1 \leq \tau' \leq \tau_2$. As is obvious from (6–74), a nonvanishing acceleration at a time $\tau < \tau_1$ is then possible, since the whole future motion contributes to $a^\mu(\tau)$. Thus, it appears as though the force at time $\tau'\epsilon\Theta$ had sent a signal along the world line (backward in time), telling the particle that it will act on it at a later time; the particle responds to this and shows $a^\mu(\tau) \neq 0$ for $\tau < \tau_1$. Such a preacceleration seems indeed to be a violation of causality (in the third meaning of causality).

As is apparent from (6–84), preacceleration effects occur at times of the order of τ_0 prior to the action of the force. Thus, in order to exhibit such an effect, one must "switch on" the force over a time interval $\lesssim \tau_0$. No classical interactions, i.e. interactions which can be described by nonquantum physics, are known that have time or space variations corresponding to such short intervals. No classical particle can enter from a free state into interaction over a time interval of order τ_0.

However, even if this is not possible, a violation of causality could still take place when both force and acceleration have been present for a long time, as long as $a^\mu(\tau)$ can be shown to be equal to $(1/m)K^\mu(\tau + \xi\tau_0)$ with $\xi > 0$ rather than to $K^\mu(\tau)$, according to (6–84). This requires measurements of two quantities with a time resolution of the order τ_0. No measurements of this accuracy are possible.

Consider the *Schott energy term*, i.e. the time component of $\dot{a}^\mu$. Clearly, the observation of $\dot{a}^\mu$ in the Lorentz-Dirac equation is equivalent to the observation of the nonlocal effect in the integrodifferential equation. The time component of (6–57) is

$$\frac{dE}{d\tau} = \gamma \mathbf{v} \cdot \mathbf{F} + \tau_0 \frac{d}{d\tau} (\gamma \mathbf{v} \cdot \mathbf{F}) - \gamma \mathcal{R}.$$

Here we used (A1–39) and the observation that, since the Schott term is small, $\dot{a}_\mu = \dot{F}_\mu/m$ is a permissible approximation in that term. $E = \gamma mc^2$ is the kinetic energy of the particle. Since $\gamma\, d\tau = dt$ and since in first approximation

$$\frac{dE}{d\tau} = \gamma \mathbf{v} \cdot \mathbf{F},$$

we have

$$\frac{dE}{dt} = \mathbf{v} \cdot \mathbf{F} + \tau_0 \frac{d}{dt} \left(\frac{E}{mc^2} \frac{dE}{dt} \right) - \mathcal{R}.$$

One of the easiest ways to detect radiation from charged particles is to observe synchrotron radiation, which was discussed in Section 5–4. In this case $\mathbf{v} \cdot \mathbf{F} = 0$, so that the radiation rate

$$\mathfrak{R}(t) = -\frac{dE}{dt} + \tau_0 \frac{d}{dt}\left(\frac{E}{mc^2}\frac{dE}{dt}\right) \doteq -\frac{dE}{dt} - \tau_0 \frac{d}{dt}\left(\frac{E\mathfrak{R}}{mc^2}\right).$$

Since $\mathfrak{R}$ is proportional to E^4 for high energies [cf. Eq. (5–46)], the last term is approximately

$$\tau_0 \frac{d}{dt}(E\mathfrak{R}) = 5\tau_0\,\mathfrak{R}\,\frac{dE}{dt}.$$

Thus

$$-\frac{dE}{dt} = \frac{\mathfrak{R}}{1 + 5\tau_0\mathfrak{R}/(mc^2)} \doteq \mathfrak{R}\left(1 - 5\tau_0\frac{\mathfrak{R}}{mc^2}\right).$$

The observation of the τ_0-term hinges on the measurements of kinetic energy loss and radiation energy. If these are not exactly equal, their difference would be due to the Schott energy, i.e. due to causality violation. Now, if these energies can be measured with an accuracy of one percent, say, the radiation energy must be $> 500(mc^2/\tau_0)$. For an electron this means energy emission of 10^{26} MeV/sec or larger, which is more than the energy rates at which pair production takes place. This rate can therefore not belong in the domain of classical physics.

Another, *a priori* more favorable system for the observation of causality violation in classical physics is uniform acceleration. In Section 6–11 we shall find that it leads to similar conclusions. Thus, the best one can hope for are quantum mechanical techniques of observation. Even if these were successful, which does not seem possible in the present framework of quantum mechanics, one could not establish causality violation of the above equations, because these are classical equations; their competence cannot be extended into the quantum domain.

The conclusion concerning causality violation is therefore the following. The equations of motion (6–82) or (6–80) are causal in the sense of predictability as well as restriction of signal velocities to $\leq c$. Causality violations in the sense of preacceleration are in principle implied but are, even in the most favorable cases, outside the domain of classical systems and classical measurements; and there is considerable doubt that effects of this size could be observable in principle even in quantum mechanics.

These conclusions about causality violation were based on the particle equations (6–57) or (6–80). These equations were obtained from the coupled system of the Maxwell-Lorentz equations and from

$$ma^\mu(\tau) = \frac{e}{c}\,\bar{F}^{\mu\nu}(z)\,v_\nu(\tau) \tag{6–51}$$

by elimination of the field variables. It is clear that (6–51) is a local equation, devoid of preacceleration, and shows *no causality violation*, as long as one does not express $\bar{F}^{\mu\nu}$ in terms of its sources, i.e. the particle variables.

As will be proven at the end of Section 6–9, the conservation laws and, in particular, energy conservation hold *exactly* (not only within time intervals of order τ_0). This result arises from the customary treatment of field energy in its own right, the conservation law referring to the sum of particle and field energy.

Thus, the occurrence of nonlocal phenomena and preacceleration is clearly dependent on the choice of variables (elimination of field variables) and the resultant form of the equations. When the fields are not eliminated, no questions of causality violation arise.

6–8 EXISTENCE, UNIQUENESS, AND PERTURBATION SERIES OF THE SOLUTIONS

It is necessary to study the equations of motion mathematically, inquiring about the existence of solutions and about their uniqueness. More precisely, we are interested in the conditions under which the integrodifferential equations of motion have a solution; furthermore, we want to know whether these solutions are unique when the usual physically reasonable initial conditions are specified.

The existence problem was solved for very weak conditions by Hale and Stokes.* Their proof requires rather advanced mathematical techniques which are beyond the level of this exposition. Consequently it will suffice to give here only the essential features of their results and omit all proofs.

The starting point is not the equation of motion in the form (6–82) but rather the Lorentz-Dirac equation together with the asymptotic conditions (6–37) and (6–38). Since we are concerned with mathematical questions, it will be convenient to choose the particle mass and the velocity of light as units (together with the centimeter, say). Thus, (6–57) can be written

$$a^\mu(\tau) = F^\mu(z, v, \tau) + \tau_0[\dot{a}^\mu(\tau) - a^\lambda(\tau)\,a_\lambda(\tau)\,v^\mu(\tau)]. \qquad (6\text{--}85)$$

Since z, v, a, F are vectors† in a space with indefinite metric, it is convenient to define a quantity which estimates the size of these vectors. Let

$$\|z\| \equiv \sqrt{z_0^2 + \mathbf{z}^2}$$

and analogously $\|v\|$, $\|a\|$, etc. For example $\|v\| = \sqrt{\gamma^2 + \gamma^2 \mathbf{v}^2} = \sqrt{2\gamma^2 - 1} \geq 1$. The asymptotic conditions are then stated in more precise form as

$$\lim_{\tau \to \infty} \|a(\tau)\| = 0, \qquad \int_0^{\pm\infty} \|a\|\,d\tau < \infty, \qquad (6\text{--}37')$$

and

$$\lim_{\tau \to \infty} \|v\| \quad \text{exists}, \qquad \lim_{\tau \to \infty} \frac{\|z\|}{\tau} \quad \text{exists}. \qquad (6\text{--}38')$$

---

* J. K. Hale and A. P. Stokes, *J. Math. Phys.* **3**, 70 (1962).
† Since we do not refer to components, the indices can be omitted.

The difficulty in studying (6–85) lies in the asymptotic conditions. If these would not have to be satisfied, the existence and uniqueness of the solutions of this equation would follow from standard and well-known theorems on ordinary differential equations.* Since (6–85) is of *third* order, a solution $z(\tau)$ would exist and be unique over any finite time interval when the *three* initial values $z(0)$, $v(0)$, and $a(0)$ are specified and certain analyticity conditions are satisfied. The latter are usually fulfilled in actual physical systems. The need to specify the initial *acceleration*, however, indicates that the Lorentz-Dirac equation by itself cannot be an equation of motion. The asymptotic conditions are exactly what is needed to eliminate this requirement: the third-order Lorentz-Dirac equation, together with the asymptotic conditions (6–37), is equivalent to a *second*-order equation, (6–57). Therefore, among all the possible initial accelerations which could be chosen, only the ones which give a solution satisfying (6–37′) are admissible. This is the much more difficult problem of the existence of solutions with specified asymptotic conditions.

The most important results† concerning the existence of solutions of the equations of motion can be stated in the following two theorems:

THEOREM 1

There exists a solution $z(\tau)$ $(0 \leq \tau < \infty)$ of the equation of motion (6–80) (or 6–82) which satisfies the asymptotic conditions (6–37′) and (6–38′) for any initial set $z(0)$ and $v(0)$, provided

(a) $\|z(0)\| < \infty$,

(b) $\|F(z, v, \tau)\| \leq \phi(\tau)$ is continuous,

(c) $\phi(\tau) \to 0$ for $\tau \to \infty$,

(d) $\int_0^\infty \phi(\tau)\, d\tau < \infty$.

THEOREM 2

There exists a solution $z(\tau)$ $(\tau_i \leq \tau < \infty)$ of the equation of motion (6–80) (or 6–82) which satisfies the asymptotic conditions (6–37′) and (6–38′) for any initial set $z(\tau_i)$, $v(\tau_i)$, provided

(a) $\|F(z, v, \tau)\| \leq \dfrac{b + c\|v\|}{\|z\|^\alpha}$ $(b \geq 0, c \geq 0, \alpha > 1)$,

(b) τ_i is such that

$$\inf_{\tau \geq \tau_i} \left\| \frac{z(\tau_i)}{\tau} + \frac{v(\tau_i)(\tau - \tau_i)}{\tau} \right\| \geq \beta(b + c\|v(\tau_i)\|) + \left(\frac{1 + c\beta}{1 - \tau_0/\tau'_0} \right)^{1/\alpha},$$

where $\beta \equiv 1/(\alpha - 1)\tau_i^{\alpha-1}$ and τ'_0, defined by

$$\tau'_0 = \frac{\tau_i^\alpha}{(b + c\|v(\tau_i)\|)(\|v(\tau_i)\| + \beta(b + c\|v(\tau_i)\|)},$$

satisfies $\tau'_0 > \tau_0 > 0$.

---

* See, for example, E. Goursat, *Differential Equations*, Dover Publications, New York, 1959.

† Hale and Stokes, loc. cit.

Theorem 1 admits all forces which are absolutely integrable and bounded. Theorem 2 admits a larger class of forces but puts restrictions on the initial time τ_i, on $z(\tau_i)$ and $v(\tau_i)$. More general theorems presumably exist; nevertheless, the theorems seem to be general enough to include all cases of physical interest. A few specific cases will be tested in Part C of this chapter.

The next general question is that of uniqueness. It may seem intuitively obvious to physicists that, if a solution of the equations of motions exists and satisfies certain initial conditions $z(\tau_i)$, $v(\tau_i)$, this solution is also unique at least for reasonable forces. Nevertheless, it is essential to have a *proof* of this uniqueness. So far, no such proof has been given. The uniqueness problem must be considered one of the most important unsolved problems of the theory.

In almost all problems in which the equation of motion is used, an exact solution is not known and one resorts to approximation methods. The most obvious of these and also the one almost always used is the *perturbation series approximation*.

This method starts with the observation that τ_0 in (6–80) is very small and a power series expansion is thus suspected to be possible. Let us therefore assume that such a series indeed exists,

$$K(\tau + \alpha\tau_0) = \sum_{n=0}^{\infty} \frac{(\alpha\tau_0)^n}{n!} K^{(n)}(\tau), \qquad (6\text{–}86)$$

where $K^{(n)}$ is the nth derivative of K. If we further assume that the integral in (6–80) is absolutely and uniformly convergent, this summation can be interchanged with the integration and we have

$$ma_\mu(\tau) = \sum_{n=0}^{\infty} \frac{\tau_0^n}{n!} K_\mu^{(n)}(\tau) \int_0^{\infty} \alpha^n e^{-\alpha} \, d\alpha = \sum_{n=0}^{\infty} \tau_0^n K_\mu^{(n)}(\tau). \qquad (6\text{–}87)$$

Written out in detail, the first few terms are

$$ma_\mu(\tau) = F_\mu(\tau) - \mathcal{R}(\tau)v_\mu(\tau) + \tau_0[\dot{F}_\mu(\tau) - \dot{\mathcal{R}}(\tau)v_\mu - \mathcal{R}(\tau)a_\mu] + O(\tau_0^2), \qquad (6\text{–}88)$$

where

$$F_\mu \equiv F_\mu^{\text{in}} + F_\mu^{\text{ext}}.$$

These equations (6–87) show that the nonlocal second-order equation, (6–80), is equivalent to the local equation, (6–87), with an infinite number of derivatives.

The commonly used approximation (zero order)

$$ma^\mu(\tau) = F^\mu(\tau) \qquad (6\text{–}89)$$

neglects radiation completely. This means that it ignores not only the recoil due to the radiation reaction momentum but also the energy lost in the form of radiation.

The next approximation seems to be

$$ma^{\mu}(\tau) \doteq F^{\mu}(\tau) - \frac{1}{c^2}\mathfrak{R}(\tau)\,v^{\mu}(\tau) = K^{\mu}(\tau), \qquad (6\text{–}90)$$

which is just the $(\xi = 0)$-approximation of (6–84). However, this approximation is in general not meaningful unless $\mathfrak{R}$ is negligible. The reason for this is twofold. Firstly, (6–90) is mathematically inconsistent unless $\mathfrak{R} = 0$, because a_{μ} and F_{μ} are orthogonal to v_{μ}, so that multiplication by v_{μ} would give an inconsistency. Secondly, it is clear that $\mathfrak{R}$ is of order τ_0, the same order as $\tau_0\dot{F}$, a term which was neglected in (6–90). Thus, the next approximation after (6–89) is not (6–90) but

$$ma^{\mu} \doteq F^{\mu} - \frac{1}{c^2}\mathfrak{R}v^{\mu} + \tau_0\dot{F}^{\mu} = F^{\mu} + \tau_0\left(\dot{F}^{\mu} - \frac{m}{c^2}a^{\lambda}a_{\lambda}v^{\mu}\right). \qquad (6\text{–}91)$$

The factor τ_0 is here displayed explicitly to emphasize the order of magnitude of the respective terms. Of course, in this approximation $\dot{F}^{\mu}$ can be replaced by its zero-order value from (6–89), $m\dot{a}^{\mu}$. This yields

$$ma^{\mu} = F^{\mu} + m\tau_0\left(\dot{a}^{\mu} - \frac{1}{c^2}a^{\lambda}a_{\lambda}v^{\mu}\right), \qquad (6\text{–}92)$$

which is *exactly* the Lorentz-Dirac equation. The two approximations made have exactly canceling errors.

A consistent successive approximation procedure following (6–91) yields a series which is a rearrangement of (6–87). Let

$$G^{\mu} \equiv \tau_0\dot{F}^{\mu} - \frac{1}{c^2}\mathfrak{R}v^{\mu}. \qquad (6\text{–}93)$$

Then the perturbation expansion, (6–87), can be written

$$ma_{\mu}(\tau) = F_{\mu}(\tau) + \sum_{n=0}^{\infty} \tau_0^n G_{\mu}^{(n)}(\tau). \qquad (6\text{–}94)$$

In classical physics this expansion is not very interesting, because there seems to be no classical process by which terms higher than the first (that is, $G_{\mu} \equiv G_{\mu}^{(0)}$) can be observed. Furthermore, it is known empirically that (6–89) and (6–91) are excellent approximations. Nevertheless, it is of importance to ascertain the convergence of (6–94) mathematically. As we shall see later, this series must be related to the perturbation expansion in *quantum* electrodynamics, whose convergence is being questioned. Does the analogous classical series converge?

We observe that, while the existence of (6–94) has not been proven, this series can be rewritten as follows. Let

$$\varphi^{\mu} \equiv F^{\mu} - ma^{\mu}$$

such that

$$G^{\mu} = \Gamma^{\mu} + \tau_0\dot{\varphi}^{\mu};$$

then (6-94) is

$$\varphi^\mu + \sum_{n=0}^\infty \tau_0^n (\Gamma^\mu + \tau_0 \dot\varphi^\mu)^{(n)} = 0$$

or, after rearrangement,

$$\sum_{n=0}^\infty \tau_0^n (\Gamma^\mu + \varphi^\mu)^{(n)} = 0.$$

This series vanishes *term by term*, because $\Gamma^\mu + \varphi^\mu = 0$ is just the Lorentz-Dirac equation. Thus, the convergence of the series is ensured when the above rearrangements are allowed.

6-9   THE ACTION PRINCIPLE*

The analysis of physical systems consisting of a charged particle, a free electromagnetic field, and an external force has led us to the conclusion that the field equations (Maxwell-Lorentz equations), the conservation laws, and certain asymptotic requirements are sufficient to establish equations of motion. Thus, the theory seems to be reduced to these three "ingredients." It is the task of the present section to provide a different reduction. We shall need only two requirements, viz. asymptotic conditions and an action principle. The latter will contain the particle equations and the field equations as the Euler-Lagrange equations of Hamilton's principle. However, it will also contain the conservation laws for the whole system as a consequence of Noether's theorems applied to the invariance of the action integral under the Lorentz group.

The origin of the action principle, historically, is the idea that particles follow the shortest path, a geodesic. The action integral should therefore be (apart from unimportant factors) an integral over the path of the particle† [cf. Eqs. (A1-43) through (A1-45)],

$$I_{\text{free particle}} = -mc^2 \int \sqrt{-dz^\mu\,dz_\mu} = -mc^2 \int \sqrt{-\frac{dz^\mu}{d\lambda}\frac{dz_\mu}{d\lambda}}\,d\lambda. \qquad (6\text{-}95)$$

The parameter λ is an arbitrary monotonically increasing labeling of the world line $z^\mu(\lambda)$. The timelike nature of the world line is anticipated by the minus sign under the square root.

The Euler-Lagrange equations associated with this free-particle Lagrangian

$$L_{\text{free particle}} = -mc^2 \sqrt{-\frac{dz^\mu}{d\lambda}\frac{dz_\mu}{d\lambda}} \qquad (6\text{-}96)$$

gives, [cf. (3-51)],

$$m\frac{d}{d\lambda}\frac{dz^\mu/d\lambda}{\sqrt{-(dz^\alpha/d\lambda)(dz_\alpha/d\lambda)}} = 0.$$

---

† The constant factor is so chosen that I has the same dimensions as in (4-103), namely energy times length.

* F. Rohrlich, *Phys. Rev. Lett.* **12**, 375 (1964).

This expression can be greatly simplified by introducing

$$c \, d\tau \equiv \sqrt{- \frac{dz^\alpha}{d\lambda} \frac{dz_\alpha}{d\lambda}} \, d\lambda. \qquad (6\text{-}97)$$

It becomes

$$m \frac{d^2 z^\mu}{d\tau^2} = 0, \qquad (6\text{-}98)$$

which is indeed the equation of motion of a free particle.

The parameter $d\tau$ introduced in (6–97) can now be identified with the proper time. It is chosen in such a way that the equation

$$v^\mu v_\mu = -c^2 \qquad (v^\mu \equiv dz^\mu/d\tau) \qquad (\text{A1-37})$$

(cf. Appendix 1) holds as an identity.

The integral I is invariant under the transformations

$$\lambda \to \lambda' = \lambda + \epsilon(\lambda), \qquad (6\text{-}99)$$

where $\epsilon(\lambda)$ is an arbitrary function which is assumed to be restricted only by the requirements of Noether's theorem. This is a transformation involving an arbitrary *function* rather than an arbitrary *parameter*. Noether's second theorem is therefore applicable (cf. footnote in Section 4–11) and gives the identity

$$\frac{d^2 z^\mu}{d\tau^2} \frac{dz_\mu}{d\tau} \equiv 0 \qquad \text{or} \qquad (v^\mu v_\mu) \equiv \text{const},$$

which also follows from the above (A1–37).

The action integral

$$I = +mc^2 \int \frac{dz^\mu}{d\lambda} \frac{dz_\mu}{d\lambda} \, d\lambda,$$

which is often used, should be compared with (6–95). This action integral does not involve integration of the path length and therefore lacks the physical and intuitive appeal of the Fermat-type principle of extremum path length. Furthermore, it is not invariant under (6–99) but only under

$$\lambda \to \lambda' = \lambda + \epsilon,$$

where ϵ is a parameter. Noether's first theorem, (3–50), then yields

$$0 = \delta\lambda \frac{d}{d\lambda} \left[L - 2mc^2 \frac{dz^\alpha}{d\lambda} \frac{dz_\alpha}{d\lambda} \right] = -\delta\lambda \frac{d}{d\lambda} L$$

as a conservation law, *provided the equation of motion*

$$m \frac{d^2 z^\mu}{d\lambda^2} = 0$$

is satisfied. One can choose $c \, d\tau = d\lambda$, so that this equation becomes identical with (6–98) and so that (A1–37) holds; however, (A1–37) now does not hold as an identity, but as a conservation law, subject to the equation of motion.*

The above action principle (6–95) for a free particle and the action principle for the free Maxwell-Lorentz field, Eqs. (4–103) and (4–111), must now be combined with suitable interaction terms in order to describe the whole system particle-electromagnetic field in interaction. So far as the *field equations* are concerned, the Lagrangian (4–106) seems satisfactory since it yields (4–51), but it remains to be seen whether this interaction Lagrangian will also provide the *particle equation* in agreement with Section 6–5. By means of the explicit expression for the current density, (4–82), let us first assume the following action integral for a closed one-particle system,

$$I = -mc^2 \int \sqrt{-\frac{dz^\mu}{d\lambda} \frac{dz_\mu}{d\lambda}} \, d\lambda + e \int A_\mu(x) \, \delta(x - z(\lambda)) \frac{dz^\mu}{d\lambda} \, d\lambda \, d^4x$$

$$- \frac{1}{16\pi} \int [\partial_\mu A_\nu(x) - \partial_\nu A_\mu(x)][\partial^\mu A^\nu(x) - \partial^\nu A^\mu(x)] \, d^4x.$$

$$(6\text{–}100)$$

The invariance of this action integral under the transformation group (6–99) should be noted. The identity associated with this invariance property via Noether's second theorem is

$$v_\mu \left(m a^\mu - \frac{e}{c} F^{\mu\alpha} v_\alpha \right) \equiv 0.$$

This can be recognized as a generalization of the identity obtained in the free-particle case.

One can now use Hamilton's principle and obtain the Euler-Lagrange equations corresponding to the variations of $A^\mu(x)$ and $z^\mu(\lambda)$, respectively.

Let us first keep z^μ fixed and vary A^μ. Because of the δ-function in the interaction term only a functional change of $A^\mu(x)$ can contribute. If $A_\mu(x)$ is moved off the world line z, the integral gives no contribution. We find, as in Section 4–9, the Maxwell-Lorentz equation

$$\partial_\mu F^{\mu\nu}(x) = -\frac{4\pi}{c} j^\nu(x) \qquad (4\text{–}51)$$

with $F^{\mu\nu}$ defined in terms of the potentials, (4–47), and j^μ defined as the point-charge current (4–82).

---

* The integral (6–95) is also invariant under the one-parameter group of (6–99). But Noether's first theorem in this case leads to a trivial identity rather than to a conservation law.

This derivation is completely gauge invariant. The explicit occurrence of the potential (rather than the field strength) in the interaction term of I does not produce gauge-variant field equations from Hamilton's principle; this principle requires that end points of line integrals and boundaries of domains be held *fixed*. A gauge transformation would change (6–100) only by a term

$$\frac{1}{c} \int \partial_\mu \Lambda(x) j^\mu(x) \, d^4x = \frac{1}{c} \int \partial_\mu (\Lambda j^\mu) \, d^4x$$

because according to (4–51) current conservation holds as an identity. By Gauss's theorem this integral is equal to a surface integral. Therefore, the extra term does not contribute to the field equations.

For the variation of z^μ only the first and second integral is relevant,

$$\delta \int L_{\text{particle}} \, d\lambda = 0,$$

$$\text{(6–101)}$$

$$L_{\text{particle}} = -mc^2 \sqrt{-\frac{dz^\mu}{d\lambda} \frac{dz_\mu}{d\lambda}} + e A_\mu(z) \frac{dz^\mu}{d\lambda}.$$

The Euler-Lagrange equations (3–51) give

$$\frac{d}{d\lambda} \left(\frac{mc^2 \, dz^\mu/d\lambda}{\sqrt{-(dz^\alpha/d\lambda)(dz_\alpha/d\lambda)}} + e A^\mu(z) \right) - e \frac{dz_\alpha}{d\lambda} \frac{\partial A^\alpha(z)}{\partial z_\mu} = 0.$$

The proper time τ can again be introduced by (6–97):

$$0 = \frac{d}{d\tau} \left(mc \frac{dz^\mu}{d\tau} + e A^\mu(z) \right) - e \frac{dz_\alpha}{d\tau} \frac{\partial A^\alpha}{\partial z_\mu},$$

$$m a^\mu = -\frac{e}{c} v_\alpha F^{\alpha\mu} = \frac{e}{c} F^{\mu\alpha} v_\alpha. \qquad \text{(6–102)}$$

The same remark about gauge invariance that was made before for the field equations is also valid for the derivation of this particle equation.

Equation (6–102) must now be compared with the results of Section 6–5. There we found from the Maxwell-Lorentz equations and from the conservation laws (6–50)

$$m a^\mu = \frac{e}{c} F^{\mu\nu} v_\nu \qquad \text{(6–51)}$$

for closed systems consisting of a charge and an incoming electromagnetic field. The field strengths in (6–102) must therefore be identified with

$$F^{\mu\nu} = F^{\mu\nu}_{\text{in}} + F^{\mu\nu}_-$$

[cf. Eqs. (6–45) and (6–46)]. The vector potential in the interaction term of the action integral (6–100) should therefore not be the total potential [cf. (6–46)]

$$A^\mu = A^\mu_{\text{in}} + A^\mu_{\text{ret}} = A^\mu_{\text{in}} + A^\mu_+ + A^\mu_- \qquad \text{(6–103)}$$

but should rather be

$$\bar{A}^\mu = A^\mu_{in} + A^\mu_-. \tag{6-104}$$

By the same token, all potentials in I must be $\bar{A}^\mu$. Therefore, the field equation resulting from Hamilton's principle should not be the inhomogeneous ones, (4–51), but the homogeneous ones; $\bar{F}^{\mu\nu}$ is a free field.

We are thus led to a dilemma: either (6–103) holds and A^μ in the action integral leads to the *in*homogeneous field equations and to the *in*correct particle equations, or (6–104) holds and $A^\mu = \bar{A}^\mu$ in I leads to the homogeneous field equations and the correct particle equations. In the latter case, the action principle does not give us all the necessary information, because it cannot describe how the charge acts as the source of a field. The action principle (6–100) is therefore unsatisfactory.

In order to find a satisfactory action principle, we can be guided by the following considerations. We observe that the derivation of the inhomogeneous Maxwell-Lorentz equation (4–51) from (6–100) leads to a mathematical inconsistency. We know from Section 6–5, Eq. (6–62), that this field equation leads to a potential $A^\mu(x)$ which diverges like $1/r$ for $x \to z$, on the world line. Correspondingly, the interaction term in I,

$$e \int A^\mu(z) \frac{dz_\mu}{d\lambda} \, d\lambda,$$

does not exist. The field strengths $F^{\mu\nu}(x) \sim 1/r^2$ for $x \to z$, so that the integrand of the last term in I diverges like $1/r^4$ and the integral diverges like $1/r$ when the domain of integration contains a world line.

But we also learned from (6–62) that the divergence of $F^{\mu\nu}_{ret}(z)$ is entirely due to $F^{\mu\nu}_+(z)$, while $F^{\mu\nu}_-(z)$ is well defined. The divergent terms in (6–100) are therefore separated by writing $A^\mu = \bar{A}^\mu + A^\mu_+$; they are

$$e \int A^\mu_+(z) \frac{dz_\mu}{d\lambda} \, d\lambda - \frac{1}{16\pi} \int F^+_{\mu\nu}(x) F^{\mu\nu}_+(x) \, d^4x.$$

Note that the cross term

$$-\frac{1}{16\pi} 2 \int \bar{F}_{\mu\nu}(x) F^{\mu\nu}_+(x) \, d^4x \qquad \text{is convergent.}$$

But we saw (end of Section 6-5) that the divergent terms give the Coulomb self-energy part of the mass. Thus, we can omit these terms and use the observed mass, m.

$$I = -mc^2 \int \sqrt{-\frac{dz^\mu}{d\lambda} \frac{dz_\mu}{d\lambda}} \, d\lambda + e \int \bar{A}_\mu(x) \, \delta(x - z \, (\lambda)) \frac{dz^\mu}{d\lambda} \, d\lambda \, d^4x$$

$$-\frac{1}{16\pi} \int [\partial_\mu \bar{A}_\nu(x) - \partial_\nu \bar{A}_\mu(x)][\partial^\mu \bar{A}^\nu(x) - \partial^\nu \bar{A}^\mu(x)] \, d^4x \tag{6-105}$$

$$-\frac{1}{8\pi} \int [\partial_\mu \bar{A}_\nu(x) - \partial_\nu \bar{A}_\mu(x)][\partial^\mu A^\nu_+(x) - \partial^\nu A^\mu_+(x)] \, d^4x.$$

The invariance property (6–99) still holds.

Let us now forget this heuristic argument and take (6–105) as *the axiomatically given basis* of our theory.* It is an action integral which involves *three* sets of variables: the particle coordinates $z^\mu(\lambda)$, the potentials $\bar{A}^\mu(x)$, and the potentials $A^\mu_+(x)$. Correspondingly, we obtain *three* sets of Euler-Lagrange equations. They are obtained, as before, by variation of these quantities:

$$\delta z: \qquad ma^\mu(\tau) = \frac{e}{c}\, \bar{F}^{\mu\nu}_{(z)} v_\nu(\tau), \qquad\qquad (6\text{–}51)$$

$$\delta\bar{A}: \qquad \partial_\mu(\bar{F}^{\mu\nu}(x) + F^{\mu\nu}_+(x)) = -\frac{4\pi}{c}\, j^\nu(x), \qquad (4\text{–}51)$$

$$\delta A_+: \qquad \partial_\mu \bar{F}^{\mu\nu}(x) = 0. \qquad\qquad\qquad (6\text{–}106)$$

All $F^{\mu\nu}$ are defined in terms of the corresponding potentials.

We are thus led to the correct particle equations and to the identification of $\bar{F}^{\mu\nu}$ as a *free* field and $F^{\mu\nu}_+$ as a field tied to the particle as its source. The former is obviously associated with acceleration: $\bar{F}^{\mu\nu} = 0$ implies $z^\mu = 0$ according to (6–51).

PROBLEM 6–5

Substitute the expansion of $v^\mu(\mp\tau)$ and $\rho(\mp\tau)$ of (6–61) into the expression (4–91) for the potential four-vector and show that on the world line

$$A^\mu_+(z) = \lim_{\rho\to 0} \frac{e}{c}\frac{v^\mu}{\rho}, \qquad A^\mu_-(z) = -\frac{e}{c^2}\, a^\mu.$$

Therefore, A^μ_+ is just the generalized Coulomb potential.

The incident field $F^{\mu\nu}_{\text{in}}$ can be identified asymptotically: at $t \to -\infty$ the total field consists only of $F^{\mu\nu}_{\text{in}}$ and the Coulomb field $F^{\mu\nu}_C$

$$\lim_{t\to-\infty} (\bar{F}^{\mu\nu} + F^{\mu\nu}_+) = \lim_{t\to-\infty} (F^{\mu\nu}_{\text{in}} + F^{\mu\nu}_C)$$

in accordance with (6–39). Analogously, we have for the distant future

$$\lim_{t\to+\infty} (\bar{F}^{\mu\nu} + F^{\mu\nu}_+) = \lim_{t\to+\infty} (F^{\mu\nu}_{\text{out}} + F^{\mu\nu}_C).$$

The known asymptotic properties of $F^{\mu\nu}_{\text{ret}}$ and $F^{\mu\nu}_{\text{adv}}$ then lead us to the identification

$$\bar{F}^{\mu\nu} + F^{\mu\nu}_+ = F^{\mu\nu}_{\text{in}} + F^{\mu\nu}_{\text{ret}} = F^{\mu\nu}_{\text{out}} + F^{\mu\nu}_{\text{adv}},$$

---

* It is interesting to observe that the following basic equations and conservation laws could also be obtained from (6–105) if the third term, $\mathcal{L}_{\text{field}}(\bar{A})$ were omitted. Its inclusion seems to be partly a matter of taste.

as in (6–43) through (6–46), or

$$\bar{F}^{\mu\nu} = F^{\mu\nu}_{in} + F^{\mu\nu}_{-} = F^{\mu\nu}_{out} - F^{\mu\nu}_{-} = \tfrac{1}{2}(F^{\mu\nu}_{in} + F^{\mu\nu}_{out}),$$

$$F^{\mu\nu}_{\pm} = \tfrac{1}{2}(F^{\mu\nu}_{ret} \pm F^{\mu\nu}_{adv}).$$

As we shall see in Section 9–2, this identification is unique when time-reversal invariance is taken into account.

With this identification the above particle equation (6–51) becomes the Lorentz-Dirac equation (6–57) with $F_{ext} = 0$, since (6–54) and (6–55) can be used. As in Section 6–6, the asymptotic conditions can be postulated and combined with the particle equation to yield the equations of motion.

The fundamental equations of the theory for a closed system have thus been obtained from the action integral (6–105) and the asymptotic requirements without encountering any divergences or other self-energy difficulties.

When we are dealing with an *open* system (e.g. one in which an externally controlled electromagnetic field $F^{\mu\nu}_{ext}$ is present) which can also be a static field, the action integral (6–105) must be amended by a term,

$$e \int A^{\mu}_{ext}(z) \frac{dz_{\mu}}{d\lambda} \, d\lambda.$$

Here A^{μ}_{ext} is given and fixed and is not to be varied together with the other potentials. It gives rise to an additional force term,

$$F^{\mu}_{ext} = \frac{e}{c} F^{\mu\nu}_{ext} v_{\nu},$$

in the particle equation (6–51); thus,

$$ma^{\mu} = F^{\mu} + F^{\mu}_{ext}$$

in agreement with the result of Section 6–5 for an open system.

Having obtained a satisfactory action principle from which the field equations and the particle equations can be obtained, we now turn to the derivation of conservation laws from this action principle. Noether's theorem assures us of 10 conservation laws, since I is invariant under the 10-parameter inhomogeneous Lorentz group.

Let us now consider momentum conservation as it follows from translation invariance of I. With $\delta z^{\mu} = \delta x^{\mu} = \epsilon^{\mu}$, Noether's theorem in the forms (3–50) and (3–58) gives us, since the Euler-Lagrange equations are assumed to be satisfied,

$$\int d\lambda \, \frac{d}{d\lambda} \frac{\partial L_{particle}[\lambda]}{\partial(dz_{\mu}/d\lambda)} + \int d^{4}x \, \partial_{\nu} \left(\eta^{\mu\nu} \mathcal{L}_{field} - \frac{\partial \mathcal{L}_{field}}{\partial \partial_{\nu} \bar{A}_{\alpha}} \partial^{\mu} \bar{A}_{\alpha} - \frac{\partial \mathcal{L}_{field}}{\partial \partial_{\nu} A^{+}_{\alpha}} \partial^{\mu} A^{+}_{\alpha} \right) = 0.$$

Here, $L_{particle}[\lambda]$ is given by (6–101) with A_{μ} replaced by $\bar{A}_{\mu}$ and $\mathcal{L}_{field}$ is the integrand of the last two terms of I, (6–105). The two integrals do not vanish separately; only their *sum* vanishes, because a change in the domain X_4 implies a corresponding change in the end points of the world line traversing X_4. This is the meaning of $\delta x^\mu = \delta z^\mu$. It is indicated in Fig. 6-2.

The first integral gives, using (6-97)

$$\int d\tau \left(\frac{dp^\mu}{d\tau} + \frac{e}{c}\, v_\alpha\, \partial^\alpha \overline{A}^\mu\right)$$

with $p^\mu = mv^\mu$. The second integral gives, in analogy with (4-113),

$$-\frac{1}{c} \int d^4x\, \partial_\nu (\overline{T}^{\nu\mu} + T_D^{\nu\mu}),$$

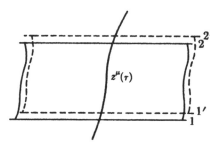

Fig. 6-2. Variation of the domain X_4.

with the following relations between the canonical and the symmetric energy tensors:

$$\overline{T}^{\nu\mu} = \overline{\Theta}^{\nu\mu} - \frac{1}{4\pi}\, \overline{F}^{\nu\alpha}\, \partial_\alpha \overline{A}^\mu,$$

$$\overline{\Theta}^{\nu\mu} = \frac{1}{4\pi}\left(\overline{F}^{\nu\alpha}\overline{F}_\alpha{}^\mu + \frac{\eta^{\nu\mu}}{4}\,\overline{F}^{\alpha\beta}\overline{F}_{\alpha\beta}\right) = \overline{\Theta}^{\mu\nu};$$

$$T_D^{\nu\mu} = \overline{\Theta}_D^{\nu\mu} - \frac{1}{4\pi}\left(F_+^{\nu\alpha}\, \partial_\alpha \overline{A}^\mu + \overline{F}^{\nu\alpha}\, \partial_\alpha A_+^\mu\right),$$

$$\Theta_D^{\nu\mu} = \frac{1}{4\pi}\left(F_+^{\nu\alpha}\overline{F}_\alpha{}^\mu + \overline{F}^{\nu\alpha}F_{+\alpha}{}^\mu + \tfrac{1}{2}\eta^{\nu\mu}F_+^{\alpha\beta}\overline{F}_{\alpha\beta}\right) = \Theta_D^{\mu\nu}.$$

Noether's theorem therefore tells us that

$$\int d\tau \left(\frac{dp^\mu}{d\tau} + \frac{e}{c}\, v_\alpha\, \partial^\alpha \overline{A}^\mu\right) - \frac{1}{c}\int d^4x\, \partial_\nu (\overline{T}^{\nu\mu} + T_D^{\nu\mu}) = 0.$$

But this can also be written as

$$\int d\tau \frac{dp^\mu}{d\tau} - \frac{1}{c}\int d^4x\, \partial_\nu (\overline{\Theta}^{\nu\mu} + \Theta_D^{\nu\mu}) = 0,$$

which exhibits gauge invariance explicitly.

The last step follows from the exact cancellation of all gauge-variant terms,

$$\int d\tau \frac{e}{c}\, v_\alpha\, \partial^\alpha \overline{A}^\mu + \frac{1}{4\pi c}\int d^4x\, \partial_\nu (\overline{F}^{\nu\alpha}\, \partial_\alpha \overline{A}^\mu + F_+^{\nu\alpha}\, \partial_\alpha \overline{A}^\mu + \overline{F}^{\nu\alpha}\, \partial_\alpha A_+^\mu) = 0.$$

The first and the last terms in the four-dimensional integral vanish on account of the field equation (6-106)

$$\partial_\nu(\overline{F}^{\nu\alpha}\, \partial_\alpha \overline{A}^\mu + \overline{F}^{\nu\alpha}\, \partial_\alpha A_+^\mu) = \partial_\nu\, \partial_\alpha [\overline{F}^{\nu\alpha}(\overline{A}^\mu + A_+^\mu)] - \partial_\nu[(\overline{A}^\mu + A_+^\mu)\, \partial_\alpha\, \overline{F}^{\nu\alpha}] = 0.$$

The remaining two terms cancel because of the field equations and charge conservation. We have

$$\frac{1}{4\pi c}\int d^4x\, \partial_\nu(F^{\nu\alpha}_+\partial_\alpha\overline{A}^\mu) = 0 - \frac{1}{4\pi c}\int d^4x\, \partial_\nu(\overline{A}^\mu\partial_\alpha F^{\nu\alpha}_+)$$

$$= -\frac{1}{c^2}\int d^4x\, \partial_\nu(\overline{A}^\mu j^\nu)$$

$$= -\frac{e}{c}\int \partial_\nu\overline{A}^\mu d^4x \int \delta(x-z)v^\nu\, d\tau$$

$$= -\frac{e}{c}\int v^\nu \partial_\nu\overline{A}^\mu(z)\, d\tau.$$

This term exactly cancels the first gauge-variant term.

The momentum conservation law is now obtained by means of the equality

$$\int d^4x\, \partial_\nu\Omega^{\mu\nu} = -\int d\tau\, \frac{d}{d\tau}\int_{\sigma(\tau)} d^3\sigma_\nu\Omega^{\mu\nu}$$

for any tensor $\Omega^{\mu\nu}$ that vanishes fast enough in spacelike directions. This is a direct consequence of Gauss's theorem, (A1-62). Thus, we find the conservation law

$$dp^\mu + d\overline{P}^\mu + dP^\mu_D = 0$$

with

$$\overline{P}^\mu \equiv \frac{1}{c}\int \overline{\Theta}^{\mu\nu}\, d^3\sigma_\nu, \qquad P^\mu_D \equiv \frac{1}{c}\int \Theta^{\mu\nu}_D\, d^3\sigma_\nu \Bigg\}. \tag{6-107}$$

The first term is the kinetic momentum of the particle. It includes the momentum of the Coulomb field, since this is an inseparable part of the physical particle. The second term is the momentum of the free fields $\overline{F}^{\mu\nu}$. Written in terms of F_{in} and F_- it becomes

$$\overline{P}^\mu = P^\mu_{\text{in}} + P^\mu_- + P^\mu_{\text{in}-}.$$

Since field momenta are bilinear functionals of the corresponding fields, (via the energy tensor) there is a cross term. All these momenta are associated with *free* fields, so that they are separately conserved,

$$dP^\mu_{\text{in}} = 0, \qquad dP^\mu_- = 0, \qquad d\overline{P}^\mu = 0. \tag{6-107}_{\text{free}}$$

This is a consequence of the free-field equations and is an application of (4-129). The conservation law (6-107) can therefore also be written

$$dp^\mu + dP^\mu_D = 0. \tag{6-107'}$$

The momentum P^μ_D comes from the interference terms between $\overline{F}^{\mu\nu}$ and $F^{\mu\nu}_+$, as is clearly evident from the form of $\Theta^{\mu\nu}_D$. This tensor is exactly the one which gives rise to $\overline{F}^\mu \, d\tau$ in Dirac's derivation of the particle equation, (6–47). In fact, if we define

$$\Theta^{\mu\nu}_C \equiv \frac{1}{4\pi}\left(F^{\mu\alpha}_+ F^\nu_{\alpha+} + \frac{\eta^{\mu\nu}}{4} F^{\alpha\beta}_+ F^+_{\alpha\beta} \right)$$

and observe that

$$\partial_\mu \Theta^{\mu\nu}_C = \frac{1}{c} F^{\nu\mu}_+ j_\mu,$$

$$\partial_\mu \Theta^{\mu\nu} = \frac{1}{c} F^{\nu\mu} j_\mu,$$

we have from $F = \overline{F} + F_+$

$$\partial_\mu \Theta^{\mu\nu}_D = \partial_\mu(\Theta^{\mu\nu} - \overline{\Theta}^{\mu\nu} - \Theta^{\mu\nu}_C) = \frac{1}{c}(F^{\nu\mu} - F^{\nu\mu}_+)j_\mu = \frac{1}{c}\overline{F}^{\nu\mu} j_\mu.$$

Therefore

$$\Delta P^\mu_D = \int d\tau \, \frac{d}{d\tau}\, \frac{1}{c}\int d\sigma_\nu\, \Theta^{\nu\mu}_D = -\frac{1}{c}\int d^4 \, \partial_\nu \Theta^{\nu\mu}_D$$

$$= -\frac{1}{c^2}\int d^4x \, \overline{F}^{\mu\nu}(x)\, j_\nu(x) = -\frac{e}{c}\int d\tau\, \overline{F}^{\mu\nu}(z)\, v_\nu$$

or

$$dP^\mu_D = -\frac{e}{c}\,\overline{F}^{\mu\nu} v_\nu \, d\tau = -\overline{F}^\mu \, d\tau. \tag{6–49'}$$

Thus, dP^μ_D is to be identified with dP^μ of (6–49) in Section 6–5.

PROBLEM 6–6

Prove that, in agreement with (6–48) and (6–49),

$$\int_{-\infty}^{\infty} dP_D = P^\mu_{out}(+\infty) - P^\mu_{in}(-\infty). \tag{6–48'}$$

For this purpose it is convenient to compute

$$\int_{-\infty}^{\infty} d(P^\mu_D + \overline{P}^\mu + P^\mu_-),$$

which equals the above integral, since $d\overline{P}^\mu = dP^\mu_- = 0$ by $(6–107)_{free}$. Note that

$$\Theta^{\mu\nu}_D + \overline{\Theta}^{\mu\nu} + \Theta^{\mu\nu}_- = \Theta^{\mu\nu}_{in} + \overline{\Theta}^{\mu\nu}_{ret} = \Theta^{\mu\nu}_{out} + \overline{\Theta}^{\mu\nu}_{adv},$$

where $\overline{\Theta}^{\mu\nu}_{ret}$ and $\overline{\Theta}^{\mu\nu}_{adv}$ are formed from $\overline{F}^{\mu\nu}$, $F^{\mu\nu}_{ret}$ and $\overline{F}^{\mu\nu}$, $F^{\mu\nu}_{adv}$, respectively, in the same way as $\Theta^{\mu\nu}_D$ is formed from $\overline{F}^{\mu\nu}$, $F^{\mu\nu}_+$. Asymptotically for $t \to \pm\infty$ the free fields and the Coulomb fields do not interfere, since the particle is free [see also Eqs. (9–55)].

Problem 6–3 above shows that the total momentum of the system is

$$P^\mu_{\text{tot}} = p^\mu + \bar{P}^\mu + P^\mu_D + P^\mu_-, \tag{6-108}$$

since it reduces to $p^\mu + P^\mu_{\text{in}}$ and to $p^\mu + P^\mu_{\text{out}}$ for $t \to -\infty$ and $t \to +\infty$, respectively. It is conserved, of course, but it is not the momentum which arises from Noether's theorem, (6–107). The free-field momentum conservation (6–107)$_{\text{free}}$ is therefore essential: only the combination of both conservation laws yields the constancy of the physical total momentum.

PROBLEM 6–7

Derive the conservation laws appropriate to the invariance of the action integral (6–105) under the homogeneous Lorentz group.

It is of some interest to compare the above law of momentum conservation with the corresponding results of the old renormalization theory. In that theory the particle momentum is separated into a bare-particle momentum and the momentum of the surrounding Coulomb field. Because of (6–69), which has its origin in $\Theta^{\mu\nu}(F_+)$, we can write this as

$$p^\mu = p^\mu_{\text{bare}} + P^\mu_+.$$

Observing the bilinearity of the energy tensor and the decomposition $F^{\mu\nu}_{\text{ret}} = F^{\mu\nu}_+ + F^{\mu\nu}_-$, we then have for the *convergent* momenta

$$p^\mu + \bar{P}^\mu + P^\mu_D = (p^\mu_{\text{bare}} + P^\mu_+) + (P^\mu_{\text{in}} + P^\mu_{\text{in},-} + P^\mu_-) + (P^\mu_{\text{in},+} + P^\mu_{-,+})$$

$$= (p^\mu_{\text{bare}} + P^\mu_+) + P^\mu_{\text{in}} + P^\mu_{\text{in,ret}} + (P^\mu_{\text{ret}} - P^\mu_+)$$

$$= p^\mu_{\text{bare}} + (P^\mu_{\text{in}} + P^\mu_{\text{in,ret}} + P^\mu_{\text{ret}}).$$

Each one of the two terms on the right are given by *divergent* integrals.

Thus, the insistence on a retarded field leads to the identification of m with the bare mass, corresponding to the action principle (6–100). The realization that one is dealing with two fields, $\bar{F}^{\mu\nu}$ and $F^{\mu\nu}_+$, leads via the action principle (6–105) to the identification of m with the experimental mass including the Coulomb field effects. The appearance of fields which are not all retarded is not a violation of causality but only a manifestation of the time symmetry of the theory. We shall return to this point in Section 7–3 and in Chapter 9.

In retrospect we can now assert that the formulation of the one-particle theory in terms of the action integral (6–105) and certain asymptotic conditions is completely satisfactory in that it provides the desired field equations and equations of motion and in that, at the same time, it is free of mathematical difficulties such as the divergent self-energy terms. The problems of stability discussed in Part A of this chapter do not arise here despite the fact that our charge is without structure.

C. Special Systems

6–10 THE FREE PARTICLE

The simplest test of the equations of motion is to see whether the law of inertia is included in them. We start with

$$a^\mu(\tau)a_\mu(\tau) = \frac{1}{m}\int_0^\infty F^\mu(\tau + \tfrac{1}{2}\alpha\tau_0)\, a_\mu(\tau + \tfrac{1}{2}\alpha\tau_0)\, e^{-\alpha}\, d\alpha \qquad (6\text{–}83)$$

and apply it to a particle without imposed force F^μ. Then $a^\mu a_\mu = 0$, which implies that $a^\mu = 0$, since $a^\mu a_\mu$ is positive definite. Thus,

$$v^\mu(\tau) = \text{const} = v^\mu(\tau_i) \qquad (6\text{–}109)$$

is the solution. It is easily verified that this is indeed a solution of (6–77): $p_{\text{in}}^\mu = p_{\text{out}}^\mu = p^\mu(\tau)$ is independent of τ. Equation (6–82) is satisfied and yields

$$z^\mu(\tau) = z^\mu(\tau_i) + v^\mu(\tau_i)(\tau - \tau_i),$$

giving the law of inertia.

It is at least of historical interest to investigate the solutions of the Lorentz-Dirac equation (6–57) if the asymptotic conditions are *not* imposed. For $F^\mu = 0$, this equation is

$$a^\mu = \tau_0\left(\dot{a}^\mu - \frac{1}{c^2}\, a^\lambda a_\lambda v^\mu\right). \qquad (6\text{–}110)$$

Clearly, the above solution $v^\mu = \text{const}$ is also a solution of this equation. However, there are other solutions.

For any initial velocity and acceleration there always exists a Lorentz frame in which these two four-vectors are both in the $(z\text{-}t)$-plane. It then follows by symmetry that the whole motion lies in this plane. The problem is therefore one-dimensional (with respect to ordinary three-space).

For one-dimensional motion we write for any vector $V^\mu = (V^0; 0, 0, V^3) \equiv (V^0, V^3)$. Let $u = \gamma v$; then

$$v^\mu = (\sqrt{c^2 + u^2},\, u), \qquad a^\mu = (u\dot{u}/\sqrt{c^2 + u^2},\, \dot{u})$$

and (6–110) becomes for the space-component

$$\dot{u} = \tau_0[\ddot{u} - u\dot{u}^2/(c^2 + u^2)].$$

Define $w(\tau)$ by

$$\frac{u(\tau)}{c} = \sinh w(\tau). \qquad (6\text{–}111)$$

This simplifies the equation greatly and reduces it to a *linear* equation,

$$\dot{w} - \tau_0\ddot{w} = 0. \qquad (6\text{–}112)$$

The solution of this equation is

$$w = Ae^{\tau/\tau_0} + B$$

so that

$$\frac{u(\tau)}{c} = \sinh\left(Ae^{\tau/\tau_0} + B\right)$$

and

$$v^\mu = \left(c\cosh\left(Ae^{\tau/\tau_0} + B\right), c\sinh\left(Ae^{\tau/\tau_0} + B\right)\right). \tag{6–113}$$

This family of solutions does not satisfy the asymptotic conditions (6–38), since $\lim v(\tau)$ for $\tau \to \infty$ does not exist unless $A = 0$. In the latter case it reduces to our previous solution (6–109).

The occurrence of solutions of the type (6–113) for *free* charged particles was the cause of much concern and led many to abandon the Lorentz-Dirac equation as physically meaningless. The implication of these solutions would be a steadily increasing velocity without the presence of an imposed force. These *self-accelerating* solutions (more picturesquely called *runaway* solutions) are indeed physically meaningless. But the fault is not the Lorentz-Dirac equation; rather, it is the lack of additional restrictions. This was already recognized by Dirac in his 1938 paper. These restrictions, in the form of asymptotic conditions, are not imposed ad hoc in order to eliminate these undesirable solutions; instead, there are other physical reasons for their presence, as was explained in Section 6–4.

The expressions (6–113) can be solutions of the equations of motion (6–77) or (6–82) only for $A = 0$.

6–11   UNIFORM ACCELERATION

Apart from the free particle solution, the equation of motion has another easily obtainable exact solution, that of uniform acceleration. In order to find it, one starts most conveniently from the definition of uniform acceleration,

$$\dot{a}^\mu = a^\lambda a_\lambda v^\mu / c^2. \tag{5–23'}$$

Because v^μ and a^μ are orthogonal, this equation implies that

$$0 = a_\mu \dot{a}^\mu = \frac{1}{2}\frac{d}{d\tau}(a_\mu a^\mu).$$

Therefore, the invariant radiation rate is a constant,

$$\mathcal{R} \equiv \frac{2}{3}\frac{e^2}{c^3} a_\mu a^\mu = \text{const.} \tag{6–114}$$

The same result was obtained previously in Eq. (5–38).

Equation (5-23′) states therefore that $\dot{a}^\mu$ is parallel to v^μ,

$$\frac{d^2 v^\mu}{d\tau^2} = \dot{a}^\mu = \frac{\mathcal{R}}{\tau_0 mc^2} v^\mu \equiv \lambda^2 v^\mu \qquad (\lambda > 0). \qquad (6\text{-}115)$$

This equation is easily integrated,

$$v^\mu(\tau) = \alpha^\mu e^{\lambda\tau} + \beta^\mu e^{-\lambda\tau}. \qquad (6\text{-}116)$$

The time-independent four-vectors α^μ and β^μ are restricted by $v_\lambda v^\lambda = -c^2$, which requires

$$\alpha_\mu \alpha^\mu = 0, \qquad \beta_\mu \beta^\mu = 0, \qquad 2\alpha_\mu \beta^\mu = -c^2. \qquad (6\text{-}117)$$

Both α^μ and β^μ are therefore null vectors.

One more integration of v^μ gives the position four-vector of the particle,

$$z^\mu(\tau) = \gamma^\mu + (1/\lambda)(\alpha^\mu e^{\lambda\tau} - \beta^\mu e^{-\lambda\tau}). \qquad (6\text{-}118)$$

We shall now show that the solution (5-32),

$$z = \sqrt{\alpha^2 + (ct)^2}, \qquad (5\text{-}32)$$

is a special case of (6-118). Let us therefore restrict ourselves to only one space dimension, $z^\mu = (ct, z)$.

As the initial condition on the velocity we can take the simplest one, $v(0) = 0$, that is

$$v^\mu(0) = (c, 0). \qquad (6\text{-}119)$$

This restricts α^μ and β^μ to [cf. (6-116)]

$$\alpha^\mu = (\alpha^0, \alpha^0) \qquad \text{and} \qquad \beta^\mu = (\beta^0, -\beta^0).$$

A convenient choice is $\alpha^0 = \beta^0$ which makes the solution time symmetric; from (6-117)

$$\alpha^\mu = c(\tfrac{1}{2}, \tfrac{1}{2}), \qquad \beta^\mu = c(\tfrac{1}{2}, -\tfrac{1}{2}). \qquad (6\text{-}120)$$

Thus,

$$v^\mu(\tau) = c(\cosh \lambda\tau, \sinh \lambda\tau), \qquad (6\text{-}121)$$

and

$$a^\mu(\tau) = \lambda c(\sinh \lambda\tau, \cosh \lambda\tau). \qquad (6\text{-}122)$$

Define g by

$$a^\mu(0) \equiv (0, g) \qquad (g > 0); \qquad (6\text{-}123)$$

then

$$\lambda = g/c. \qquad (6\text{-}124)$$

So far, only one initial condition was used, namely (6-119). The second one can be taken to be

$$z^\mu(0) = (0, \alpha) \qquad (\alpha > 0). \qquad (6\text{-}125)$$

From (6–120) and (6–118) follows that

$$z^{\mu}(\tau) = \gamma^{\mu} + \frac{c}{\lambda}\,(\sinh \lambda\tau, \cosh \lambda\tau)$$

so that the initial condition requires

$$z^{\mu}(0) = \gamma^{\mu} + \frac{c}{\lambda}\,(0, 1).$$

If we take $\gamma^{\mu} = 0$,

$$\alpha = \frac{c}{\lambda} = \frac{c^2}{g}. \tag{6–126}$$

The solution is therefore

$$(ct, z) = \alpha(\sinh \lambda\tau, \cosh \lambda\tau). \tag{6–127}$$

Elimination of $\lambda\tau$ gives

$$z = \sqrt{\alpha^2 + (ct)^2}, \tag{5–32}$$

which was to be established.

The reader should modify the various assumptions made above and investigate more general solutions. It should be noted, however, that a time translation can always produce the choice (6–120) and the appearance of the hyperbolic functions as above. This is another reason for the characterization of this system as "hyperbolic motion."*

The Lorentz-Dirac equation, (6–57), reduces to

$$ma^{\mu}(\tau) = F^{\mu}(\tau)$$

for hyperbolic motion because of (5–32). As remarked earlier, this motion has therefore $\Gamma^{\mu} = 0$ for all τ despite the fact that $\mathfrak{R} \neq 0$.

Let us now return to Eq. (6–80),

$$ma^{\mu}(\tau) = \int_{0}^{\infty} (F^{\mu} - \mathfrak{R}v^{\mu})_{\tau + \alpha\tau_0}e^{-\alpha}\,d\alpha, \tag{6–80}$$

and assume a force of the form

$$F^{\mu} = F_{\text{ext}}^{\mu} = m\,\lambda(\alpha^{\mu}e^{\lambda\tau} - \beta^{\mu}e^{-\lambda\tau}), \qquad F_{\text{in}}^{\mu} = 0, \qquad \lambda\tau_0 < 1, \tag{6–128}$$

with α^{μ} and β^{μ} satisfying (6–117). It then follows that (6–118) is an *exact* solution of (6–80). This can be verified by an elementary integration.

The force (6–128) does not vanish asymptotically. Correspondingly, the asymptotic conditions (6–37) and (6–38) are not satisfied. This can be seen explicitly from (6–116). Thus, if the force is acting over an infinite time interval, $0 \leq \tau < \infty$ say, the solution (6–118) of the equation (6–80) will hold for that same time interval, but the equation of motion (6–82) *cannot* be satisfied.

---

* Compare with the remarks made in Section 5–3 following Eq. (5–32).

One concludes that uniform acceleration is possible only over a finite time interval. The physical reason for this situation was already discussed in Chapter 5 (following Problem 5–5).

Uniform acceleration seems to offer an ideal way to observe the violation of causality in classical physics predicted by the nonlocal nature of the equation of motion (6–82). As we saw in Section 5–7, it is sufficient for this purpose to observe the Schott term $\dot{a}^\mu$. For uniform acceleration,

$$\dot{a}^\mu = 1/(m\tau_0 c^2)\Re v^\mu,$$

as follows from (5–23'). The time component of this equality states that

$$m\tau_0 \frac{da^0}{dt} = \frac{1}{c}\,\Re. \tag{6-129}$$

Observation of the noncausal behavior is therefore synonymous with observation of radiation from uniformly accelerated charges. Since a qualitative measurement (i.e. the verification of the existence of such radiation) would suffice, this particular physical system appears indeed to be promising in this respect. In fact, it is the most promising system in the following sense.

In Section 6–7 we derived an equation for the energy loss,

$$\frac{dE}{dt} = \mathbf{v}\cdot\mathbf{F} + \tau_0\frac{d}{dt}\left(\frac{E}{mc^2}\frac{dE}{dt}\right) - \Re.$$

The approximation used in deriving it happens to be exactly valid for the present system, so that this equation is exact for uniform acceleration. Now for the observation of the τ_0-term, it is desirable that this term involve as large a fraction as possible of the imbalance of energy rate: the work by the external force does not all go into kinetic energy increase and radiation energy. In the case of synchrotron radiation, we saw in Section 6–7 that, since $\mathbf{v}\cdot\mathbf{F} = 0$, almost all the radiation energy comes from the kinetic energy; the τ_0-term contributes only a small fraction. One can ask under what conditions the τ_0-term (Schott term) contributes most to $\Re$. These are clearly the conditions for uniform acceleration since in that motion *all* of the radiation energy comes from the Schott term. The kinetic energy increase is *exactly* equal to the work done by the applied force. The above equation consists of two separate equalities,

$$\frac{dE}{dt} = \mathbf{v}\cdot\mathbf{F} \quad \text{and} \quad \tau_0\frac{d}{dt}\left(\frac{E}{mc^2}\frac{dE}{dt}\right) = \Re.$$

The last equation is of course just (6–129). The observability of causality violation is thus reduced to the problem of producing a uniform acceleration field g so large that

$$\Re = m\tau_0 g^2$$

is measurable within classical physics. This does not seem to be possible.

6-12 THE INTRINSIC ORBIT GEOMETRY

Both the free-field case and the problem of uniform acceleration led to exact solutions of the Lorentz-Dirac equation. A comparison between Eqs. (6–111) and (6–121) shows that in both cases very similar transformations appeared. We shall now explore the deeper reason behind this apparently accidental success. We shall see that all rectilinear one-dimensional problems admit exact solutions in closed form in terms of quadrature if the force $F_{\text{ext}}^{\mu}(\tau)$ is known.

The trajectory of a point particle is a curve in three-dimensional space. Such a curve can be characterized by certain intrinsic geometrical quantities, namely the first and the second curvatures, κ_1 and κ_2. At each point of the trajectory there exists a set of three mutually orthogonal, dimensionless unit vectors $\hat{v}$, $\hat{n}$, and $\hat{b}$. As one proceeds along the curve, these unit vectors change in direction but remain mutually orthogonal. Their dependence on the proper time τ can be expressed in terms of a set of simultaneous first-order differential equations called *Frenet equations,**

$$\frac{d\hat{v}}{d\tau} = \kappa\hat{n}, \qquad \frac{d\hat{n}}{d\tau} = -\kappa\hat{v} + \kappa'\hat{b}, \qquad \frac{d\hat{b}}{d\tau} = -\kappa'\hat{n}. \qquad (6\text{--}130)$$

These equations differ from the ones usually encountered in differential geometry in that the proper time τ is used here rather than the path length s. As a consequence, the invariants κ and κ' do not represent the curvature κ_1 and κ_2, but differ from them by a factor $ds/d\tau = v\gamma$, the magnitude of the velocity. For example, $\kappa = \gamma v\kappa_1 = \gamma v/r$, where r is the radius of curvature in the osculating plane (Fig. 6–3).

The second curvature $\kappa_2 = \kappa'/v\gamma$, which is often referred to as *torsion*, is related to $\hat{b}$ and $\hat{n}$ in exactly the same way as κ_1 is related to $\hat{v}$ and $\hat{n}$. The unit vectors are so chosen that $\hat{v}$ points in the direction of motion,

$$\hat{b} = \hat{v} \times \hat{n},$$

and $\hat{n}$ points toward the center of curvature (Fig. 6–3).

We shall try to represent the equations of a point charge in terms of this moving triad as the basis system. This will permit us to obtain explicit equations for $\kappa(\tau)$ and $\kappa'(\tau)$.

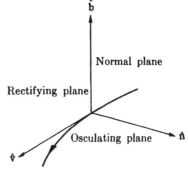

FIG. 6-3. The unit triad accompanying a space curve.

The velocity four-vector $v^{\mu}(\tau)$ depends on three arbitrary functions, since it is restricted by

$$v_{\mu}v^{\mu} = -c^2. \qquad (\text{A1--37})$$

---

*See, e.g., D. J. Struick, *Differential Geometry*, Second Edition, Addison-Wesley Publishing Co., Inc., Reading, Mass., 1961.

It can therefore be written in the form

$$v^\mu = c(\cosh w, \hat{v} \sinh w). \tag{6-131}$$

The three functions which specify w and the direction of the unit vector $\hat{v}$ are arbitrary. Equation (6–131) obviously satisfies (A1–37).

In order to introduce the variables w and $\hat{v}$ into the Lorentz-Dirac equation we must compute, by means of (6–130),

$$a^\mu/c = (\dot{w} \sinh w, \hat{v}\dot{w} \cosh w + \kappa\hat{n} \sinh w), \tag{6-132}$$

$$\begin{aligned} d^\mu/c = (\ddot{w} \sinh w + \dot{w}^2 \cosh w, \hat{v}[\ddot{w} \cosh w + (\dot{w}^2 - \kappa^2) \sinh w] \\ + \hat{n}(2\kappa\dot{w} \cosh w + \dot{\kappa} \sinh w) + \kappa\kappa'\hat{b} \sinh w). \end{aligned} \tag{6-133}$$

From this it follows that

$$a^\mu a_\mu/c^2 = \dot{w}^2 + \kappa^2 \sinh^2 w. \tag{6-134}$$

The force four-vector $F^\mu_{ext} + F^\mu_{in}$ can also be expressed in terms of w. Using

$$f^\mu \equiv \frac{1}{mc} F^\mu = \frac{1}{mc}\left(\gamma \frac{\hat{v}}{c} \cdot \mathbf{F}, \gamma\mathbf{F}\right), \tag{A1-49}$$

we have

$$f^\mu = (\sinh w \, \hat{v}\cdot\mathbf{f}, \cosh w \, \mathbf{f}), \quad \mathbf{f} = \mathbf{F}/mc. \tag{6-135}$$

These expressions can now be substituted into the Lorentz-Dirac equation (6–57). The result is a four-vector equation of which the space component is a three-vector equation, involving a linear combination of $\hat{v}$, $\hat{n}$, and $\hat{b}$. Since these are orthogonal, the coefficients of each of these vectors must vanish separately. This yields the following three linearly independent equations:

$$\begin{aligned} \dot{w} - \tau_0(\ddot{w} - \kappa^2 \sinh w \cosh w) &= \hat{v}\cdot\mathbf{f}, \\ \kappa - \tau_0(\dot{\kappa} + 2k\dot{w} \coth w) &= \coth w \, \hat{n}\cdot\mathbf{f}, \\ -\tau_0\kappa\kappa' &= \coth w \, \hat{b}\cdot\mathbf{f}. \end{aligned} \tag{6-136}$$

The time component gives an equation which is identical with the first equation (6–136).

In the coordinate system $\hat{v}$, $\hat{n}$, $\hat{b}$, the components of $\mathbf{f}$ are known, and (6–136) constitutes three simultaneous first-order equations for the three functions $w(\tau)$, $\kappa(\tau)$, and $\kappa'(\tau)$.

For $\kappa = \kappa' = 0$ we have rectilinear motion. In this case, (6–136) reduces to only one equation,

$$\dot{w} - \tau_0\ddot{w} = \hat{v}\cdot\mathbf{f}. \tag{6-137}$$

This equation can be solved easily in closed form,

$$\dot{w}(\tau) = \int_0^\infty (\hat{\mathbf{v}} \cdot \mathbf{f})_{\tau+\alpha\tau_0} e^{-\alpha} \, d\alpha + \frac{A}{\tau_0} e^{\tau/\tau_0}, \tag{6-138}$$

$$w(\tau) = w(0) + \int_0^\tau d\tau' \int_0^\infty (\hat{\mathbf{v}} \cdot \mathbf{f})_{\tau'+\alpha\tau_0} e^{-\alpha} \, d\alpha + A(e^{\tau/\tau_0} - 1). \tag{6-139}$$

Substitution into (6-131) results in the explicit solution $v(\tau)$ of the equation. It follows that *the one-dimensional Lorentz-Dirac equation can be solved exactly in closed form* provided that the above integrals converge and that the force is known as a function of proper time. If the latter is not the case, the above furnishes an integral equation.

PROBLEM 6-8

Show that the previously obtained solutions for the free particle and the uniformly accelerated particle follow from (6-138) and (6-139).

The equations (6-136) do not contain the asymptotic conditions (6-37) and (6-38). These can be incorporated, however, by changing the equations into integrodifferential equations, as was done in Section 6-6. In complete analogy with the procedure used there, one has for the first two equations (6-136)

$$\dot{w}(\tau) = \int_0^\infty (\hat{\mathbf{v}} \cdot \mathbf{f} - \tau_0 \kappa^2 \sinh w \cosh w)_{\tau+\alpha\tau_0} e^{-\alpha} \, d\alpha, \tag{6-140}$$

and

$$\kappa(\tau) = \int_0^\infty (\cosh w \, \hat{\mathbf{n}} \cdot \mathbf{f} - 2\tau_0 \kappa \dot{w})_{\tau+\alpha\tau_0} e^{-\alpha} \, d\alpha.$$

These equations correspond to (6-80) and must still be combined with the asymptotic conditions (6-38). The asymptotic conditions are

$$\lim_{|\tau| \to \infty} \dot{w}(\tau) = 0, \qquad \lim_{|\tau| \to \infty} \kappa(\tau) = 0, \qquad \lim_{|\tau| \to \infty} \kappa'(\tau) = 0. \tag{6-141}$$

Equations (6-140) and (6-141) together are equivalent to the equations of motion (6-82) if suitable initial conditions are specified.

Comparison of (6-140) with (6-138) shows that the constant A in the latter must be zero if the asymptotic conditions are to hold. This eliminates the self-accelerating solution for $\mathbf{f} = 0$.

In the following two sections we shall make use of the above general exact solution for one-dimensional rectilinear problems. But we shall also attack two-dimensional problems by means of the intrinsic formulation (6-140) and (6-136).

One might think it accidental that the use of the Frenet equations led us to a general solution for one-dimensional problems. The success of this approach should not be surprising. In fact, the use of intrinsic properties in the description of physical systems is much deeper. It is the minimum description because it permits the characterization of the observable features in a way which holds not only for any one particular observer but for all observers: it is independent of the coordinate system. For example, the characterization of any curve in three-dimensional space by $\kappa(\tau)$ and $\kappa'(\tau)$ is complete. At the same time, it is the minimum description in the sense that it cannot be given by fewer than two functions.

The intrinsic characterization therefore is the fundamental description. The intrinsic parameters and functions can be identified with the "observables" in a much more basic sense than coordinate dependent quantities. This is not always recognized. It is also not used to full advantage in solving physical problems, but we shall see that it is of great help in the applications of the theory.

6-13 THE PULSE

As the first application of the solution for rectilinear problems, let us consider a constant force acting in the direction of the velocity of a uniformly moving particle. Since we assume a force of finite duration, the asymptotic conditions will be satisfied by the solution*

$$v^{\mu} = c(\cosh w, \sinh w), \qquad\qquad (6\text{-}131')$$

$$\dot{w}(\tau) = \int_{0}^{\infty} f(\tau + \alpha\tau_0)\, e^{-\alpha}\, d\alpha. \qquad\qquad (6\text{-}140')$$

If the pulse is very short, it can be expressed by a δ-function,

$$f(\tau) = A\,\delta(\tau). \qquad\qquad (6\text{-}142)$$

We choose the origin of τ to coincide with the application of the force. Substitution of the integral representation of the δ-function yields

$$\dot{w}(\tau) = \int_{0}^{\infty} e^{-\alpha}\, d\alpha\, \frac{A}{2\pi} \int_{-\infty}^{\infty} e^{i\omega(\tau+\alpha\tau_0)}\, d\tau$$

$$= \frac{A}{2\pi} \int_{-\infty}^{\infty} e^{i\omega\tau}\, d\tau\, \frac{1}{1 - i\omega\tau_0} = -\frac{A}{2\pi i\tau_0} \int_{-\infty}^{\infty} \frac{e^{i\omega\tau}}{\omega - (1/i\tau_0)}\, d\omega.$$

This integral has a simple pole in the complex ω-plane at $\omega = -i/\tau_0$. It can be evaluated by contour integration as follows. For $\tau > 0$, the contour can be closed by a large semicircle in the upper half-plane. Since the pole is in the

---

* These equations differ from the corresponding unprimed equations only in that $\hat{v}$ is eliminated by expressing the direction relative to $\hat{v}$ through the sign of $w(\tau)$ and $f(\tau)$.

lower half-plane, the integral vanishes. For $\tau < 0$, the completion of the contour in the lower half-plane yields the negative residue, since the contour is followed in the negative (clockwise) direction. Thus,

$$\dot{w}(\tau) = \frac{A}{\tau_0} e^{\tau/\tau_0} \quad (\tau < 0), \qquad \dot{w}(\tau) = 0 \quad (\tau > 0). \qquad (6\text{-}143)$$

Therefore,

$$w(\tau) = w(-\infty) + A e^{\tau/\tau_0} \qquad (\tau \leq 0),$$

$$w(\tau) = w(-\infty) + A = w(0) \qquad (\tau \geq 0). \qquad (6\text{-}144)$$

The velocity follows from (6-131'),

$$\frac{dz}{dt} = v[t(\tau)] = c \tanh w(\tau) \qquad (6\text{-}145)$$

with $w(\tau)$ given by (6-144). The corresponding acceleration is

$$\frac{d^2z}{dt^2} = a[t(\tau)] = \frac{1}{\gamma} \frac{dv}{d\tau} = \frac{c\dot{w}(\tau)}{\cosh^3 w(\tau)}, \qquad (6\text{-}146)$$

since $\gamma = \cosh w$ according to (6-131'). These results are shown graphically in Fig. 6-4.

The preacceleration phenomenon is clearly visible in Fig. 6-4(a). The particle experiences an acceleration at arbitrarily long time intervals prior to the action of the force. But the effect has an exponential time behavior, so that in the classical domain where any measurement has an accuracy $\Delta\tau \gg \tau_0$, preacceleration is not observable. This was the conclusion we drew also earlier in Section 6-7.

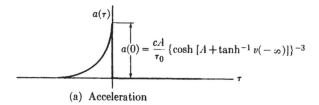

(a) Acceleration

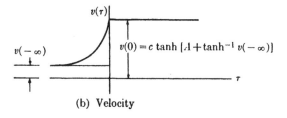

(b) Velocity

FIG. 6-4. Acceleration and velocity of a point charge due to a delta-function pulse.

The preacceleration effect also causes a rounding off of the onset of the velocity [Fig. 6–4(b)]. Comparison with the velocity of a neutral particle subject to the same force shows a difference only in this rounding off: $v_{\text{neutral}}(\tau)$ remains equal to $v(-\infty)$ until $\tau = 0$ and then rises to *exactly* the same value $v(0)$ which is maintained also by the charged particle for all times $\tau > 0$. This value, indicated in the figure, is of course the velocity obtained by the relativistic addition of velocities from $v(-\infty)$ and the velocity acquired by the force, $c \tanh A$.

A more general situation is encountered when the pulse lasts for a finite time τ_1, say.

$$f(\tau) = f \quad \text{for } 0 < \tau < \tau_1,$$
$$f(\tau) = 0 \quad \text{otherwise.} \tag{6–147}$$

Integration of (6–140′) yields

$$\dot{w}(\tau) = \begin{cases} f e^{\tau/\tau_0}(1 - e^{-\tau_1/\tau_0}) & (\tau < 0), \\ f(1 - e^{-(\tau_1 - \tau)/\tau_0}) & (0 < \tau < \tau_1), \\ 0 & (\tau_1 < \tau). \end{cases} \tag{6–148}$$

For convenience let us assume that $v(-\infty) = 0$; any other choice leads to the complication of relativistic velocity addition but does not otherwise affect the problem in any way. Then the integration of (6–148) gives

$$w(\tau) = \begin{cases} \tau_0 f e^{\tau/\tau_0}(1 - e^{-\tau_1/\tau_0}) & (\tau \le 0) \\ f\tau + \tau_0 f(1 - e^{-(\tau_1 - \tau)/\tau_0}) & (0 \le \tau \le \tau_1) \\ f\tau_1 & (\tau_1 \le \tau). \end{cases} \tag{6–149}$$

Velocity and acceleration are again given by substitution into (6–145) and (6–146). The results are shown in Fig. 6–5.

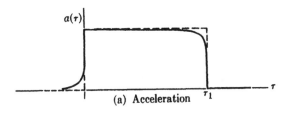

(a) Acceleration τ_1

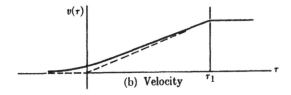

(b) Velocity τ_1

FIG. 6–5. Acceleration and velocity of a point charge due to a constant force of finite duration.

The figure shows the "rounding off" of the discontinuities because of pre-acceleration for $\tau \lesssim 0$ and *predeceleration* for $\tau \lesssim \tau_1$. Again, the final velocity $v(\tau > \tau_1)$ is *exactly* the same as that of a neutral particle subjected to the same force.

The acceleration is continuous for all τ while a_{neutral} is discontinuous at $\tau = 0$ and at $\tau = \tau_1$. The velocity is correspondingly differentiable every-where while v_{neutral} is not differentiable (although continuous) at $\tau = 0$ and $\tau = \tau_1$. The smallness of the preacceleration and predeceleration effects is clearly evident from the factors τ_0 multiplying f in (6-149). If the $(\tau_0 f)$-terms in $w(\tau)$ were omitted, the neutral particle results would follow. In fact, in the limit $\tau_0 \to 0$, these terms vanish, providing an example of the important relationship between charged and neutral particles. This will be elaborated further in Section 8-2.

The present example of a force of the type (6-147) emphasizes the transient effects of switching on and off. The action of a constant force over a long period of time, so that the transient effects are negligible, is already well known to us. It is the case of uniform acceleration. In fact, for $0 < \tau < \tau_1$, (6-149) gives for $\tau_1 - \tau \gg \tau_0$

$$w(\tau) = f\tau,$$

where we have chosen the initial condition $w(0) = 0$. The velocity is therefore

$$v^\mu = c(\cosh f\tau, \sinh f\tau)$$

according to (6-131'). Comparison with the solution for uniformly accelerated charges, (6-121), shows that they are identical because

$$f \equiv \frac{F}{mc} = \frac{mg}{mc} = \frac{g}{c} = \lambda$$

according to (6-124).

Another special case of the extended pulse (6-147) is the δ-function pulse (6-142). Indeed, in the limit $\tau_1 \to 0$ with

$$\int_{-\infty}^{\infty} f(\tau)\, d\tau = f\tau_1 = A = \text{const}$$

the solution (6-149) for $w(\tau)$ becomes exactly the solution (6-144) with $w(-\infty) = 0$. The latter is the initial condition which was imposed on (6-149).

6-14 THE OSCILLATOR

So far we have considered only problems in which the force is a known function of proper time τ. When the force is given as a function of the particle position (and possibly also of its velocity), even the linear problems can no longer be solved in closed forms. But a few exceptions do exist. An interesting one is given by the force

$$F(z) = -k\zeta, \tag{6-150}$$

$$\zeta = c \tanh^{-1} \frac{v}{c}. \tag{6-151}$$

A charged particle in this force field can be considered a relativistic harmonic oscillator. Its nonrelativistic limit is the usual harmonic oscillator.

With the usual definition of circular frequency,

$$\omega_0 = \sqrt{k/m}, \tag{6-152}$$

the quantity f defined in (6–135) becomes

$$f = -\frac{k\zeta}{mc} = -\frac{\omega_0^2 \zeta}{c}. \tag{6-150'}$$

The Lorentz-Dirac equation for linear motion, (6–137), now reads

$$\ddot{\zeta} - \tau_0 \dddot{\zeta} + \omega_0^2 \zeta = 0, \tag{6-153}$$

since (6–145) tells us that

$$w = \dot{\zeta}/c. \tag{6-154}$$

Equation (6–153) can be solved easily by means of exponentials. Assume that

$$\zeta(\tau) = e^{-i\alpha\tau}. \tag{6-155}$$

Substitution into (6–153) then requires that only those values of α are permitted which satisfy the cubic equation

$$-\alpha^2 - i\tau_0\alpha^3 + \omega_0^2 = 0. \tag{6-156}$$

This cubic can be solved by standard methods. It has the three solutions

$$\alpha_1 = \omega - i\gamma, \qquad \alpha_2 = -\omega - i\gamma, \qquad \alpha_3 = i\left(\frac{1}{\tau_0} + 2\gamma\right), \tag{6-157}$$

where ω and γ are given by

$$\omega\tau_0 = \tfrac{1}{2}\sqrt{3}\,(A - B) \qquad \text{and} \qquad \gamma\tau_0 = \tfrac{1}{2}(A + B) - \tfrac{1}{3}. \tag{6-158}$$

The constants A and B follow from the solution of a cubic,

$$\left.\begin{matrix} A \\ B \end{matrix}\right\} = \left[\frac{(\omega_0\tau_0)^2}{2} + \frac{1}{27} \pm \sqrt{\left(\frac{(\omega_0\tau_0)^2}{2} + \frac{1}{27}\right)^2 - \frac{1}{27^2}}\,\right]^{1/3}. \tag{6-159}$$

These relations imply that $\omega \geq 0$, $\gamma \geq 0$, and

$$\omega^2 = \gamma(3\gamma + 2/\tau_0). \tag{6-160}$$

The first two solutions (6–157) are located symmetrically with respect to the imaginary axis of the complex α-plane in the negative half-plane. The third solution is on the positive imaginary axis. Since the equation is linear, the general solution is a superposition of the three solutions corresponding to (6–157),

$$\zeta(\tau) = e^{-\gamma\tau}(Ae^{-i\omega\tau} + Be^{i\omega\tau}) + Ce^{(1/\tau_0+2\gamma)\tau}. \qquad (6\text{–}161)$$

The asymptotic conditions require that v^μ remain finite in the limit $\tau \to \infty$, so that $C = 0$. Furthermore, if we choose $\zeta(0) = 0$, we have $A + B = 0$, which results in

$$\zeta(\tau) = \text{const } e^{-\gamma\tau} \sin \omega\tau. \qquad (6\text{–}162)$$

The velocity of the charged particle is therefore

$$v(\tau) = c \tanh \frac{\dot\zeta(\tau)}{c},$$

and the position is

$$z(\tau) = z(0) + c\int_0^\tau d\tau' \tanh \frac{\dot\zeta(\tau')}{c}, \qquad (6\text{–}163)$$

with $\zeta(\tau)$ given by (6–162). If $\dot\zeta \ll c$, we have approximately

$$z(\tau) = z(0) + \zeta(\tau) = z(0) + \text{const } e^{-\gamma\tau} \sin \omega\tau. \qquad (6\text{–}164)$$

The motion is that of a damped harmonic oscillator. If τ_0 is very small, so that $\omega_0\tau_0 \ll 1$, it follows from (6–158) and (6–159) that

$$\omega \to \omega_0 \qquad \text{and} \qquad \gamma \to \tfrac{1}{2}\omega_0^2\tau_0. \qquad (6\text{–}165)$$

The frequency is therefore almost the harmonic frequency (6–152) and the damping is very small, of order $\omega_0\tau_0$.

It is interesting to compute the radiation rate of this oscillator. According to (5–13) and (6–134),

$$\Re = \frac{2}{3}\frac{e^2}{c^3}\dot w^2. \qquad (6\text{–}166)$$

Differentiation of (6–162) and use of (6–154) yield

$$\Re = \frac{2}{3}\frac{e^2}{c^3}\ddot\zeta^2 = \frac{2}{3}\frac{e^2}{c^3}(\text{const})^2 e^{-2\gamma\tau}[(\gamma^2 - \omega^2)\sin\omega\tau - 2\gamma\omega\cos\omega\tau]^2, \qquad (6\text{–}167)$$

$$\overline{\Re} = \frac{1}{3}\frac{e^2}{c^3}(\text{const})^2 e^{-2\gamma\tau}\omega^4. \qquad (6\text{–}168)$$

The last expression is the time average over one period of oscillation, assuming $\omega \gg \gamma$ as is always the case because of the smallness of τ_0.

6–15 COULOMB FORCE PROBLEMS

The previously studied systems all involved rectilinear orbits, characterized by the vanishing of both curvatures, $\kappa_1 = \kappa_2 = 0$.

The simplest more general case is

$$\kappa_2 = 0, \qquad \kappa_1 = \text{const},$$

which is obviously a circle. The straight line is a special case: a circle with vanishing curvature. On the basis of our Eqs. (6–136), $\kappa' = 0$ requires $\mathfrak{b} \cdot \mathbf{f} = 0$ at all times; i.e. the orbit lies entirely in the osculating plane spanned by $\mathbf{\hat{t}}$ and $\mathbf{\hat{n}}$. The two remaining equations (6–136) then determine all possible forces which lead to circular motion.

Since the orbit is a circle, we might ask for the force which will give this orbit as a motion with constant magnitude of velocity. The latter is suggested by the symmetry of the problem. Thus, we try to find $\mathbf{f}$ from (6–136), assuming that

$$w = \text{const}, \qquad \kappa = \gamma\,\frac{v}{r} = \text{const}, \qquad \kappa' = 0. \tag{6–169}$$

One finds easily

$$f_v \equiv \mathbf{\hat{t}} \cdot \mathbf{f} = \tau_0 \kappa^2 \sinh w \cosh w, \qquad f_n \equiv \mathbf{\hat{n}} \cdot \mathbf{f} = \kappa \tanh w.$$

The relation (6–131) permits us to write

$$\mathbf{f} = -\kappa \tanh w\, \mathbf{\hat{t}} + \tau_0 \kappa^2\, \frac{\tanh w}{1 - \tanh^2 w}\, \mathbf{\hat{t}},$$

$$\mathbf{F} \equiv mc\,\mathbf{f} = -m\gamma\,\frac{v^2}{r}\,\mathbf{\hat{t}} + \tau_0\,\frac{\gamma^2 v^2}{r^2}\,\gamma^2 \mathbf{v} = -m\gamma\,\frac{v^2}{r}\,\mathbf{\hat{t}} + \frac{2}{3}\,\frac{e^2}{r^2}\,\gamma^2(\gamma^2 - 1)\,\frac{\mathbf{v}}{c}. \tag{6–170}$$

Here we took into account that $\mathbf{\hat{n}}$ points toward the center of curvature so that $\mathbf{\hat{n}} = -\mathbf{\hat{t}}$. The force consists of an attractive radial force (centripetal force) and a tangential force in the direction of motion $\mathbf{\hat{t}}$.

If we desire the attractive radial part to be a static Coulomb force due to a charge Ze at the origin,*

$$\mathbf{F} \cdot \mathbf{\hat{t}} = -m\,\frac{v^2}{r}\,\gamma = -\frac{Ze^2}{r^2}, \tag{6–171}$$

we must require that

$$\gamma v^2 r = Ze^2/m \qquad \text{or} \qquad \gamma \left(\frac{v}{c}\right)^2 \frac{r}{r_0} = Z. \tag{6–172}$$

---

* For circular motion $v = \omega r$ and $\boldsymbol{\omega} \times \mathbf{v} = -a\mathbf{\hat{r}} = (d\mathbf{v}/dt)$ are also true relativistically. Equation (6–171) then follows from $\mathbf{p} = m\gamma\mathbf{v}$ and $(d\mathbf{p}/dt) = e\mathbf{E}$ if one notes that $\gamma = \text{const}$. for circular motion.

where r_0 is the classical electron radius (6–10). This relation is exactly the one obtained in elementary Newtonian physics when the centripetal force mv^2/r of circular motion is identified with the Coulomb force, (6–171).

The result (6–170) of course also admits other kinds of forces. For example, the centripetal force could be of gravitational origin due to a very heavy mass M at the origin,

$$\gamma \frac{mv^2}{r} = G \frac{Mm}{r^2}, \tag{6–171′}$$

$$\gamma v^2 r = GM. \tag{6–172′}$$

The nature of the force is seen to be quite secondary; we started with a prescribed orbit, and only (6–170) needs to be satisfied. In particular, the radiation rate is independent of the force, involving only the acceleration. From (6–134) and (5–14), we have

$$\mathcal{R} = mc^2 \tau_0 (\dot{w}^2 + \kappa^2 \sinh^2 w). \tag{6–173}$$

In our case, (6–169) gives

$$\mathcal{R} = mc^2 \tau_0 \left(\frac{\gamma v}{r}\right)^2 \left(\frac{\gamma v}{c}\right)^2 = \frac{2c}{3} \frac{e^2}{r^2} (\gamma^2 - 1)^2 \tag{6–174}$$

in agreement with (5–45). For small velocities this rate is proportional to v^4/r^2. If the centripetal force is a Coulomb force so that (6–172) holds,

$$\mathcal{R} = mc^2 \tau_0 (\gamma^2 - 1)^3 \left(\frac{v}{Zr_0}\right)^2 = \frac{2c}{3} \frac{e^2}{(Zr_0)^2} \left(\frac{v}{c}\right)^8 \gamma^6. \tag{6–175}$$

This is the radiation rate of a particle of charge $\pm e$ and mass m which is tangentially accelerated in the Coulomb field of a charge $\pm Ze$ [the $\hat{v}$-component in (6–170)] so that its orbit is exactly a circle. The corresponding circular frequency is

$$\omega = v/r. \tag{6–176}$$

The energy radiated is exactly accounted for by the tangential force. Equation (6–170) gives

$$\mathbf{F} \cdot \mathbf{v} = \frac{2}{3} \frac{e^2}{r^2} \gamma^2 \frac{v^2}{c} (\gamma^2 - 1),$$

which is just (6–174).

The tangential force is evidently necessary to ensure circular motion, but we notice that it is small, of order $\omega \tau_0$, compared with the centripetal force. What would be the motion and the radiation rate if it were completely absent?

Since $\mathbf{F} \cdot \hat{v}$ in (6–170) is small, its absence will clearly cause deviations from circular motion of only that small order, $O(\omega \tau_0)$. The radiation rate $\mathcal{R}$ will, in zeroth order, be unaffected by this change. But it can now no longer be accounted for by the tangential force. The particle is therefore expected to lose kinetic energy; its velocity would decrease and its motion would correspondingly be a spiral motion toward the origin. However, since the relationship (6–172) would still hold approximately in a Coulomb field, we are now led to a contradiction: (6–172) requires an *increased* radius for a smaller velocity.

The resolution of this apparent difficulty lies in the incorrect assumption that the radiation energy loss is accounted for by the loss of kinetic energy. Rather, it is a loss of *potential* energy due to the fact that the particle is very slowly falling toward the center.

Let us verify this quantitatively. If the particle is slowly spiraling toward the center, the vectors $\hat{v}$ and $-\hat{n}$ will not exactly coincide with the basis vectors of the radial coordinate system fixed at the center, $\hat{\varphi}$ and $\hat{r}$. They will be rotated by an angle of order $\omega\tau_0$,

$$-\hat{n} = \hat{r} + \alpha\hat{\varphi}, \qquad \hat{v} = \hat{\varphi} - \alpha\hat{r}. \tag{6–177}$$

This simple transformation can be used to obtain the solution for the particle spiraling toward the center of a pure Coulomb field, to first order, from the solution of exact circular motion. In fact, the forces transform as

$$F_n = -F_r^0 - \alpha F_\varphi^0, \qquad F_r = -F_n^0 - \alpha F_v^0,$$
$$F_v = F_\varphi^0 - \alpha F_r^0, \qquad F_\varphi = F_v^0 - \alpha F_n^0, \tag{6–178}$$

where the superscript zero indicates the quantities referring to the circular motion. Since we require that

$$F_\varphi = 0,$$

the last equation (6–178) determines α:

$$\alpha = F_v^0/F_n^0 = f_v^0/f_n^0 = \kappa_0\tau_0 \cosh^2 w_0. \tag{6–179}$$

Correspondingly,

$$F_r = -F_n^0 - \alpha F_v^0 = -F_n^0 - \alpha(\alpha F_n^0) \doteq -F_n^0$$

to first order, and

$$F_r = -mc\kappa_0 \tanh w_0 = -m\gamma v_0^2/r.$$

If we accept (6–172) as the restriction on v_0,

$$\gamma \left(\frac{v_0}{c}\right)^2 \frac{r}{r_0} = Z, \tag{6–172'}$$

this equation means that in the rotated coordinate system we have exactly the desired force to first order,

$$\mathbf{F} = F_r\hat{r} + F_\varphi\hat{\varphi} = -(Ze^2/r^2)\hat{r}. \tag{6–180}$$

The two functions $v(\tau)$ and $\kappa(\tau)$ which characterize the motion are now also easily found from the rotation (6–177):

$$\mathbf{v} = v\hat{\mathbf{v}} = v\hat{\boldsymbol{\varphi}} - \alpha v\,\hat{\mathbf{r}}.$$

Thus, using (6–179), we have

$$v_\varphi = v_0, \qquad v_r = -\alpha v_0 = -\omega_0\gamma_0\tau_0 v_0\gamma_0^2. \tag{6–181}$$

The magnitude of the velocity remains unchanged,

$$v = |v_0\hat{\boldsymbol{\varphi}} - \alpha v_0\hat{\mathbf{r}}| = v_0$$

to first order. Similarly, the curvature vector is

$$\kappa = \kappa\hat{\mathbf{n}} = -\kappa\hat{\mathbf{r}} - \alpha\kappa\hat{\boldsymbol{\varphi}} = -\kappa_0\hat{\mathbf{r}} - \alpha\kappa_0\hat{\boldsymbol{\varphi}},$$
$$\kappa = |\kappa_0\hat{\mathbf{r}} + \alpha\kappa_0\hat{\boldsymbol{\varphi}}| = \kappa_0. \tag{6–182}$$

It also remains unchanged to first order.

These results tell us that the radius of curvature ρ remains unchanged to first order,

$$\rho = \gamma\frac{v}{\kappa} = \gamma_0\frac{v_0}{\kappa_0} = \rho_0. \tag{6–183}$$

Equation (6–181) also gives us the time dependence of the particle distance from the center of attractions,

$$r(\tau) = r(0) + v_r\tau = r(0) - \omega_0\tau_0 v_0\gamma_0^3\tau. \tag{6–184}$$

The radius decreases at a constant rate v_r. The azimuth angle is clearly

$$\varphi = \omega\tau = \omega_0\tau. \tag{6–185}$$

It can be used in (6–184) to obtain the trajectory of the particle in polar coordinates,

$$r(\varphi) = r(0) - (\gamma_0^3 v_0\tau_0)\varphi. \tag{6–186}$$

The work done by the Coulomb potential per unit time is

$$\mathbf{F}\cdot\mathbf{v} = F_r v_r = +(Ze^2/r^2)\cdot\omega_0\tau_0 v_0\gamma_0^3. \tag{6–187}$$

It is seen to be identical with the radiation rate $\mathcal{R}$, as we found it in (6–174). One needs only to observe (6–176) and (6–171). The radiation rate is of course the same as for the circular orbit, in first approximation.

This proves that the energy emitted by the spiraling particle is due to the loss of potential energy, i.e. the work gained from the Coulomb field as the particle draws nearer to the center. The kinetic energy is unchanged to first order, according to (6–181).

PROBLEM 6-9

Obtain the above results for the spiraling particle by means of perturbation theory. Use the circular orbit as zeroth-order approximation and find the corrections to w_0 and κ_0 from the intrinsic equations (6–136).

In addition to the above bound state problem, there is the scattering problem: a charged particle moves toward a Coulomb field that is produced by a very heavy charge so that it remains at rest despite the impact. Two cases must be distinguished, the case of elastic scattering (Coulomb scattering), and the case of scattering with the emission of radiation (Bremsstrahlung*). In both cases, the domain of applicability of classical physics is very limited.

The case of Coulomb scattering is limited to high enough incident energy so that the binding energy of the electrons surrounding the atomic nucleus can be neglected. On the other hand, that energy cannot be so high that the impacted particle moves: the Coulomb field approximation (elastic scattering) would otherwise no longer hold. Also, the higher the energy, the more likely the scattering becomes inelastic: radiation emitted during the process can no longer be ignored; otherwise, the process becomes bremsstrahlung.

In the case of bremsstrahlung, the emission of radiation during the scattering process is taken into account. This is of course an approximation: any acceleration of a charge will cause radiation emission. It is simply ignored in so-called elastic collisions (Coulomb scattering). However, when such radiation emission is taken into account, the process requires a quantum mechanical treatment of the process, i.e. quantum electrodynamics. It follows that the classical treatment of both, Coulomb scattering and bremsstrahlung are very restricted indeed. Detailed discussions on Coulomb scattering and bremsstrahlung can be found in Jackson's text.**

---

*'Bremsstrahlung' is a German word that has become generally accepted for this process. It means literally 'deceleration radiation' and refers to the loss of kinetic energy of the incident charge due to its emission of radiation during the scattering process.

**J.D. Jackson, *Classical Electrodynamics*, John Wiley, New York, 3rd edition, 1999.

CHAPTER 7

Generalizations

The classical theory of systems of one point charge can be generalized in various ways. In the present chapter, systems of an arbitrary but finite number of point charges will be studied and their fundamental equations will be derived from an action principle. This generalization of the one-particle system will permit us to deduce what a test charge would measure, since such a charge is a limiting case of a physical charged particle. Another generalization of the considerations of the previous chapter refers to the structure of the particle. Instead of a structureless point charge one can study extended charges of various characteristics. This generalization is of minor importance for the classical description of elementary particles, because their structure lies outside the classical domain. Correspondingly, our discussion of particles with structure will be only brief.

7–1 SYSTEMS OF MORE THAN ONE CHARGED PARTICLE[*]

The one-particle theory developed in Section 6–9 on the basis of an action principle can be extended to an arbitrary number n of charged particles. The procedure will be the same: the action is a functional of the two potentials $\overline{A}$ and A_+ and the particle positions. Variation of the former with fixed boundaries produces an inhomogeneous and a homogeneous field equation. Then the particle positions can be varied as in Hamilton's principle, leading to particle equations consistent with the Maxwell-Lorentz field equations. General variations of the field and the particle variables yield conservation laws, provided the Euler-Lagrange equations of Noether's theorem (i.e. the field and the particle equations) are satisfied. Additional conservation laws are obtained for the free fields.

Let us again consider a closed system of particles and fields. The n charged particles have world lines $z_k^\mu(\lambda_k)$ $(k = 1, 2, \ldots, n)$. The parameter λ_k labels the world line of the kth particle and is related to the kth proper time τ_k by

$$c\,d\tau_k \equiv \sqrt{-\frac{dz_k^\mu}{d\lambda_k}\frac{dz_\mu^k}{d\lambda_k}}\,d\lambda_k, \tag{7–1}$$

an obvious generalization of (6–97). The fact that we shall be led to n different proper times is a necessity. The proper time is the time recorded in the instantaneous rest frame of the particle; it is in general not possible to be instantaneously at rest with respect to more than one particle. The proper times τ_k can easily

[*] F. Rohrlich, *Phys. Rev. Lett.* **12**, 375 (1964).

be reduced to a "common denominator" by relating them to some "observer time" t_0,

$$d\tau_k = (d\tau_k/dt_0)\, dt_0. \tag{7-2}$$

When the world lines are known, the functions $\tau_k = \tau_k(t_0)$ can be determined.

There are again two different fields $\bar{F}^{\mu\nu}$ and $F^{\mu\nu}_+$, as in (6–105). They will be related to the incident field $F^{\mu\nu}_{\text{in}}$ by the asymptotic condition. The n particles act as sources of fields, since they have charges e_k $(k = 1, 2, \ldots, n)$. We shall further assume that they have masses m_k $(k = 1, 2, \ldots, n)$ which are not necessarily equal.

The action integral (6–105) has the obvious generalization

$$I = -\sum_{k=1}^{n} m_k c^2 \int \sqrt{-\frac{dz^{\mu}_k}{d\lambda_k}\frac{dz^k_{\mu}}{\lambda_k}}\, d\lambda_k + \sum_{k=1}^{n} e_k \int \bar{A}_{\mu}(x)\, d^4 x\, \delta[x - z_k(\lambda_k)]\, \frac{dz^{\mu}_k}{d\lambda_k}\, d\lambda_k$$

$$- \frac{1}{16\pi}\int \bar{F}_{\mu\nu}(x)\bar{F}^{\mu\nu}(x)\, d^4 x - \frac{1}{8\pi}\int \bar{F}_{\mu\nu}(x)F^{\mu\nu}_+(x)\, d^4 x. \tag{7-3}$$

The variations of $\bar{A}_{\mu}(x)$ and A^+_{μ} with fixed boundaries yield, exactly as in Section 6–9,

$$\partial_{\mu}(\bar{F}^{\mu\nu} + F^{\mu\nu}_+) = -\frac{4\pi}{c}\, j^{\nu}(x) \tag{4-51}$$

and $\partial_{\mu}\bar{F}^{\mu\nu} = 0$. But in contradistinction to the one-particle problem, we now have

$$j^{\mu}(x) = \sum_{k=1}^{n} e_k c \int_{-\infty}^{\infty} \delta(x - z_k)\, \frac{dz^{\mu}_k}{d\lambda_k}\, d\lambda_k = \sum_{k=1}^{n} j^{\mu}_k(x). \tag{7-4}$$

These are the field equations for the n-particle system. The identification of $\bar{F}^{\mu\nu}$ and $F^{\mu\nu}_+$ with the expressions (6–46) also carries through via the asymptotic conditions exactly as in the one-particle case. One finds

$$F^{\mu\nu}_+ = \frac{1}{2}\sum_i (F^{\mu\nu}_{i,\text{ret}} + F^{\mu\nu}_{i,\text{adv}}) \quad\text{and}\quad \bar{F}^{\mu\nu} = F^{\mu\nu}_{\text{in}} + \frac{1}{2}\sum_i (F^{\mu\ \nu}_{i,\text{ret}} - F^{\mu\nu}_{i,\text{adv}}),$$

where $F^{\mu\nu}_i$ is the field produced by the current density j^{μ}_i.

The particle equations based on (7–3), however, lead to difficulties. Variation with fixed end points of z^{μ}_k gives, as before [cf. Eq. (6–51)],

$$m_k a^{\mu}_k = \frac{e_k}{c}\, \bar{F}^{\mu\nu}(z_k)v_{\nu} = \frac{e_k}{c}\, F^{\mu\nu}_{\text{in}} v^k_{\nu} + \sum_{i=1}^{n} \frac{e_k}{c}\, F^{\mu\nu}_{i,-} v^k_{\nu}. \tag{7-5}$$

This equation is empirically incorrect, because it does not give the retarded interaction of particle k with all the other particles. It contains only the interaction with the incident field, the radiation reaction term $(e_k/c)F^{\mu\nu}_{k,-}v^k_{\nu}$, and mutual interactions via $F^{\mu\nu}_-$. The action integral (7–3) must therefore be amended.

The missing term in (7–3) can be found easily. We shall proceed to verify that the following action integral provides a satisfactory basis for a system of n point charges:*

$$I = \sum_{k=1}^{n} \left[-m_k c^2 \int \sqrt{-\frac{dz_k^\mu}{d\lambda_k} \frac{dz_\mu^k}{d\lambda_k}} \, d\lambda_k + e_k \int \overline{A}_\mu(x) \, d^4x \, \delta(x - z_k) \frac{dz_k^\mu}{d\lambda_k} \, d\lambda_k \right]$$

$$+ \frac{1}{2} \sum_{\substack{i=1 \\ (i \neq j)}}^{n} \sum_{j=1}^{n} e_i e_j \iint \frac{dz_i^\nu}{d\lambda_i} \frac{dz_\nu^j}{d\lambda_j} \, \delta[(z_i - z_j)^2] d\lambda_i \, d\lambda_j$$

$$- \frac{1}{16\pi} \int (\partial_\mu \overline{A}_\nu - \partial_\nu \overline{A}_\mu)(\partial_\mu \overline{A}^\nu - \partial^\nu \overline{A}^\mu) \, d^4x$$

$$- \frac{1}{8\pi} \int (\partial_\mu \overline{A}_\nu - \partial_\nu \overline{A}_\mu)(\partial^\mu A^\nu_+ - \partial^\nu A^\mu_+) \, d^4x. \tag{7–6}$$

It is obvious that this action integral provides the same field equations as (7–3), differing from (7–3) only by the double integral which is independent of the fields.

The variation of the z_k^μ can now be carried out in accordance with Hamilton's principle with

$$\sum_k \delta \int L_{\text{particle}}^k \, d\lambda_k = 0, \tag{7–7}$$

where

$$\frac{\delta I}{\delta \lambda_k} \equiv L_{\text{particle}}^k = -m_k c^2 \sqrt{-\frac{dz_k^\mu}{d\lambda_k} \frac{dz_\mu^k}{d\lambda_k}} + e_k \overline{A}_\mu(z_k) \frac{dz_k^\mu}{d\lambda_k}$$

$$+ e_k \frac{dz_\nu^k}{d\lambda_k} \sum_{\substack{i=1 \\ (i \neq k)}}^{n} e_i \int \frac{dz_i^\nu}{d\lambda_i} \, \delta[(z_i - z_k)^2] \, d\lambda_i. \tag{7–8}$$

Note that the factor $\frac{1}{2}$ in the double integral term of (7–6) is absent in (7–8). The result of this variation is the Euler-Lagrange equation (3–51),

$$\frac{d}{d\lambda_k} \frac{\partial L_{\text{particle}}^k}{\partial(\partial z_k^\mu / \partial \lambda_k)} - \frac{\partial L_{\text{particle}}^k}{\partial z_k^\mu} = 0.$$

The first two terms in (7–8) yield, exactly as in (6–102),

$$m_k \frac{d^2 z_k^\mu}{d\tau_k^2} - \frac{e_k}{c} \overline{F}^{\mu\alpha} (z_k) v_\alpha^k, \tag{7–9}$$

provided we divide by $\sqrt{-(dz_k^\mu/d\lambda_k)(dz_\mu^k/d\lambda_k)}$ and use (7–1). Without the new

---

*See also J. D. Hamilton, *Am. J. Phys.* **39**, 1172 (1971).

term this would give (7–5). But now the third term in (7–8) gives, with $\partial_k^\mu \equiv \partial/\partial z_\mu^k$,

$$e_k \sum_{\substack{i=1 \\ i \neq k}}^{n} e_i \int d\lambda_i \left[\frac{dz_i^\mu}{d\lambda_i} \frac{d}{d\lambda_k} - \frac{dz_i^\nu}{d\lambda_i} \frac{dz_\nu^k}{d\lambda_k} \partial_k^\mu \right] \delta[(z_i - z_k)^2]$$

$$= e_k \sum_{\substack{i=1 \\ i \neq k}}^{n} e_i \frac{dz_\nu^k}{d\lambda_k} \int d\lambda_i \left[\frac{dz_i^\mu}{d\lambda_i} \partial_k^\nu - \frac{dz_i^\nu}{d\lambda_i} \partial_k^\mu \right] \delta[(z_i - z_k)^2]. \qquad (7\text{–}10)$$

This expression can be greatly simplified. From Eq. (4–84') we learn that

$$A_+^\mu(x) = e \int_{-\infty}^{\infty} \delta[(x - z)^2] \, v^\mu(\tau) \, d\tau.$$

The ith particle will therefore produce a potential $A_{i,+}^\mu$ which, at the position of particle k, is

$$A_{i,+}^\mu(z_k) = e_i \int_{-\infty}^{\infty} \delta[(z_k - z_i)^2] \frac{dz_i^\mu}{d\lambda_i} \, d\lambda_i.$$

This potential can now be introduced into (7–10). The above expression therefore becomes

$$e_k \frac{dz_\nu^k}{d\lambda_k} \sum_{\substack{i=1 \\ i \neq k}}^{n} \left(\partial_k^\nu A_{i,+}^\mu(z_k) - \partial_k^\mu A_{i,+}^\nu(z_k) \right).$$

Division by $\sqrt{-(dz_k^\mu/d\lambda_k)(dz_\mu^k/d\lambda_k)}$ and use of (7–1) yields

$$-\frac{e_k}{c} \sum_{\substack{i=1 \\ i \neq k}}^{n} F_{i,+}^{\mu\nu}(z_k) v_\nu^k. \qquad (7\text{–}11)$$

The terms (7–9) and (7–11) now combine to give the particle equation

$$m_k a_k^\mu = \frac{e_k}{c} \left(F^{\mu\alpha} v_\alpha^k + \sum_{\substack{i=1 \\ i \neq k}}^{n} F_{i,+}^{\mu\alpha} v_\alpha^k \right) = \frac{e_k}{c} \left(F_{\text{in}}^{\mu\alpha} v_\alpha^k + F_{k,-}^{\mu\alpha} v_\alpha^k + \sum_{\substack{i=1 \\ i \neq k}}^{n} F_{i,\text{ret}}^{\mu\alpha} v_\alpha^k \right).$$

We define*

$$F_{k,\text{in}}^\mu \equiv \frac{e_k}{c} F_{\text{in}}^{\mu\alpha}(z_k) v_\alpha^k, \qquad (7\text{–}12)_{\text{in}}$$

$$\Gamma_k^\mu \equiv \frac{e_k}{c} F_{k,-}^{\mu\alpha}(z_k) v_\alpha^k, \qquad (7\text{–}12)_{-}$$

$$F_{(k)\text{ret}}^\mu \equiv \frac{e_k}{c} \sum_{\substack{i=1 \\ i \neq k}}^{n} F_{i,\text{ret}}^{\mu\alpha}(z_k) v_\alpha^k, \qquad (7\text{–}12)_{\text{ret}}$$

---

* The notation $F_{(k)\text{ret}}^\mu$ was chosen in order to avoid possible confusion with the force due to the self-interaction via $F_{k,\text{ret}}^{\mu\nu}$, namely $(e_k/c)F_{k,\text{ret}}^{\mu\nu} v_k^\nu$. The latter is divergent.

and obtain the final form of the particle equations

$$m_k a_k^\mu = F_{k,\text{in}}^\mu + \Gamma_k^\mu + F_{(k)\text{ret}}^\mu \qquad (k = 1, 2, \ldots, n). \qquad (7\text{-}13)$$

The acceleration of the kth charged particle is due to the sum of three forces: the force of the incident radiation field, the Abraham force or "radiation re-action" force Γ_k^μ, and the force due to all the other charged particles. The latter is a purely retarded interaction.

As was proven in Section 6-5 [cf. Eq. (6-63)], Γ_k^μ is related to the kinematical properties of the kth charge by (6-55),

$$\Gamma_k^\mu = \frac{2}{3} \frac{e_k^2}{c^3} \left(\dot{a}_k^\mu - \frac{1}{c^2} a_k^\lambda a_\lambda^k v_k^\mu \right), \qquad (7\text{-}14)$$

with $\dot{a}_\mu^k \equiv d a_\mu^k / d\tau_k$. Therefore, the particle equations (7-13) are just the Lorentz-Dirac equations for each of the particles. They are now coupled to each other through the term $F_{(k),\text{ret}}^\mu$, which represents the only mutual inter-action of the charged particles. This term disappears if one has a system of only one charged particle and (7-13) reduces exactly to the one-particle Lorentz-Dirac equation (6-57).

The absence of the term with $i = j$ in the double integral of (7-6) is respon-sible for the absence of a self-interaction via $F_+^{\mu\nu}$. It is exactly this feature which ensures a well-defined theory free of self-energy divergences. Writing $F_{k,\text{ret}}^{\mu\nu} = F_{k+}^{\mu\nu} + F_{k-}^{\mu\nu}$, we see that $F_{k+}^{\mu\nu}$ due to the kth particle can interact only with *another* particle. The free fields $F_{k-}^{\mu\nu}$, on the other hand, do not behave in this fashion. They give rise to self-interactions (explicit occurrence of $F_{k-}^{\mu\nu}$) as well as to mutual interactions (via $F_{k,\text{ret}}^{\mu\nu}$); the self-interaction, however, is finite and is just the Abraham term. It occurs also for a single particle. The fields $F_+^{\mu\nu}$ do not occur in the particle equations of a one-particle system.

The equations of motion for a system of n particles can be derived in exactly the same way as for a single particle. One adjoins the asymptotic and initial conditions to the Lorentz-Dirac equations. Let $z_k^\mu(0)$ and $v_k^\mu(0)$ be the position and velocity vectors of particle k at the initial time $\tau_k = 0$. Then the simple generalization of (6-82) is

$$z_k^\mu(\tau_k) = z_k^\mu(0) + v_k^\mu(0)\tau_k + \frac{1}{m_k} \int_0^{\tau_k} d\tau' \int_{\tau_1}^{\tau'} d\tau'' \int_0^\infty K_k^\mu(\tau' + \alpha\tau_0^k) e^{-\alpha} \, d\alpha$$

$$(k = 1, 2, \ldots, n), \qquad (7\text{-}15)$$

where

$$K_k^\mu \equiv F_{k,\text{in}}^\mu + F_{(k),\text{ret}}^\mu - \mathfrak{R}_k v_k^\mu, \qquad (7\text{-}16)$$

$$\mathfrak{R}_k \equiv \frac{2}{3} \frac{e_k^2}{c^3} a_\mu^k a_k^\mu, \qquad \tau_0^k \equiv \frac{2}{3} \frac{e_k^2}{m_k c^3}. \qquad (7\text{-}17)$$

The closed system of n particles, possibly including a radiation field $F_{\text{in}}^{\mu\nu}$ is now determined by the simultaneous solution of the n equations of motion (7-15) and the Maxwell-Lorentz equations (6-106), (4-51), and (7-4). This poses a very complicated mathematical question. In fact, the precise formulation as an initial-value problem poses a problem by itself.

At first one might think that the z_k^μ and v_k^μ of each particle can be specified for some arbitrary set of initial times $\tau_{k,\text{in}}$ ($k = 1, \ldots, n$), which, say, all lie on a given spacelike plane σ_{in}. But the fields must also be given on σ_{in}. This raises a consistency problem: can these radiation fields *arbitrarily* specified on σ_{in} be produced by the n particles at suitable retarded points? The z_k^μ and v_k^μ are also chosen arbitrarily on σ_{in} and they obviously depend on the amount of radiation emitted in the past.

Our formulation of the n-particle problem suggests, in fact, that there may not be a consistent arbitrary set of data of this sort on a spacelike plane of finite time.*

A way out of this difficulty is to specify the initial state of the system in the infinite past. The particles are then all free and noninteracting. The incident radiation field vanishes by assumption at the positions of the particles. There is therefore no difficulty in specifying a consistent set of initial positions and velocities. Alternatively, one can specify initial velocities at $t \to -\infty$ and final velocities at $t \to +\infty$, and use the equation of motion (6-81). This is the type of formulation used in quantum field theory.

We can now turn to the conservation laws. The Lorentz invariance of the action integral (7-6) assures us, via Noether's theorem, of ten conservation laws. As an example, invariance of I under translations yields the momentum conservation law

$$\sum_{k=1}^{n} (dp_k^\mu + dp_{(k)+}^\mu) + d\overline{P}^\mu + dP_D^\mu = 0. \tag{7-18}$$

Here

$$p_k^\mu = m_k v_k^\mu \tag{7-19}$$

is the kinetic momentum (including the Coulomb field) of the kth particle, $\overline{P}^\mu$ and P_D^μ have the same meaning as in (6-107), and

$$dp_{(k)+}^\mu = d\tau_k \frac{d}{d\tau_k} \sum_{\substack{i=1 \\ (i \neq k)}}^{n} e_k e_i \int_{-\infty}^{\infty} \delta[(z_i - z_k)^2] v_i^\mu \, d\tau_i \tag{7-20}$$

is the four-momentum due to the mutual interactions of the particles via $F_+^{\mu\nu}$. Using (7-10) and (4-91), we can also write this latter momentum as

$$p_{(k)+}^\mu = \sum_{\substack{i=1 \\ i \neq k}}^{n} e_k A_{i,+}^\mu(z_k) = \frac{1}{2} \sum_{\substack{i=1 \\ i \neq k}}^{n} e_k \frac{e_i}{c} \left[\frac{v_{i,\text{ret}}^\mu}{\rho_{ik,\text{ret}}} + \frac{v_{i,\text{adv}}^\mu}{\rho_{ik,\text{adv}}} \right].$$

---

* The mathematical problem here is that of formulating an initial-value problem for a system of delay-differential equations; cf. R. D. Driver, *Ann. Phys.* (N.Y.), **21**, 122 (1963).

In the infinite past and infinite future this mutual interaction vanishes, if we assume that the particles are then free, i.e. infinitely far apart from one another. In these limits only the kinetic momenta and the free-field momenta contribute to the conservation law.

Equation (7–18) reduces to the one-particle case (6–107) for $n = 1$, since the mutual interaction (7–20) is then absent.

PROBLEM 7-1

Prove the conservation law (7–18).

PROBLEM 7-2

Extend the action integral (7–6) to open systems and find the corresponding particle equations.

7-2 ACTION AT A DISTANCE

In the last section it was proven that a consistent action integral for an electromagnetic n-particle system ($n > 1$) requires the introduction of a direct interaction between the charges, not mediated by a field, (7–7). One could ask whether it might not be possible to dispense with the electromagnetic field completely. Such a theory has indeed been proposed by Fokker* and was studied in great detail by Wheeler and Feynman†. It is known as the *action-at-a-distance* theory, because charges can interact with one another only over finite distances without a mediating entity (a field) that "carries" this action from one particle to the other.

Their starting point is the action integral

$$-\sum_{k=1}^{n} m_k c^2 \int \sqrt{-\frac{dz_k^\mu}{d\lambda_k} \frac{dz_\mu^k}{d\lambda_k}} \, d\lambda_k + \frac{1}{2} \sum_{\substack{i=1 \\ i \neq j}}^{n} \sum_{j=1}^{n} e_i e_j \int \frac{dz_i^\mu}{d\lambda_i} \frac{dz_\mu^j}{d\lambda_j} \delta[(z_i - z_j)^2] \, d\lambda_i \, d\lambda_j.$$

$$(7-21)$$

This is exactly the action integral (7–7) without the terms which involve the electromagnetic field. We can therefore take over the results of the previous section. Hamilton's principle applied to (7–21) yields, according to (7–9) and (7–11),

$$m_k a_k^\mu = \frac{e_k}{c} v_\nu^k \sum_{\substack{i=1 \\ i \neq k}}^{n} F_{i,+}^{\mu\nu}(z_k) \equiv F_{(k)+}^\mu \qquad (k = 1, \ldots, n), \qquad (7-22)$$

---

* A. D. Fokker, Z. *Physik*, **58**, 386 (1929).

† J. A. Wheeler and R. P. Feynman, *Revs. Modern Phys.* **17**, 157 (1945) and **21**, 425 (1949).

where, according to (7–10), $F_{i,+}^{\mu\nu}$ is *defined* by

$$F_{i,+}^{\mu\nu}(x) \equiv \partial^\mu A_{i,+}^\nu(x) - \partial^\nu A_{i,+}^\mu(x), \qquad A_{i,+}^\mu(x) \equiv e_i \int d\tau_i\, v_i^\mu\, \delta[(z_i - x)^2].$$

$$(7\text{–}23)$$

The particle equations (7–22) of the action-at-a-distance approach involve only the Coulomblike forces $F_{(k)+}^\mu$ due to all the other charges.

The first question here concerns the relationship between (7–22) and the particle equations (7–13). These equations become identical when

$$F_{k,\text{in}}^\mu + \Gamma_k^\mu + F_{(k),-}^\mu = 0.$$

According to (7–12) this can be written

$$F_{k,\text{in}}^\mu = -\frac{e_k}{c} \sum_{i=1}^{n} F_{i,-}^{\mu\nu} v_\nu^k \qquad (7\text{–}24)$$

or

$$F_{k,\text{in}}^\mu = \tfrac{1}{2}(F_{k,\text{adv}}^\mu - F_{k,\text{ret}}^\mu), \qquad (7\text{–}24)'$$

where

$$F_{k,\text{ret}}^\mu \equiv \frac{e_k}{c} v_\nu^k \sum_{i=1}^{n} F_{i,\text{ret}}^{\mu\nu}.$$
$$\quad\;\;\text{adv} \qquad\qquad\qquad\qquad\quad \text{adv}$$

The requirement (7–24) is mathematically consistent, because the fields on both sides of this equation satisfy the free-field equations.

In order to understand the physical interpretation of (7–24') let us first assume that the system consists only of n charged particles and that $F_{\text{in}}^{\mu\nu} = 0$. Then (7–24') requires

$$\sum_{i=1}^{n} F_{i,\text{ret}}^{\mu\nu}(x) = \sum_{i=1}^{n} F_{i,\text{adv}}^{\mu\nu}(x) \qquad \text{for all } x, \qquad (7\text{–}25)$$

or from (7–24)

$$\sum_{i=1}^{n} F_{i,-}^{\mu\nu}(x) = 0 \qquad \text{for all } x. \qquad (7\text{–}25)'$$

Since the $F_{i,-}^{\mu\nu}$ fields are asymptotically responsible for the four-momentum of radiation, (7–25') means that there is no radiation present asymptotically and that there is no net free field present anywhere in space-time.

This result is fully expected: the interaction term in (7–6) which is neglected in (7–21) is exactly the interaction of the charged particles with the free fields. But it is interesting that the neglect of this term still gives exactly the same results as (7–6), provided only that *all the particles in the system absorb exactly whatever radiation they all emit*. And this must hold at every space-time point through which a particle world line passes. The equality (7–25) is therefore often referred to as the *complete absorber* assumption.

When this assumption is satisfied and $F_{in}^{\mu\nu} = 0$, the action integral (7–21) gives exactly the equations (7–13), i.e. radiation reaction and effectively only retarded fields from the other particles.

When $F_{in}^{\mu\nu} \neq 0$, the condition (7–24) can now be interpreted as *incomplete absorption:* not all the radiation emitted by the system is also absorbed, the difference being felt as the presence of an extraneous radiation field, $F_{in}^{\mu\nu}$.

It is clear from these considerations that according to the action-at-a-distance theory radiation is only a spurious effect, almost a way of speaking. The only process taking place is that of interaction between charges. Radiation is a by-product of mutual interaction between charges.

The interest in the action-at-a-distance theory as stated by Wheeler and Feynman was threefold: the complete elimination of the field, the avoidance of self-energy difficulties, and the explicit symmetry in past and future, evident from (7–21). These reasons for seeking the action-at-a-distance theory are today no longer very cogent: the previous sections have shown that a consistent theory of the interaction of point particles with the electromagnetic field can be constructed which is free of self-energy difficulties; the time symmetry is also satisfied by Eq. (7–6), as will be demonstrated in Section 9–2; and the autonomous nature of the radiation field (as also evidenced by the existence of photons) makes the elimination of *all* electromagnetic fields somewhat arbitrary and not justified. Finally, the absorber conditions, (7–24) and (7–25), do not seem to lend themselves easily to inclusion in a set of basic assumptions of a theory. Nevertheless, the action-at-a-distance approach should not go unnoticed. It represents an interesting chapter in the history of the theory of point charges.

7–3 THE TEST CHARGE

Consider an open system of two charged particles in an external force field, $F_{in}^{\mu\nu} = 0$, $F_{ext}^{\mu} \neq 0$. The particle equations are, according to (7–13),

$$ma^{\mu} = \frac{e}{c} F_{ret}'^{\mu\nu} v_{\nu} + \Gamma^{\mu} + F_{ext}^{\mu},$$

$$m'a'^{\mu} = \frac{e'}{c} F_{ret}^{\mu\nu} v_{\nu}' + \Gamma'^{\mu} + F_{ext}'^{\mu}. \tag{7–26}$$

If the primed particle were absent, the system would consist of one particle (m, e) subject to the external force F_{ext}^{μ},

$$ma^{\mu} = \Gamma^{\mu} + F_{ext}^{\mu}, \tag{6–57}$$

which is just the Lorentz-Dirac equation with $F_{in}^{\mu\nu} = 0$. If we want to measure the electromagnetic field $F^{\mu\nu}$ due to the charge e whose equation is (6–57), we must introduce into this field a *test charge*, a charge which is affected by this field but which does not affect the motion of e. For any finite charge e' this is not exactly possible, but in the limit $e' \rightarrow 0$ such a measurement is expected to give the desired result; the lack of influence of e' on e can be approached as closely as the ingenuity of the experimenter will allow.

In order to see how this can be done, we consider the two-body problem (7–26). The test charge e' is influenced by the external force as well as by the field $F^{\mu\nu}$ of the charge e. The observer, in trying to minimize the effect of F'^{μ} on e', may shield the test charge or place it in a suitable position. If this is accomplished so that the effects of F'^{μ} are not observable within the error limits in question, the field $F^{\mu\nu}$ is the only cause for the test charge to deviate from uniform motion. The measurement of $F^{\mu\nu}$ consists in the inference of $F^{\mu\nu}$ from the observed behavior of the test charge e'.

In practice it is often advantageous to subject the test charge e' to a *known* external force which does not influence the charge e. One then observes the deviation of the behavior of e' from its motion under the known force.

As an example, consider a free electron in the conduction band of a metal which is shaped into a wire. An oscillating voltage represents F^{μ}_{ext} and causes the charge e to oscillate. The resultant field $F^{\mu\nu}$ is to be measured. Let the test charge be another electron e' located in a different wire which is not subject to F^{μ}_{ext}. The field $F^{\mu\nu}$ will cause the test charge to oscillate. These oscillations can be recorded and $F^{\mu\nu}$ can be inferred from it.

The importance of describing a measurement of $F^{\mu\nu}$ as a two-body problem, (7–26), lies in the following:

(1) Despite the absence of Coulomb effects in the one-particle theory, any measurement will record exactly $F^{\mu\nu}_{ret}$ and not $F^{\mu\nu}_{+}$ or $F^{\mu\nu}_{-}$, as might be expected on the basis of the one-particle equations (6–51) and (4–51) and the action principle (6–105).

(2) There are in general *two* forces acting on each of the two particles. The measurement will be the more accurate, the closer one can approach the ideal situation:

(a) the field $F'^{\mu\nu}_{ret}$ does not affect e,
(b) the force F'^{μ}_{ext} which affects e' is known (e.g. zero),
(c) the recorded motion of e' (or force on e') due to $F^{\mu\nu}$ is observed with maximum precision.

7–4 CHARGED PARTICLES WITH STRUCTURE

Ever since the Abraham electron (cf. Section 2–2) there has been interest in the description of extended charged particles. The self-energy and stability problems of the point charge, aggravated by the runaway solutions of the Lorentz-Dirac equation, induced many to seek a theory for charged particles of finite extension. These attempts were made considerably more complicated

after the special theory of relativity became established and Lorentz invariance
became a necessity.

Today these difficulties of point charge theory are overcome, as we have seen
in Sections 6–10 and 7–1, and there is no longer any need to introduce extended particles into the theory on these grounds. The only remaining reason
for a theory of extended charges therefore lies in its use as a model of actual
particle structure. Obviously, nuclei and the other elementary charged particles belong in the quantum domain. Nevertheless, it is felt by some that a
classical theory may offer either a crude phenomenological model or at least
a better understanding of the quantum theory of elementary particles which,
after all, is yet to be constructed.

The basic guidelines of a classical theory of extended charges are twofold:
the continued validity of the field equations (Maxwell-Lorentz equations) and
the invariance of the theory under Lorentz transformations. As mentioned
before, the latter requirement proved to be far more difficult to satisfy.

The first requirement makes it clear that the main modification of the point
charge theory will lie in a generalization of the point charge current density

$$j^\mu(x) = ec\int_{-\infty}^{\infty} \delta(x - z(\tau))v^\mu(\tau)\, d\tau \tag{4-82}$$

to that of an extended charge. This must be done in a covariant way.

Assume that we are dealing with a spherically symmetric charge distribution.
It will be described by a charge density ρ which is spherically symmetric around
the *charge center*. If the charge center has position vector z, the charge density
of the point charge,

$$\rho(x, z) = e\, \delta(\mathbf{r} - \mathbf{z}), \tag{7-27}$$

will be generalized to

$$\rho(x, z) = e\, f[(\mathbf{r} - \mathbf{z})^2]. \tag{7-28}$$

The *form factor* or *shape factor* f has dimensions of a reciprocal three-dimensional
volume. It is of course normalized in the same way as the δ-function,

$$\int f[(\mathbf{r} - \mathbf{z})^2]\, d^3r = 1.$$

These static considerations must now be generalized to a moving charge
within the framework of the special theory of relativity. This requires first
the characterization of the particle as *rigid*: a Lorentz observer instantaneously
at rest with the charge center (instantaneous inertial system) will see exactly
the charge distribution (7–28), irrespective of his velocity relative to some
initial Lorentz frame. Thus, the charge density (7–28) must be referred to
the three-dimensional spacelike "now-plane" of the instantaneous rest system.
If $v^\mu(\tau)$ is the velocity of the charge center, this plane is given by all points
x^μ satisfying

$$\sigma: v^\mu(x_\mu - z_\mu) = 0. \tag{7-29}$$

This is exactly the equation (A1–55) for a spacelike plane of normal (timelike) unit vector n^μ if we identify

$$n^\mu = v^\mu/c \quad \text{and} \quad \tau = -n_\mu z^\mu.$$

Equations (7–27) and (7–28) must correspondingly be brought into a covariant form. The covariant form of the three-dimensional δ-function $\delta(\mathbf{r} - \mathbf{z})$ is $\delta_\sigma(x - z)$, which is a function of the four-vector $x^\mu - z^\mu$. It is characterized by [see also Eq. (4–60)]

$$\int \delta_\sigma(x - z)\, d\sigma = 1, \tag{7–30}$$

the integration being over the surface (7–29). The *surface δ-function* $\delta_\sigma(x - z)$ is an invariant. Its relation to $\delta(\mathbf{r} - \mathbf{z})$ follows by comparison of (7–30) with

$$\int \delta(\mathbf{r} - \mathbf{z})\, d^3r.$$

It is

$$\delta_\sigma(x - z) = \delta(\mathbf{r} - \mathbf{z}) \frac{d\tau}{dt}. \tag{7–31}$$

Since $\delta(\mathbf{r} - \mathbf{z})$ transforms like a reciprocal volume (in ordinary space) which is contracted under a Lorentz transformation by a factor $1/\gamma$, $\delta(\mathbf{r} - \mathbf{z})$ receives a factor γ. The factor $d\tau/dt = 1/\gamma$ exactly compensates for this.

The generalization (7–27) $\rightarrow$ (7–28) can thus be put into covariant form: the equation

$$\rho_{\text{cov}}(x, z) = e\, \delta_\sigma(x - z) \tag{7–27'}$$

is generalized to

$$\rho_{\text{cov}}(x, z) = e\, f[(x - z)^2]|_\sigma. \tag{7–28'}$$

The generalization of the current density (4–82) is now easily made, starting with the form

$$j^\mu(x) = e\, \delta(\mathbf{r} - \mathbf{z}) \frac{dz^\mu}{dt}, \tag{4–82'}$$

which was derived in Section 4. In covariant form this can be written

$$j^\mu(x) = e\, \delta_\sigma(x - z) \frac{dz^\mu}{d\tau} = e\, \delta_\sigma(x - z)\, v^\mu(\tau). \tag{4–82''}$$

Thus, the extended charge current density is

$$j^\mu(x) = e\, f[(x - z)^2]\, v^\mu(\tau)|_\sigma. \tag{7–32}$$

The subscript σ in (7–28') and (7–32) refers to the determination of τ for a given x: τ specifies that point on the world line of the charge center $z^\mu(\tau)$ which is intersected by a spacelike plane orthogonal to it that passes through the point x.

The essential point here is the observation that for a given x there may exist more than one τ. This is a characteristic feature of special relativity and is closely related to the concept of simultaneity. The reason for this is best understood in geometrical terms.

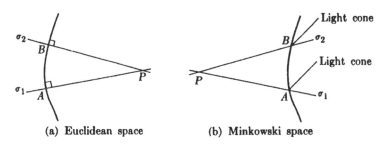

FIG. 7–1. Planes orthogonal to a curve at points A and B, respectively, are not parallel.

If Fig. 7–1(a) represents a curve in the plane of the paper and P a point in it, there are, in general, several lines through P which are orthogonal to the curve (i.e. orthogonal to the tangent at the point of intersection). Two such lines are drawn. If the curve is a space curve in Euclidean space, there will generally be several planes through P which will have an orthogonal intersection with the curve.

In Fig. 7–1(b) the curve is a world line in Minkowski space. The three-dimensional "now-planes" are drawn for the two points A and B, following the geometrical representation of moving frames in this space* (Fig. A1–1). These two three-dimensional planes, σ_1 and σ_2, intersect in a two-dimensional plane on which the point P lies. There may exist one or more other spacelike planes, $\sigma_3, \sigma_4, \ldots$, which pass through P and are orthogonal (in the Minkowski sense) to the world line.

What is the current density at the space-time point P in Fig. 7–1(b)? This point is simultaneous with *two* points on the world line, A and B. And in case there exist additional orthogonal planes σ_k, it will be simultaneous also with other points on that world line. Correspondingly, the current $j_\mu(x)$ at P is the sum of all the currents associated with each of these points as charge centers. This situation must now be allowed for in (7–32).

Let $\tau_1, \tau_2, \ldots$ be the proper time instants of $A, B, \ldots$; then (7–32) must be written

$$j^\mu(x) = e \sum_{k=1}^{n} \delta_k f[(x - z(\tau_k))^2] \, v^\mu(\tau_k) \Big|_{v^\alpha(\tau_k)(x_\alpha - z_\alpha(\tau_k)) = 0}. \qquad (7\text{–}33)$$

---

* Note that the hyperbolic nature of this space makes the planes appear not orthogonal to the world line.

Obviously, the τ_k are the n zeros of the equation

$$v^\alpha(\tau_k)(x_\alpha - z_\alpha(\tau_k)) = 0 \qquad (k = 1, 2, \ldots, n), \qquad (7\text{–}34)$$

since this is the equation of the spacelike plane through x which is orthogonal to the world line $z(\tau)$.

The $\delta_k = \pm 1$ in (7–33) occurs because not all the contributions to $j^\mu(x)$ are positive: as τ_1 at A is increased to $\tau_1 + d\tau$, the plane σ_1 will move slightly and will no longer pass through P. But will it have swept through P in the positive or negative time-direction? This depends on the curvature of the world line at A and the location of P. If σ_1 sweeps through P in the positive sense, $\delta_1 = +1$; otherwise, $\delta_1 = -1$. Two examples are drawn in Fig. 7–2. Thus, with $x^0 = ct$,

$$\delta_k = \operatorname{sgn} \left. \frac{d\tau_k}{dt} \right|_{\mathbf{r}=\text{const}}. \qquad (7\text{–}35)$$

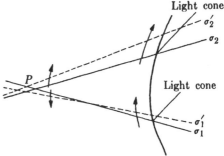

The consistent application of the simultaneity concept in special relativity thus leads to the conclusion that the motion of an extended positive charge forward in time may appear to another, suitably stationed, observer as a motion backward in time. He will interpret this as a *negative* charge moving forward in time, since these two descriptions are locally indistinguishable.

FIG. 7–2. The plane σ_2 sweeps through P in the positive sense; the plane σ_1 sweeps through it in the negative sense.

The result (7–33) can be written in a more elegant fashion as follows. Differentiation of (7–34) with respect to t, keeping $\mathbf{r}$ fixed, yields

$$\frac{d\tau_k}{dt} a^\alpha(\tau_k)(x_\alpha - z_\alpha(\tau_k)) + v^\alpha \left[\eta_{\alpha 0} c - \frac{d\tau_k}{dt} v_\alpha(\tau_k) \right] = 0$$

or

$$\frac{d\tau_k}{dt} \left[1 + \frac{a^\alpha(\tau_k)}{c^2}(x_\alpha - z_\alpha(\tau_k)) \right] = -\frac{v_0}{c} = \gamma(\tau_k) > 0.$$

Thus, Eq. (7–35) tells us that

$$\delta_k = \operatorname{sgn} \left[1 + \frac{a^\alpha}{c^2}(x_\alpha - z_\alpha) \right]_{\tau=\tau_k} \qquad (7\text{–}36)$$

Equation (7–33) can therefore be written

$$j^\mu(x) = e \sum_{k=1}^{n} \int \frac{1 + a^\alpha(x_\alpha - z_\alpha)/c^2}{|1 + a^\alpha(x_\alpha - z_\alpha)/c^2|} f[(x - z(\tau))^2] v^\mu(\tau) \, \delta(\tau - \tau_k) \, d\tau.$$

$$(7\text{–}37)$$

The identity (4-85) can be applied with $g(\tau)$ given by (7-29),

$$\int f(\tau)\, \delta[v^\alpha(x_\alpha - z_\alpha(\tau))]\, d\tau = \sum_{k=1}^{n} \frac{f(\tau_k)}{c^2 \left|1 + \dfrac{a^\alpha}{c^2}(x_\alpha - z_\alpha)\right|_{\tau = \tau_k}}$$

$$= \sum_{k=1}^{n} \int f(\tau) \frac{\delta(\tau - \tau_k)\, d\tau}{|c^2 + a^\alpha(x_\alpha - z_\alpha)|},$$

since

$$\frac{d}{d\tau}[v^\alpha(x_\alpha - z_\alpha)] = a^\alpha(x_\alpha - z_\alpha) - v^\alpha v'_\alpha.$$

When this identity is used in (7-37), the current density takes on its final form,

$$j^\mu(x) = e c^2 \int [1 + a^\alpha(x_\alpha - z_\alpha)/c^2]\, f[(x - z)^2]\, v^\mu(\tau)\, \delta[v^\alpha(x_\alpha - z_\alpha)]\, d\tau. \quad (7\text{-}38)$$

PROBLEM 7-3

Show that the current density (7-38) satisfies charge conservation, $\partial_\mu j^\mu = 0$.

A further generalization can now be made to *spinning* extended charges. One can assume that the charge distribution is rotating around the charge center. For this purpose one defines the angular velocity tensor $\omega^{\mu\nu} = -\omega^{\nu\mu}$ which has the following structure in the instantaneous rest system:

$$\omega_{(0)}^{\mu\nu} = \begin{pmatrix} 0 & 0 & 0 & 0 \\ 0 & 0 & \omega_z & -\omega_y \\ 0 & -\omega_z & 0 & \omega_x \\ 0 & \omega_y & -\omega_x & 0 \end{pmatrix}. \quad (7\text{-}39)$$

The angular velocity vector in the rest system is then $\omega = (\omega_x, \omega_y, \omega_z)$. It can then be shown* that a spinning, spherically symmetric charge distribution has the current density

$$j^\mu(x) = e c^2 \int_{-\infty}^{\infty} \{[1 + a^\alpha(x_\alpha - z_\alpha)/c^2]\, f[(x - z)^2]\, v^\mu$$

$$+ \omega^{\mu\nu}(x_\nu - z_\nu)\, g[(x - z)^2]\}\, \delta[v^\alpha(x_\alpha - z_\alpha)]\, d\tau. \quad (7\text{-}40)$$

This expression contains two form factors, one for the charge distribution and one for the spin.

---

* John S. Nodvik, *Ann. Phys.* (N.Y.) **28**, 225 (1964).

It is interesting to observe that the δ-function in (7–38) permits us to use the following effective equality:

$$v^\mu a^\alpha (x_\alpha - z_\alpha)/c^2 \doteq \omega_T^{\mu\alpha}(x_\alpha - z_\alpha),$$

where

$$\omega_T^{\mu\alpha} \equiv \frac{1}{c^2}(v^\mu a^\alpha - v^\alpha a^\mu). \tag{7-41}$$

The current density (7–38) can therefore also be written as

$$j^\mu(x) = e\,c^2 \int_{-\infty}^{\infty} [v^\mu + \omega_T^{\mu\nu}(x_\nu - z_\nu)]\,f[(x - z)^2]\,\delta[v^\alpha(x_\alpha - z_\alpha)]\,d\tau. \tag{7-38'}$$

The angular velocity tensor $\omega_T^{\mu\nu}$ is just the covariant form of the angular velocity of the Thomas precession. The presence of acceleration is thus closely related to the existence of a Thomas precession term in the extended current density. Equation (7–40) is then an obvious generalization of (7–38').

Further generalizations to charges lacking spherical symmetry are also possible.

How is the theory of a point charge, studied in Chapter 6, to be modified when we generalize the current to (7–38) or (7–40)? The action principle (6–105) can be generalized by replacing the interaction term between the particle and the field by

$$e \int \overline{A}_\mu(x)\,d^4x\,j^\mu(x)$$

with $j^\mu(x)$ given in (7–38). One can then derive the field equations

$$\partial_\mu F^{\mu\nu} = -\frac{4\pi}{c}j^\nu \tag{4-51}$$

(where $F^{\mu\nu}$ is the total field) and the particle equations

$$ma^\mu(\tau) = e \int F^{\mu\nu}(x)\,v_\nu\,[1 + a^\alpha(x_\alpha - z_\alpha)]\,f[(x - z)^2]\,\delta[v^\alpha(x_\alpha - z_\alpha)]\,d^4x, \tag{7-42}$$

which are really the equations for the charge center. The charge center and the mass center are assumed to be identical. The electrostatic self-energy of the extended charge is fully contained in the mass and occurs nowhere explicitly.

The electromagnetic field inside the extended charge, as it would be measured by a test charge, is not described in this way. The possibility of such a measurement presupposes the extended charge as nonelementary. It must consequently be described as a superposition of elementary charge distributions. For example, it would have to be given by a stream of many point charges, which are approximated by a smooth form factor f in (7-38). One must then use the many-particle theory of Section 7-1.

Such a classical description of subatomic particles as nonelementary is not completely unrealistic. When the particle can interact with forces that are strong compared with the electromagnetic forces, it will have a structure due to these strong interactions. This structure is in first approximation unaffected by electromagnetic interactions, so that these can be considered as superimposed on it. In the classical limit, it is reasonable to expect this structure to take the form of a distribution of matter which, when electromagnetic interactions are present, results in a charge distribution, a magnetic moment distribution, and possibly higher multipole distributions.

It is not surprising that in those cases the particles occur in both uncharged and charged varieties, corresponding to charge distributions whose integrals do and do not vanish, respectively. If the classical limit of the appropriate higher-level theory of strong interactions provides us with suitable form factors such as $f[(x - z)^2]$, at least in the nonrelativistic case, the static self-interaction via $F_+^{\mu\nu}$ can be computed. The corresponding interaction energy,

$$e \int A_\mu^+(x) \; j^\mu(x) \; d^4x,$$

constitutes a *finite* contribution to the self-energy of the particle. This energy can be measured as the difference between the masses of the charged and the neutral variety of the particle in case both these varieties exist in nature (proton-neutron, $\pi^\pm - \pi^0$ meson, etc.).

This calculation is separable in a covariant manner from the above dynamical theory based on Eq. (7-42). The mass in the latter contains all electromagnetic self-energy contributions.

7-5 POINT PARTICLES WITH SPIN

We consider a charged particle that has an intrinsic angular momentum $\vec{S}$, (a spin), and a corresponding magnetic dipole moment, μ, which is related to $\vec{S}$ by

$$\vec{\mu} = g \frac{e}{2mc} \vec{S}.$$

$$(7\text{-}43)$$

The coefficient g is called the *gyromagnetic ratio* or g-factor. It is known that for orbital motion (orbital angular momentum $\vec{L}$ instead of $\vec{S}$) the g-factor is unity.

There are several equivalent ways in which $\vec{S}$ can be generalized to a relativisitically covariant quantity. One way is to regard the three components of $\vec{S}$ as the space-space parts of an antisymmetric tensor $S^{\mu\nu}$,

$$\vec{S} = (S^{23}, S^{31}, S^{12}).$$

The components S^{ok} of this tensor can be fixed by requiring

$$S^{\mu\alpha}v_\alpha = 0$$

$$(7\text{-}44)$$

or equivalently, in the notation of (A1-36)

$$S^{ok} = -\left(\frac{\vec{v}}{c} \times \vec{S}\right)^k \cdot$$

$$(7\text{-}44')$$

The relativistic equation of motion of the spin $\vec{S}$ will now be obtained from its form in the rest frame. A magnetic dipole moment μ in a magnetic field $\vec{B}$ experiences a torque $\vec{\mu} \times \vec{B}$. Since a torque is the time rate of change of an angular momentum the equation of motion for $\vec{S}$ is*

$$\frac{d\vec{S}}{dt} = \vec{\mu} \times \vec{B} = \frac{ge}{2mc} \vec{S} \times \vec{B}. \qquad \text{(rest frame)} \qquad (7\text{-}45)$$

The coordinate time in the rest frame is denoted here by τ. The covariant relativistic generalization of this equation would seem to be

$$\frac{dS^{\mu\nu}}{dt} = \frac{ge}{2mc} \left(F^{\mu\nu} S^\nu_\alpha - F^{\nu\alpha} S^\mu_\alpha \right) \qquad (7\text{-}46)$$

since this reduces to (7–45) for the space-space components in the rest frame. However, the time-space components of (7–46) are in the rest frame

$$\frac{dS^{ok}}{dt} = \frac{ge}{2mc} (\vec{S} \times \vec{E})^k$$

or, in the view of (7–44′), they lead to the equation of motion for $\vec{v}$,

$$m\frac{d\vec{v}}{dt} = \frac{ge}{2} \vec{E}.$$

Therefore, (7–46) cannot be the correct generalization of (7–45) unless g = 2. Equation (7–46) was used by H. A. Kramers (1934) who later recognized this fallacy.

The correct generalization of (7–45) is

$$\frac{dS^{\mu\nu}}{dt} = \frac{ge}{2mc} \left[\left(\eta^{\mu\alpha} + \frac{v^\mu v^\alpha}{c^2} \right) F_{\alpha\beta} S^{\beta\nu} - \left(\eta^{\nu\alpha} + \frac{v^\nu v^\alpha}{c^2} \right) F_{\alpha\beta} S^{\beta\mu} \right]$$
$$+ \left(v^\mu a_\alpha S^{\alpha\mu} - v^\nu a_\alpha S^{\alpha\mu} \right) \frac{1}{c^2} \qquad (7\text{-}47)$$

The extra terms are all v-dependent and vanish for the space-space part of the equation in the rest frame. For the time-space part they lead to the rest frame derivative of (7–44′), and they do not imply an equation of motion for $\vec{v}$.

An immediate consequence of (7–47) is the constancy of the spin angular momentum $|\vec{S}|$ in the rest frame; covariantly this means

$$\frac{1}{2} \frac{d}{dt} \left(S^{\mu\nu} S_{\mu\nu} \right) = S_{\mu\nu} \frac{dS^{\mu\nu}}{dt} = 0 \qquad (7\text{-}48)$$

---

*See for example, J. D. Jackson, *Classical Electrodynamics*, Chapter 5, John Wiley and Sons, Inc., New York; second edition, 1975.

The equation (7–47) was first obtained by J. Frenkel in 1926.

Another way to generalize $\vec{S}$ to a relativistically covariant quantity involves the Pauli-Lubanksi fourvector w^μ. This is a space-like vector whose time component vanishes in the rest-frame,

$$w \cdot v = 0 \qquad \text{or} \qquad w^0 = \vec{w} \cdot \vec{v}/c \tag{7–49}$$

One now identifies $\vec{S}$ with the space-part of w^μ. In the rest frame one thus has again two equations; one is (7–45) with $\vec{S} = \vec{w}$, the other is

$$\frac{dw^0}{dt} = \vec{w} \cdot \frac{d\vec{v}}{dt} \frac{1}{c} \qquad \text{(rest–frame)} \tag{7–50}$$

which follows from (7–49). The unique covariant generalization of these two equations is

$$\frac{dw^\mu}{dt} = \frac{ge}{2mc} \left(\eta^{\mu\alpha} + \frac{v^\mu v^\alpha}{c^2} \right) F_{\alpha\beta} w^\beta + v^\mu a_\alpha w^\alpha \frac{1}{c^2} \tag{7–51}$$

This equation of motion for the spin was derived by Bargmann, Michel, and Telegdi in 1959 (BMT equation).

The conservation of the magnitude of the vector w^μ, $\sqrt{w_\mu w^\mu}$, which is just $|\vec{S}|$ in the rest frame, is a consequence of (7–51)

$$\frac{1}{2} \frac{d}{dt} \left(w_\mu w^\mu \right) = w_\mu \frac{dw^\mu}{dt} = 0. \tag{7–52}$$

The two generalizations of $\vec{S}$, $S^{\mu\nu}$ and w^μ are related by

$$w^\mu = +\frac{1}{2c} \, \varepsilon^{\mu\nu\alpha\beta} v_\nu S_{\alpha\beta} = S_*^{\mu\nu} \frac{v_\nu}{c} \tag{7–53}$$

or equivalently by

$$S_*^{\mu\nu} = -\frac{1}{c} \left(w^\mu v^\nu - w^\nu v^\mu \right) \tag{7–53'}$$

where $S_*^{\mu\nu}$ is the dual tensor to $S^{\mu\nu}$,

$$S_*^{\mu\nu} \equiv \frac{1}{2} \varepsilon^{\mu\nu\alpha\beta} S_{\alpha\beta} \tag{7–54}$$

with $\varepsilon^{\mu\nu\alpha\beta} = +1 \, (-1)$ for $\mu\nu\alpha\beta$ an even (odd) permutation of 0123, and zero otherwise.

PROBLEM 7–4

Show that the two equations (7–51) and (7–47) can be derived from one another and are therefore equivalent. Hint: Show first that (7–47) implies an equation for $S_*^{\mu\nu}$ and then use (7–53).

PROBLEM 7-5

Show that the conditions (7–44) and (7–49) can be imposed as initial conditions to the equations for $S^{\mu\nu}$ and w^μ, respectively; they are preserved by the motion.

In 1926 L. H. Thomas derived the spin equation of motion (7–51) for the first time. He also derived at that time equation (7–47) independently of Frenkel.

The two equivalent relativistic equations (7–51) and (7–47) for the motion of a magnetic moment in an external electromagnetic field should not be put parallel with the Lorentz-Dirac equation for a charged monopole. Firstly, the latter includes radiation reaction while the Frenkel-Thomas equations do not; secondly, the Lorentz-Dirac equation is no longer valid when the charged particle has also a magnetic dipole moment. Such a moment *modifies* the Lorentz-Dirac equation, as Frenkel has shown. One finds

$$\frac{d}{dt}\left[mv^\mu + \frac{1}{c^2} S^{\mu\alpha}a_\alpha - \frac{g}{2}\frac{e}{mc^3}\left(S^{\mu\alpha}F_{\alpha\beta}v^\beta - S^{\alpha\beta}F_{\alpha\beta}v^\mu\right)\right] =$$
$$+ \frac{e}{c}F^{\mu\alpha}v_\mu + \frac{1}{2}\frac{g}{2}\frac{e}{mc}\left(\delta^\mu F_{\alpha\beta}\right)S_{\cdot}^{\alpha\beta}$$

(7–55)

This is not an entirely satisfactory equation of motion since it contains $da^\mu/d\tau$ (like the Lorentz-Dirac equation). When $S^{\mu\nu}$ vanishes (charged monopole) one obtains the Lorentz force equation

$$ma^\mu = \frac{e}{c}F^{\mu\nu}v_\nu.$$

(7–56)

A charged magnetic dipole is thus governed by the two simultaneous equations (7–55) and (7–47). Only if one can ignore terms not linear in $S^{\mu\nu}$ is it permitted to use (7–56) instead of (7–55) to eliminate the acceleration from the spin dynamics (7–47). In that approximation, one obtains for the spin equation of motion from (7–47):

$$\frac{dS^{\mu\nu}}{dt} = \frac{g}{2}\frac{e}{mc}F^{\mu\alpha}S_\alpha^\nu + \left(\frac{g}{2}-1\right)\frac{e}{mc^3}v^\mu v^\alpha F_{\alpha\beta}S^{\beta\nu} - (\mu \rightleftarrows \nu)$$

(7–57)

and from (7–51):

$$\frac{dw^\mu}{dt} = \frac{g}{2}\frac{e}{mc}F^{\mu\alpha}w_\alpha + \left(\frac{g}{2}-1\right)\frac{e}{mc^3}v^\mu v^\alpha F_{\alpha\beta}w^\beta.$$

(7–57′)

The above considerations become considerably more complicated when one wants to include radiation reaction. However, one can show that the effect of radiation reaction can be completely expressed by a modification of the electromagnetic field tensor $F^{\mu\nu}$ and of the intrinsic angular momentum tensor $S^{\mu\nu}$ to $F^{\mu\nu}_{eff}$ and $S^{\mu\nu}_{eff}$. These effective tensors differ from $F^{\mu\nu}$ and $S^{\mu\nu}$ only by the addition of local functions of the world-line variables. They then satisfy exactly the same equations (7–47) and (7–55) as those without radiation reaction. Correspondingly, the properties (7–43), (7–44), and (7–48) now hold for $S_{eff}^{\mu\nu}$ rather than for $S^{\mu\nu}$.

This result is not entirely surprising. The Lorentz-Dirac equation can also be written in the form of (7–56) which ignores radiation reaction,

$$ma^{\mu} = \frac{e}{c} F_{eff}^{\mu\alpha} v_{\alpha} ,$$

(7–58)

provided*

$$F_{eff}^{\mu\nu} = F^{\mu\nu} + F_{-}^{\mu\nu},$$

(7–59)

as we have seen in Section 6–5. The relation between $F_{-}^{\mu\nu}$ and the Thomas precession tensor $\omega_{T}^{\mu\nu}$ of (7–41)

$$F_{-}^{\mu\nu} = m\tau_{0} \frac{d\omega_{T}^{\mu\nu}}{d\tau}$$

(7–60)

shows that $F_{-}^{\mu\nu}$ is indeed a local function of the world line variables.

The detailed and lengthy relations between $F^{\mu\nu}$ and $F_{eff}^{\mu\nu}$, and between $S^{\mu\nu}$ and $S_{eff}^{\mu\nu}$ can be obtained from the literature.

---

*Note that throughout this section $F^{\mu\nu}$ refers to an external field while $F_{-}^{\mu\nu}$ refers to the self-field of the charge.

CHAPTER 8

The Relation of the Classical Lorentz-Invariant Charged-Particle Theory to Other Levels of Theory

In accordance with the outline concerning the interrelation of physical theories, as given in Section 1–2, it is essential to establish that a theory of a given level contain the corresponding lower-level theories in suitable approximation and that it be itself an approximation of the corresponding higher-level theories. In the present Chapter this interrelationship will be studied for the charged point-particle theory of the preceding chapters. It will be convenient to separate the theories to be considered into classical and quantum theories.

A. Classical Theories

8–1 THE NONRELATIVISTIC APPROXIMATION

In Chapter 3 we saw that Newtonian dynamics is an approximation to the dynamics of special relativity. This approximation is characterized by the inequality

$$\left(\frac{v}{c}\right)^2 \ll 1. \tag{8-1}$$

This means that terms of order $(v/c)^2$ are neglected while terms of order v/c are still taken into account. It would be incorrect to characterize the limit to the nonrelativistic (Newtonian) theory by $c \to \infty$. Firstly, a quantity with dimensions cannot indicate a degree of approximation; otherwise, a change of units could give a better approximation. Secondly, this limit would rule out all systems in which particles interact with electromagnetic fields, because the latter will involve factors of $1/c$ (depending on the units used).

The electromagnetic field has no nonrelativistic approximation. It propagates with the velocity of light or close to it, depending on the properties of the medium. Its equations are always characterized by Lorentz invariance. A theory of charged particles in interaction with electromagnetic fields can therefore have a nonrelativistic approximation only so far as the particles are concerned. These would satisfy Galilean-invariant equations were it not for their interaction with the fields which satisfy Lorentz-invariant equations. From the point of view of symmetry properties, the nonrelativistic approximation of charged-particle theory is therefore inconsistent: the theory is neither Galilean nor Lorentz invariant.

The space part of the Lorentz transformations reduces to the Galilean transformations in the approximation (8–1), as is seen in Appendix 1, Eqs. (A1–14) and (A1–7). But the Lorentz transformation on time is thereby not reduced to the identity transformation, unless the space intervals Δx are restricted to $\Delta x \lesssim v \, \Delta t$. This restriction can be ensured only for the particles, not for the fields. Correspondingly, the theory will in general be invariant only under the four-dimensional translation group and the space rotations, still yielding conservation of linear and angular momentum as well as energy.* But the *center-of-mass conservation law* for a closed system of particles and electromagnetic fields can no longer be obtained on the basis of invariance.

The basic equations of relativistic kinematics are given in the Appendix (Section A1–4). In order to apply the approximation (8–1) we note that

$$ d\tau = dt + O\left[\left(\frac{v}{c}\right)^2\right], \qquad v^\mu = (c, \mathbf{v}) + O\left[\left(\frac{v}{c}\right)^2\right]. \tag{8–2} $$

Similarly, from Eq. (5–26) we find for the acceleration

$$ a^\mu = \left(\frac{\mathbf{v} \cdot \mathbf{a}}{c}, \, \mathbf{a}\right) + O\left[\left(\frac{v}{c}\right)^2\right] \tag{8–3} $$

and from Eq. (5–27) for the derivative

$$ \dot{a}^\mu = \left(\frac{\mathbf{v} \cdot \mathbf{b} + \mathbf{a}^2}{c}, \, \mathbf{b} + \frac{1}{c^2}\left(3\mathbf{v} \cdot \mathbf{a} \, \mathbf{a} + \mathbf{a}^2 \mathbf{v}\right)\right) + O\left[\left(\frac{v}{c}\right)^2\right]. \tag{8–4} $$

As before [cf. Eq. (5–21)], $\mathbf{b} = d\mathbf{a}/dt = d^2\mathbf{v}/dt^2$. Note that these equations do not follow from each other by differentiation with respect to time t, because this derivative is not interchangeable with the limit (8–1). For example,

$$ \frac{d}{dt} \lim_{(v/c)^2 \to 0} v^0 = \frac{d}{dt} \lim_{(v/c)^2 \to 0} \gamma c = \frac{dc}{dt} = 0, $$

but

$$ \lim_{(v/c)^2 \to 0} \frac{d}{dt} v^0 = \lim_{(v/c)^2 \to 0} \frac{1}{\gamma} a^0 = \lim_{(v/c)^2 \to 0} \gamma^3 \frac{\mathbf{v} \cdot \mathbf{a}}{c} = \frac{\mathbf{v} \cdot \mathbf{a}}{c}. $$

Finally, the force four-vector, (A1–49), is

$$ F^\mu = \left(\frac{\mathbf{v}}{c} \cdot \mathbf{F}, \, \mathbf{F}\right) + O\left[\left(\frac{v}{c}\right)^2\right]. \tag{8–5} $$

---

* The space- and time-inversion invariance which will be discussed in Section 9–2 is also still present.

With these preliminaries we can now compute the nonrelativistic approximation of the Lorentz-Dirac equation (6–57),

$$ma^\mu = F_{\text{ext}}^\mu + \frac{e}{c} F_{\text{in}}^{\mu\nu} v_\nu + \Gamma^\mu. \qquad (6\text{–}57)$$

The second term on the right is the interaction with the incident electromagnetic field. By means of (8–5), its space components give the force

$$\mathbf{F}_{\text{in}} = e\left(\mathbf{E}_{\text{in}} + \frac{\mathbf{v}}{c} \times \mathbf{B}_{\text{in}}\right) + O\left[\left(\frac{v}{c}\right)^2\right]. \qquad (8\text{–}6)$$

The nonrelativistic limit is therefore just the Lorentz force density, (2–8), integrated over the point charge distribution $\rho = e\delta(\mathbf{r} - \mathbf{z})$.

The nonrelativistic limit of the Abraham four-vector follows from (8–2) through (8–4),

$$\Gamma^\mu = \frac{2e^2}{3c^3}\left(\frac{\mathbf{v}\cdot\mathbf{b}}{c},\ \mathbf{b} + \frac{3\mathbf{v}\cdot\mathbf{a}\,\mathbf{a}}{c^2}\right) + O\left[\left(\frac{v}{c}\right)^2\right]. \qquad (8\text{–}7)$$

For the space part of the Lorentz-Dirac equation, substitutions into (6–57) yield

$$m\mathbf{a} = \mathbf{F}_{\text{ext}} + e\left(\mathbf{E}_{\text{in}} + \frac{\mathbf{v}}{c} \times \mathbf{B}_{\text{in}}\right) + \frac{2e^2}{3c^3}\left(\mathbf{b} + \frac{3\mathbf{v}\cdot\mathbf{a}\,\mathbf{a}}{c^2}\right). \qquad (8\text{–}8)$$

The time component obtains, similarly,

$$m\mathbf{a}\cdot\frac{\mathbf{v}}{c} = \frac{\mathbf{v}}{c}\cdot\mathbf{F}_{\text{ext}} + \frac{e}{c}\mathbf{E}_{\text{in}}\cdot\mathbf{v} + \frac{2e^2}{3c^3}\frac{\mathbf{v}\cdot\mathbf{b}}{c}.$$

It is obvious that this equation is just (8–8) multiplied by $\mathbf{v}/c$ in the approximation (8–1). The time component does not contain any new information.

In many systems of interest

$$\frac{3\mathbf{v}\cdot\mathbf{a}\,\mathbf{a}}{c^2} \ll \mathbf{b} \qquad (8\text{–}9)$$

in the approximation (8–1). This inequality is often true only in a time-average sense. For example, in a periodic motion in which the maxima of $\mathbf{v}$ and $\mathbf{a}$ are of the same order as their average values, $\bar{v}$ and $\bar{a}$, one has

$$\frac{\bar{v}\bar{a}^2}{c^2} \sim \frac{\bar{v}^2}{c^2}\,\bar{b},$$

so that (8–9) follows from (8–1). When this is the case, (8–8) can be written

$$\frac{d\mathbf{p}}{dt} = \mathbf{F}_{\text{ext}} + \mathbf{F}_{\text{in}} + \frac{2e^2}{3c^3}\frac{d\mathbf{a}}{dt}, \qquad (8\text{–}8')$$

with $\mathbf{p} = m\mathbf{v}$. This equation is exactly the Lorentz equation (2–24), provided one goes to the point-particle limit (the radius of the electron goes to zero) and one identifies **p** with the electromagnetic momentum $\mathbf{p}_{\text{elm}}$. The latter is in the spirit of a purely electromagnetic particle.*

In connection with the approximation (8-9), which is apparently essential in obtaining agreement with Lorentz's result, we need to remember how his damping force was derived: a periodic time dependence of **a** and **v** was assumed [cf. the derivation of Eq. (2-10) in Section 2-2]. It is clear that in general and over sufficiently short time intervals, Δt, one can have accelerations so large that (8-9) is not satisfied but $(a\,\Delta t)^2 \ll c^2$ still holds according to (8-1). For example, in uniformly accelerated motion **b** is not negligible, but the last term in (8-8) vanishes exactly.

The Maxwell-Lorentz field equations remain unchanged in form. The careful distinction between the various asymptotic conditions, retarded and advanced solutions, etc., must still be retained also for slowly moving particles. The asymptotic conditions, too, are essentially unaltered. Equations (6-37) and (6-38) are to be stated, of course, in their nonrelativistic form.

The equation of motion, obtained from the particle equation in combination with the asymptotic conditions, deserves special mention. In analogy to the derivation of Eq. (6-76), we can derive from (8-8′) the following integro-differential equation† (we omit $\mathbf{F}_{\text{in}}$ for convenience):

$$m\mathbf{a}(t) = \frac{1}{\tau_0} \int_t^{\infty} e^{(t-t')/\tau_0} \mathbf{F}_{\text{ext}}(t')\,dt'. \qquad (8\text{-}10)$$

PROBLEM 8-1

Derive Eq. (8-10).

This equation must be compared with the nonrelativistic approximation of (6-76),

$$m\mathbf{a}(t) = \frac{1}{\tau_0} \int_t^{\infty} e^{(t-t')/\tau_0}\left[F_{\text{ext}}(t') - m\tau_0\,\frac{\mathbf{a}^2(t')\mathbf{v}(t')}{c^2}\right]dt'. \qquad (8\text{-}11)$$

It is clear that the radiation reaction term is missing in (8-10). Thus, Eq. (8-8′) contains the small nonlocality in time but not the radiation reaction force; the space part of $\mathfrak{R}v^{\mu}$ is missing. On the other hand, the energy loss by radiation *is* contained in (8-8′). This energy loss is the work done by the damping force, which is proportional to **b**. In fact, that is just how this force was constructed in Section 2-2. Multiplication of (8-8′) by **v** yields the energy conservation law (we have omitted $\mathbf{F}_{\text{in}}$)

$$\frac{d}{dt}\left(\frac{m\mathbf{v}^2}{2}\right) = \mathbf{F}_{\text{ext}}\cdot\mathbf{v} - \mathfrak{R} + m\tau_0\,\frac{d}{dt}\,(\mathbf{v}\cdot\mathbf{a}), \qquad (8\text{-}12)$$

---

* The factor $\tfrac{4}{3}$ would be absent in our expression for p_{elm} because of its covariant definition (cf. Section 6-3).

† Equation (8-10) was first given by Haag, (loc. cit. in Section 6-5), and by Iwanenko and Sokolow, *Klassische Feldtheorie*, Akademie-Verlag, Berlin 1953.

with the radiation rate

$$\Re = m\tau_0 \mathbf{a}^2, \tag{8-13}$$

the nonrelativistic limit of (5-14).* The periodicity assumption will eliminate the last term in (8-12) when this equation is integrated over any multiple of the period. The conservation law then states that the rate of work done by the external force accounts exactly for the rate of kinetic-energy increase and the rate of radiation-energy loss. This statement must also be understood in a time-average sense.

Thus, we conclude that the nonrelativistic approximation together with the validity of (8-9) take account of the energy loss but not of the reaction force (three-vector) due to radiation emission, yielding (8-10) instead of (8-11).

The result (8-12) can also be obtained by taking the nonrelativistic approximation of the time component of the integrodifferential equation (6-76):

$$\frac{d}{dt} E_{\text{kin}}(t) = \frac{1}{\tau_0} \int_t^{\infty} e^{(t-t')/\tau_0}[\mathbf{v}(t') \cdot \mathbf{F}_{\text{ext}}(t') - \Re(t')] \, dt',$$

$$= \mathbf{v}(t) \cdot \mathbf{F}_{\text{ext}}(t) - \Re(t) + \tau_0 \frac{d^2}{dt^2} E_{\text{kin}} + O(\tau_0^2), \tag{8-14}$$

as in (6-88). Here no periodicity assumptions were made and (8-9) was not used.

Equation (8-10) permits the complete integration of the equations of motion by quadrature if $\mathbf{F}_{\text{ext}}$ is independent of the particle variables. It is interesting to observe that this equation is mathematically equivalent to the relativistic equation for the one-dimensional case. This is obvious from the mathematical equivalence of (8-8') and (6-137). The reader is referred to the review article by Plass† for applications of the nonrelativistic equations of motion.

8-2 THE LIMIT TO NEUTRAL PARTICLES

The classical theory of charged particles gives no indication of a quantization of charge. There is no smallest amount of charge and the theory carries through for any arbitrary choice of charge on a particle.

It is intuitively obvious that if one can vary the charge of a particle in a continuous fashion, this charge becomes undetectable in a given experiment, provided it is made sufficiently small. How, then, would a particle behave if its charge is made so small? It will clearly appear as a neutral particle.

This trivial observation has the far-reaching consequence that *the equation of motion of a charged particle must approach that of a neutral particle in the limit e → 0*. Such a requirement is far from trivial mathematically. But physically it is so basic that it might well be postulated as an a priori require-

---

* Note that (8-8) also yields (8-12).

† Literature Selections for further study: Chapter 6.

ment on any charged-particle theory as *the principle of the undetectability of small charges.*

We must now check whether this principle holds for our theory. The Lorentz-Dirac equation,

$$ma^\mu = F^\mu_{ext} + \frac{e}{c} F^{\mu\nu}_{in} v_\nu + \frac{2}{3} \frac{e^2}{c^3} (\dot{a}^\mu - a^\lambda a_\lambda v^\mu), \qquad (6\text{-}57)$$

indeed reduces to

$$ma^\mu = F^\mu_{ext} \qquad (8\text{-}15)$$

in the limit $e \to 0$, provided F^μ_{ext} is of a nonelectromagnetic nature. This implies that the *solutions* of the Lorentz-Dirac equation also must reduce to the solutions of the neutral-particle equation (8-15). It should be noted, however, that the limit (6-57) → (8-15) is not so trivial: a third-order differential equation becomes a second-order differential equation. In fact, this cannot hold in general, since the former has a larger family of solutions than the latter, requiring three initial conditions instead of only two.

This extra freedom lies exactly in the runaway solutions. Consider the simple case of a free particle. The free charged particle solution (6-113) in the limit $\tau_0 \to 0$ (implied by $e \to 0$) does not approach the neutral particle solution, $v^\mu = const$, but actually diverges and has no limit. Only when $A = 0$ does one obtain the correct limit.

Since we have seen that the asymptotic postulate guarantees the absence of runaway solutions, we might expect that this postulate also guarantees the correct limit to neutral particles. As it stands, the *solutions* of (6-57) obviously do *not* approach those of (8-15) in the limit.

We are thus forced to investigate the integrodifferential equations (6-80) or (6-82) which combine the asymptotic postulate with the Lorentz-Dirac equation:

$$ma^\mu(\tau) = \int_0^\infty e^{-\alpha} K^\mu(\tau + \alpha\tau_0) \, d\alpha, \qquad (6\text{-}80)$$

$$K^\mu = F^\mu_{ext} + F^\mu_{in} - \frac{1}{c^2} \Re v^\mu. \qquad (6\text{-}71)$$

Since both F^μ_{in} and $\Re$ vanish for $e \to 0$, (6-80) becomes

$$ma^\mu(\tau) = \int_0^\infty e^{-\alpha} \, d\alpha F^\mu_{ext}(\tau) = F^\mu_{ext},$$

and we recover (8-15). Of course, we assumed here that it is permissible to interchange the limit and the integration.

The limit (6-80) → (8-15) involves no change in the order of the highest derivative. We can therefore expect that the limit would also carry through for the solutions of the respective equations.

This expectation is trivially established when the perturbation expansion (6-87) is valid. Conversely, it is reasonable to expect that the requirement of the existence of this limit implies that the solutions are analytic at the origin of the complex e-plane and that therefore all physical solutions actually do have a perturbation expansion of nonzero convergence radius. A proof of this assertion is so far outstanding. (See, however, the end of Section 6–8).

The reader should check the special systems in Part C of Chapter 6 and verify that all solutions of the equations of motion (6–80) do, in fact, reduce to the corresponding neutral-particle solutions.

One suspects that the principle of the undetectability of small charges could be used to establish that part of the asymptotic postulate which is sufficient to eliminate the runaway solutions. This is indeed the case.* However, the asymptotic postulate is basic to the theory and an integral part of it, in any case, because it is a field theory (Section 6–4); the neutral particle limit is therefore not needed for its structure. But it is essential that this limit carry through in a satisfactory manner.

8-3 IS THE PRINCIPLE OF EQUIVALENCE VALID FOR CHARGED PARTICLES?

In Part C of Chapter 3 two different principles were stated, both under the name of "principle of equivalence." The weaker one is Einstein's principle which states the equivalence between gravitational fields and accelerated frames of reference. It is valid only for apparent gravitational fields or, in the language of general relativity, for flat space.† The stronger principle, which states a property of the equations of motion of a test particle in free fall [statement (E) of Section 3–8], is to be valid in any kind of gravitational field. The stronger principle implies the weaker one.

We must now inquire whether these formulations of the principle of equivalence remain valid for charged particles. The difference between the weak and the strong formulation is very important here, because the strong formulation makes a nontrivial statement about particles in general gravitational fields, while the weak formulation is a statement about the *simulation* of certain gravitational fields by acceleration. Since the latter is known to be valid in Newtonian physics as well as in special relativity, it must also be valid for any relativistic theory of gravitation. The validity of the weak principle must be a consequence of the fact that such a theory is a covering theory of special relativity and Newtonian gravitation theory. Thus, if the weak principle were not also valid for charged particles, there would be an inconsistency in the theories, while the strong principle need not necessarily hold for charged particles. In order to investigate the latter we shall need to know the equations of motion of a charged test particle in an arbitrary gravitational field. This problem will be deferred to the following section. In the present section we want to study the validity of Einstein's principle of equivalence for charged particles.

---

* F. Rohrlich, *Ann. Phys.*(N.Y.) **13**, 93 (1961).

† This includes regions of curved space which are so small that they appear as flat-space regions within the approximation considered.

The difficulties encountered in this validity question are fully apparent when we restrict our consideration to a constant, static, homogeneous gravitational field (SHGF). This is a field whose lines of force are equidistant parallels, such as the gravitational field in the laboratory. It is known that this type of gravitational field can be simulated by uniform acceleration of a neutral particle in Newtonian mechanics (Section 3–9) and in special relativity. Is this also true for the motion of a charged particle?

At first the answer seems to be "no," because it would require that a neutral and a charged particle fall equally fast in an SHGF. But this cannot be the case, because the charged particle will radiate (being accelerated), will thereby lose energy, and will consequently fall more slowly than the neutral particle.

We shall now proceed to show that this argument is fallacious and we shall prove that a charged and a neutral particle in an SHGF fall equally fast, despite the fact that the charged one loses energy by radiation.* The Einstein form of the principle of equivalence will therefore be satisfied also for a charged particle.

In studying the problem of free fall of a charged particle, in an SHGF, we imagine an observer such as a man in the laboratory on the surface of the earth. Such an observer is *not an inertial observer*. We recall from Section 3–5 that an inertial frame of reference is one for which a particle at rest remains at rest when no forces are acting; i.e. Galileo's law of inertia is valid. A man in the laboratory is not in such a frame because objects drop to the ground despite the fact that no force is acting: *a constant gravitational field provides only an apparent force* that can always be transformed away by suitable motion. In this case, the suitable motion is free fall. A freely falling observer sees no gravitational field† and is consequently in force-free surroundings; he is an inertial observer. The observer on the surface of the earth is constantly supported in the gravitational field.

Maxwell's equations are valid only for inertial observers. We can therefore *not* apply them to the description of a falling charge relative to a supported observer. We *can* apply them relative to a falling observer. But this makes the problem trivial: the falling observer, I, is inertial and the falling charge will remain at rest with respect to I, since no forces are present. Thus, a neutral and a charged particle will fall equally fast relative to I.

In order to find out how a supported observer S would see this free fall, a transformation must be carried out from I to S. This is not a Lorentz transformation, because S and I are in relative acceleration. It must be determined from the description of the SHGF as a fictitious force field.

---

* The details of the following proof are contained in the author's paper, *Ann. Phys.* (N.Y.) **22**, 169 (1963).

† This is true for a constant homogeneous field. If the curvature of the surface of the earth becomes important, the lines of force will no longer be parallel, and this fact *can* be observed in free fall. (See p. 44, Problem 3–5.)

This description is offered to us by general relativity. Being a fictitious field, it is characterized by flat space, i.e. a vanishing curvature tensor $R_{\kappa\lambda\mu\nu}$. Its static nature and its symmetry (it is independent of the directions x and y, and dependent only on z) put severe restrictions on the field. One finds* that the most general (fictitious) gravitational field which satisfies these requirements of an SHGF is given by the following timelike line element:

$$c^2\, d\tau^2 = u^2(z)c^2\, dt^2 - [c^2 u'(z)/g]^2\, dz^2 - dy^2 - dx^2. \qquad (8\text{-}16)$$

The function $u(z)$ must approach

$$u(z) = 1 + gz/c^2 \qquad (8\text{-}17)$$

in the nonrelativistic approximation in order to ensure the inclusion of that approximation in the more general theory. The z-axis is so chosen that a falling object would move in the negative direction of this axis.

Equation (8-16) means that the supported observer S who uses the coordinates $x^\mu = (ct, x, y, z)$ sees an SHGF whose strength is characterized by the acceleration of gravity g. The motion of a freely falling neutral particle is given by the equation of motion (3-33),

$$\frac{d^2 x^\mu}{d\tau^2} + \Gamma^\mu_{\alpha\beta}\frac{dx^\alpha}{d\tau}\frac{dx^\beta}{d\tau} = 0. \qquad (3\text{-}33)$$

It is easy to verify that this motion is in general *not* uniformly accelerated (hyperbolic) motion. Rather, $z(t)$ is given by the solution of

$$1/u(z) = \cosh\,[g(t - t_0)/c^2]. \qquad (8\text{-}18)$$

This follows by substitution of (8-16) into Eq. (3-33). The initial time t_0 is the time at which $z = 0$.

PROBLEM 8-2

Show that the particular choice of $u(z)$,

$$u(z) = \{\cosh\sqrt{[1 - (gz/c^2)]^2 - 1}\}^{-1},$$

does yield hyperbolic motion. Compare this result with (5-32) and note the difference in initial conditions.

PROBLEM 8-3

Show that the nonrelativistic result

$$z = -\tfrac{1}{2}g(t - t_0)^2$$

follows from (8-18), irrespective of the choice of $u(z)$, so long as (8-17) is satisfied.

---

* F. Rohrlich, *loc. cit.*

The transformation from the reference frame S of the supported observer to the inertial frame I is that transformation which changes $d\tau$ given by (8–16) into

$$c^2\, d\tau^2 = c^2\, dt'^2 - dz'^2 - dy'^2 - dx'^2. \qquad (8\text{–}19)$$

The primed coordinates refer to I. One verifies that this transformation is

$$x' = x, \qquad y' = y,$$

$$\frac{1}{c^2}\, g(z' - z'_0) = u(z)\cosh g(t - t_0) - 1, \qquad (8\text{–}20)$$

$$\frac{1}{c^2}\, g(t' - t_0) = u(z)\sinh g(t - t_0).$$

We are now in a position to solve electrodynamic problems in the noninertial coordinate system S. All we need to do is to transform the Maxwell-Lorentz equations from I to S by means of (8–20). Furthermore, if we know the solutions of these equations in I, we also know them in S from (8–20).

A charged particle at rest at the origin of I (i.e. freely falling with I) produces in I the fields

$$\mathbf{B}' = 0, \qquad \mathbf{E}' = \frac{e}{r'^2}\, \hat{\mathbf{r}}'. \qquad (8\text{–}21)$$

Transformation via (8–20) shows that the observer sees this charge radiating. Nevertheless, the charged particle remains at rest in I, just as a falling neutral particle would. It is, of course, clear that the radiation rate $\mathcal{R}$ is Lorentz invariant, but it is not invariant under (8–20).*

PROBLEM 8–4

Compute the radiation rate seen by S. Using (8–20), transform (8–21) to the reference frame of the supported observer S, and compute $\mathcal{R}$.

The principle of equivalence is thus satisfied. Our equations are all consistent with it. But another problem arises now which was trivial previously: how do we know that a charged particle at rest relative to S does *not* radiate? Having established that the physicist in his laboratory is not an inertial observer and cannot apply the Maxwell-Lorentz equations in their usual form, we have no longer any assurance that the trivial problem of a charge at rest (relative to S) leads to no radiation.

In order to investigate this question, we must again start with the inertial observer I, i.e. the falling one. A supported charge (at rest on the table in the laboratory) will appear to I to be *uniformly accelerated* upward. This can

---

* A corollary of this result states that a (classical) radiation detector which moves in uniform acceleration past a Coulomb field will record radiation. This can be demonstrated by means of a detector model: R. A. Mould, Ann. Phys. (N.Y.) **27**, 1 (1964).

be most easily seen from (8–20). The charge has coordinates $x = y = z = $ const for all t in S. Elimination of these coordinates yields

$$\left[\frac{c}{g} + (z' - z'_0)\right]^2 - (t' - t'_0)^2 = \left[\frac{cu(z)}{z}\right]^2 = \text{const.} \qquad (8-22)$$

This is just the equation for hyperbolic motion, $z'(t')$ being a hyperbola in Minkowski space.

Since I is an inertial system, all the results of Section 5–3 are applicable and we find the fields (5–34) and (5–35) in the primed system, I. The supported charge as seen by the falling observer therefore radiates at the constant rate (5–38).

Again, the transformation (8–20) will tell us how the noninertial supported observer sees the supported charge. One finds *no radiation*, because it is found that $\mathbf{B} \equiv 0$ in S. However, the presence of the SHGF does make a difference when compared with field-free space. Instead of the pure Coulomb field (8–21) we now find a *distorted Coulomb* field,

$$E_r \equiv \hat{\mathbf{r}} \cdot \mathbf{E} = \frac{e}{x^2 + y^2 + (c^2/2g)^2[1 - (g/c^2)^2(x^2 + y^2) - u^2(z)]^2} \cdot \qquad (8-23)$$

When the gravitational field is sufficiently weak, $(gl/c^2) \ll 1$, l being a typical length,

$$E_r = \frac{e}{r^2} + O\left[\left(\frac{gl}{c^2}\right)\right]. \qquad (8-24)$$

Unfortunately, this distortion of the Coulomb field by the SHGF is too small to be observable.

We conclude that the freely falling charge does not pose a paradox and that it is completely consistent with Einstein's formulation of the principle of equivalence; the charge at rest in a noninertial system is also satisfactorily accounted for.

The question of energy conservation was discussed previously (end of Section 6–11) and need not be repeated here. The reader should clearly understand why the emission of radiation from a uniformly accelerated charge does not require a loss of kinetic energy.

8–4   GENERAL RELATIVITY

Newton's theory of gravitation is inconsistent with the special theory of relativity. In the attempt at constructing a theory of gravitation consistent with special relativity, it became necessary to go beyond special relativity in an essential way. The general theory of relativity emerged as a covering theory of the special theory. But this theory is meant to fulfill a double purpose: it is to describe relativistic gravitational phenomena and to establish the relativity of acceleration. How well the latter aim is accomplished was briefly mentioned in Part D of Chapter 3. In the present section we are interested in general relativity primarily as a theory of gravitation.

According to general relativity, true gravitational fields are described as deviations from flatness of the Minkowski space-time manifold. Einstein's principle of equivalence and the discussion of the last section do not deal with true gravitational fields but with flat space.

The principle of equivalence (E) stated in Section 3–8 is, however, also applicable to true gravitational fields. It is known to be satisfied in general relativity for neutral test particles. Is it also valid for charged ones?

In order to answer this question we must first find the equation of motion of a charged test particle in a true gravitational field. This problem was solved in analogy to Dirac's method of special relativity by DeWitt and Brehme.*

The simplest way to generalize the theory of one charged particle (Section 6–9) from Minkowski space to Riemann space (cf. Appendix 2) is to generalize the action integral and proceed from there in a generally covariant way. The action integral (6–105) generalized to Riemannian space with metric tensor $g_{\mu\nu}(x)$ reads

$$I = -mc^2 \int \sqrt{-g_{\mu\nu} \frac{dz^\mu}{d\lambda} \frac{dz^\nu}{d\lambda}} \, d\lambda + e \int \bar{A}_\mu(z) \frac{dz^\mu}{d\lambda} \, d\lambda$$

$$- \frac{1}{16\pi} \int F_{\mu\nu}(x) F^{\mu\nu}(x) \sqrt{-g} \, d^4x - \frac{1}{8\pi} \int F_{\mu\nu}(x) F_+^{\mu\nu}(x) \sqrt{-g} \, d^4x. \qquad (8\text{-}25)$$

The field strengths are again assumed to be defined in terms of the potentials equation (4–47). The independent quantities are $z^\mu(\tau)$, $\bar{A}^\mu(x)$, and $A_+^\mu(x)$. The variation of I with fixed domain leads to the Euler-Lagrange equations

$$ma^\mu = \frac{e}{c} F^{\mu\nu} v_\nu, \qquad (8\text{-}26)$$

$$\partial_\mu [\sqrt{-g} \, (F^{\mu\nu} + F_+^{\mu\nu})] = -\frac{4\pi}{c} j^\nu(x) \sqrt{-g}, \qquad (8\text{-}27)$$

$$\partial_\mu (\sqrt{-g} \, F^{\mu\nu}) = 0. \qquad (8\text{-}28)$$

The four-vector of acceleration a^μ is the second *covariant* derivative [cf. Eq. (A2–36)] of the particle position $z^\mu(\tau)$,

$$a^\mu \equiv \frac{\delta^2 z^\mu}{\delta \tau^2} = \frac{\delta v^\mu}{\delta \tau} = \frac{dv^\mu}{d\tau} + \Gamma^\mu_{\alpha\beta} v^\alpha v^\beta, \qquad (A2\text{-}36)$$

$\Gamma^\mu_{\alpha\beta}$ being the affine connection of the Riemann space.†

---

* B. S. DeWitt and R. W. Brehme, *Ann. Phys.* (N.Y.) **9**, 220 (1960); J. M. Hobbs, *Ann. Phys.* **47**, 141 (1968).

† Note that a^μ is a four-vector in Riemann space, while $d^2z^\mu/d\tau^2$ is not. In our discussion following Eq. (3–36) this four-vector was temporarily denoted by A^μ.

It should be noted that the field equations (8-27) and (8-28) are tensor equations and are therefore covariant in the sense of general relativity.

PROBLEM 8-5

Show that Eqs. (8-27) and (8-28) can also be written by means of the covariant differentiation operator defined in (A2-19) and (A2-24), when the special form (A2-40) for the linear connection in Riemann space is used,

$$\nabla_\mu(\overline{F}^{\mu\nu} + F_+^{\mu\nu}) = -\frac{4\pi}{c}\,j^\nu, \tag{8-27'}$$

$$\nabla_\mu \overline{F}^{\mu\nu} = 0. \tag{8-28'}$$

Hint: Note first that the field strengths $F_{\mu\nu}$ are tensor components because $F_{\mu\nu} \equiv \partial_\mu A_\nu - \partial_\nu A_\mu = \nabla_\mu A_\nu - \nabla_\nu A_\mu$; then prove that $\Gamma^\alpha_{\alpha\mu} = \partial_\mu \ln\sqrt{-g}$ and derive (8-27) and (8-28) from (8-27') and (8-28').

The similarity of Eqs. (8-26) through (8-28) to those obtained in Section 6-9 from the action integral of special relativity, (6-105), is evident. In particular, the flat space approximation in Minkowski space is obtained by setting $\Gamma^\mu_{\alpha\beta} = 0$, $g = -1$, $g_{\mu\nu} = \eta_{\mu\nu}$. The results of special relativity then follow exactly.

The work by DeWitt and Brehme explains how the inhomogeneous Dirac equation (8-27) with the point charge current*

$$j^\mu(x) = ec \int \delta_4(x, z)v^\mu(\tau)\,d\tau$$

can be solved in terms of suitable generalizations of the Green functions of Minkowski space. If this is done and the field strengths are computed on the world line of the particle, one obtains

$$\tfrac{1}{2}(F^{\mu\nu}_{\text{ret}} - F^{\mu\nu}_{\text{adv}}) = \frac{2e}{3c^4}\,(v^\mu\dot{a}^\nu - v^\nu\dot{a}^\mu) + \frac{1}{3}\frac{e}{c}\,v^\alpha\,(R^\mu_\alpha v^\nu - R^\nu_\alpha v^\mu) + \phi^{\mu\nu}_-. \tag{8-29}$$

The first term is identical with (6-63), except that

$$\dot{a}^\mu = \frac{\delta a^\mu}{\delta\tau} = \frac{da^\mu}{d\tau} + \Gamma^\mu_{\alpha\beta}a^\alpha v^\beta \tag{8-30}$$

is the covariant derivative of a^μ. The second term is a contribution from the true gravitational field: it contains the contraction of the curvature tensor $R^\mu_{\nu\alpha\beta}$ defined in equation (A2-23),

$$R_{\mu\nu} \equiv R^\alpha_{\mu\alpha\nu} = R_{\mu\alpha\nu\beta}\,g^{\alpha\beta} \tag{8-31}$$

The last term in (8-29) is the antisymmetric tensor

$$\phi^{\mu\nu}_-\big(z(\tau)\big) = \frac{e}{2}\int_{-\infty}^{\infty} \epsilon(\tau - \tau')f^{\mu\nu}_\alpha\,[z(\tau), z(\tau')]v^\alpha(\tau')\,d\tau'; \tag{8-32}$$

---

* Note that $\delta_4(x, z)$ is the generalization to Riemann space of the Dirac δ-function $\delta_4(x - z)$ of Minkowski space.

$\varepsilon(\tau) = \tau/|\tau|$, and the tensor of third rank $f_\alpha^{\mu\nu}$ owes its presence also to the true gravitational fields, i.e. to the space curvature. It vanishes for flat space.

At this point, it becomes necessary to relate $\overline{F}^{\mu\nu}$ and $F_+^{\mu\nu}$ of the action integral (8–25) to $F_{in}^{\mu\nu}$, $F_{ret}^{\mu\nu}$, and $F_{adv}^{\mu\nu}$. In flat space this was accomplished by two requirements; the sum of $\overline{F}^{\mu\nu}$ and $F_+^{\mu\nu}$ must be the total physical field,

$$\overline{F}^{\mu\nu} + F_+^{\mu\nu} = F_{in}^{\mu\nu} + F_{ret}^{\mu\nu}, \tag{8–33}$$

and time reversal invariance must be satisfied, $\overline{F}^{\mu\nu}$ and $F_+^{\mu\nu}$ must have definite time parity (see Section 9–2). The first requirement can also be made to hold for curved space, but the second one cannot because the theory is no longer time reversal invariant. Therefore, the separation of the total field into $\overline{F}^{\mu\nu}$ and $F_+^{\mu\nu}$ using only (8–33) is not unique. We shall consider two choices.

One possibility is to keep the flat space identification

$$\overline{F}^{\mu\nu} = F_{in}^{\mu\nu} + \frac{1}{2}(F_{ret}^{\mu\nu} - F_{adv}^{\mu\nu}) \equiv F_{in}^{\mu\nu} + F_-^{\mu\nu},$$

and therefore

$$F_+^{\mu\nu} = \frac{1}{2}(F_{ret}^{\mu\nu} + F_{adv}^{\mu\nu}).$$

Equation (8–29) can then be used in the Euler-Lagrange equation (8–26) and results in

$$ma^\mu = F_{in}^\mu + \Gamma^\mu - \frac{1}{3}e^2 \frac{v^\alpha}{c}\left(R_\alpha^\mu + R_{\alpha\beta}\frac{v^\beta}{c}\frac{v^\mu}{c}\right) + \Phi_-^\mu \tag{8–34}$$

where

$$\Phi_-^\mu \equiv \frac{e}{c}\Phi_-^{\mu\nu}v_\nu. \tag{8–35}$$

This equation is formally identical to the Lorentz-Dirac equation with two extra terms due to the gravitational field. However, Γ^μ is here constructed from the generally covariant fourvectors v^μ, a^μ, and $\dot{a}^\mu$.

Another possibility for the identification of $\overline{F}^{\mu\nu}$ is to choose $F_+^{\mu\nu}$ to be just the divergent self-energy contribution which in Dirac's calculation leads to (6–69). In flat space this is the same as identifying $F_+^{\mu\nu}$ with the average of retarded and advanced field strengths, (6–46)$_+$. In curved space this average contains in addition a contribution from the gravitational field. But since that contribution, $\Phi_+^{\mu\nu}$, satisfies the homogeneous equation (8–28) it can be taken as part of $\overline{F}^{\mu\nu}$,

$$\overline{F}^{\mu\nu} = F_{in}^{\mu\nu} + F_-^{\mu\nu} + \Phi_+^{\mu\nu}. \tag{8–36}$$

Substitution into (8–26) leads to

$$ma^\mu = F_{in}^\mu + \Gamma^\mu - \frac{1}{3}e^2\frac{v^\alpha}{c}(R_\alpha^\mu + R_{\alpha\beta}\frac{v^\beta}{c}\frac{v^\mu}{c}) + \Phi_{ret}^\mu \tag{8–37}$$

where

$$\Phi_-^\mu + \Phi_+^\mu = \Phi_{ret}^\mu = \frac{e}{c}\Phi_{ret}^{\mu\nu}v_\nu \tag{8–38}$$

and

$$\Phi_{ret}^{\mu\nu}(\zeta(\tau)) = e\int_{-\infty}^\tau f_\alpha^{\mu\nu}[\zeta(\tau), \zeta(\tau')]v^\alpha(\tau')\,d\tau'. \tag{8–39}$$

The difference between (8–34) and (8–37) is in the last term, the "tail term". In both cases, this term depends on the worldline of the particle and is nonlocal. But in (8–34) it depends on the future as well as the past of the worldline.

The physical interpretation of the tail term can be expressed by the following graphical picture. The "bumps" of the nonflat space cause the electromagnetic field to be "scattered". The reaction on the particle by the field $\overline{F}^{\mu\nu}$ (which is produced by the particle) is modified in a way that depends on the path the particle takes over the bumps. It is interesting that this effect, the tail term, is additive to the flat-space reaction $(e/c)F_-^{\mu\nu}$. This interpretation regards the tail term as a radiation reaction effect rather than as a direct interaction between the particle and the gravitational field. The latter type of interaction is given by the third term in (8–34) and in (8–37).

For neutral particles (8-34) reduces to the geodesic equation.

For charged particles, $a^\mu = 0$ is not a solution of (8-37) when $F^\mu_{in} = 0$ unless the true gravitational field vanishes. Thus a charged particle will *not* follow the same trajectory as a neutral particle in a *true* gravitational field. The reason for this is the nonlocal interaction of its field with the gravitational field, as explained above. This situation is therefore essentially different from free fall in the *apparent* gravitational field studied in the previous section.

Beyond that, it also follows that the equation of motion for free fall, obtained by combining

$$ma^\mu = \Gamma^\mu - \frac{1}{3} e^2 \frac{v^\alpha}{c} \left(R^\mu_\alpha + R_{\alpha\beta} \frac{v^\beta}{c} \frac{v^\mu}{c} \right) + \varphi^\mu$$

(8-40)

(where φ^μ stands for φ^μ_- or for φ^μ_{ret}) with the asymptotic condition

$$\lim_{\tau \to \pm\infty} a^\mu = 0,$$

(8-41)

corresponding to asymptotically flat space and no interaction, will depend on the mass of the particle. Thus the principle of equivalence (E) of Section 3-8 is *not* satisfied for charged particles in true gravitational fields. For apparent gravitational fields, however, $\varphi^\mu = 0$ and $R_{\mu\nu} = 0$ and the only solution of (8-40) consistent with the asymptotic condition (6-37) is $a^\mu = 0$. Thus the principle of equivalence (E) *holds for charged particles only in apparent gravitational fields.*

One can speculate about the consequence of this result. As is demonstrated in Appendix 2, the assumption that particles follow geodesics (geodesic postulate) is basic to a geometrization of gravitational interactions. This assumption, in turn, is suggested by the principle of equivalence (E). The latter requires, in particular, that a free test particle cannot obey an equation with a mass-dependent term which vanishes only for vanishing curvature of the space. Einstein's principle of equivalence is not strong enough to prohibit such an equation. But Einstein's genius realized the importance of the geodesic postulate on the basis of the weaker equivalence principle. Now we are faced with the fact that charged particles satisfy Einstein's principle of equivalence but not the principle (E). Does this imply that a geometrization of electromagnetic interactions including charged particles is not possible? This conclusion is certainly suggested. The consistent lack of success of the many attempts at a fully unified field theory seems to point in the same direction. It may well be that electromagnetic interactions have much more in common with strong and weak interactions than with gravitational ones and will correspondingly find their next level of physical theory aligned with those interactions.

The above treatment of electromagnetic interactions in the presence of true gravitational fields describes the motion of a test particle.   It must be the limit of the gravitational interaction of a physical charged particle in which the particle's own energy acts as active gravitational mass and contributes to the gravitational field actually present.   The dynamics of the gravitational field as given by Einstein's field equations must be linked to the presence of the particle and the corresponding electromagnetic and gravitational interaction energy.   This requires a theory which combines fully gravitational and electromagnetic interactions.   Such a theory has so far been carried through successfully only for electromagnetic fields in the absence of particles.*

At this time we must therefore be content with the above treatment.   It does represent a covering theory of the classical theory of charged particles based on special relativity.   This was demonstrated above by showing that in the Minkowski space limit the usual results are recovered.   The consistency and logical coherence with the covering theory as far as it is developed is therefore established.

B. Quantum Theorles

8–5 NONRELATIVISTIC QUANTUM MECHANICS

Nonrelativistic classical mechanics is an approximation to nonrelativistic quantum mechanics.   The nature of this approximation and the corresponding validity limit of Newtonian mechanics is best expressed in terms of observables, i.e. physically measurable quantities, particularly those, like energy, linear and angular momenta, etc., that enter conservation laws.   The value of an observable in the domain of validity of classical mechanics, O_{CM}, must be the same (to extremely good approximation) as the value of that same observable computed by means of quantum mechanics, O_{QM}.   The latter, as is often the case, involves a factor $\hbar$ (Planck's constant h divided by 2π) and a quantum number q (i.e. a number characteristic of O_{QM}),

$$O_{CM} = O_{QM} = q\hbar a. \tag{8–36}$$

The factor a does not depend on $\hbar$.   The classical approximation is now characterized by the inequality†

$$\hbar a / O_{CM} \ll 1. \tag{8–37}$$

-----

* Credit for this formulation of the theory is due to Rainich and to Misner and his collaborators. See J. A. Wheeler, *Geometrodynamics*, Academic Press, New York, 1962.

† Note the use of a dimensionless quantity to indicate this approximation.   It is not meaningful to speak of the limit $\hbar \to 0$, just as it is not meaningful to speak of $c \to \infty$ (cf. Section 8–1).

The validity of (8–36) thus requires *large quantum numbers q*. One can state this situation as follows: the classical approximation of a quantum mechanical observable O is that value which O_{QM} would take on for very large quantum numbers.

For example, the orbital angular momentum **L** can take on the values $\mathbf{L}^2 = l(l+1)\hbar^2$ (l = integer); a classical approximation will exist whenever l is very large. A photon has energy $\hbar\omega$. An ensemble of monochromatic radiation will have an energy E which can be described classically when it contains a large number n of such photons, $E = n\hbar\omega$.

It follows that the validity of the classical approximation of quantum mechanics is much less clear-cut than the nonrelativistic approximation of special relativity (Section 8–1). It depends on the observable under consideration, the state of the system, and the measurement intended. Thus it would be incorrect to suppose, for example, that classical mechanics is valid at distances of the order of $r_0 = e^2/(mc^2)$ just because this quantity happens to be independent of $\hbar$.

The *correspondence principle* introduced in 1923 by Niels Bohr (1885–1962) can be stated as the requirement that quantum mechanics contain classical mechanics as an approximation (a limit, in a certain sense). Moreover, Bohr thought of quantum mechanics as the minimum modification of classical mechanics necessary to account for all experiments in a consistent way. In the following we shall be concerned with the way in which the quantum mechanical description of charged particles yields the classical description in suitable approximation. The correspondence principle will thereby guide our lines of thought.

Let us consider the way in which the classical approximation emerges from a quantum mechanical one by examining first the case of a free particle. Such a particle will be characterized by a wave function $\psi(\mathbf{r}, t)$ which satisfies the Schrödinger equation

$$-\frac{\hbar^2}{2m}\nabla^2\psi = i\hbar\frac{\partial\psi}{\partial t}. \tag{8–38}$$

The probability of finding the particle inside a volume d^3x at **r** during the time interval dt at time t is

$$\rho(\mathbf{r}, t)\,d^3x\,dt = |\psi(\mathbf{r}, t)|^2\,d^3x\,dt.$$

Therefore $\rho(\mathbf{r}, t)$ is the *probability density*. This interpretation requires the normalization

$$\int\rho(\mathbf{r}, t)\,d^3x = 1. \tag{8–39}$$

The motion of the particle is characterized by the *probability current density*

$$\mathbf{s}(\mathbf{r}, t) = \frac{\hbar}{2m}(\psi^*\nabla\psi - \psi\nabla\psi^*). \tag{8–40}$$

Conservation of probability will be expressed by the continuity equation

$$\frac{\partial \rho}{\partial t} + \nabla \cdot \mathbf{s} = 0, \tag{8–41}$$

which holds as a consequence of (8–38).

These equations follow from the fundamental principles of quantum mechanics, which are incorporated into a conceptual framework that can be summarized very briefly as follows: Quantum mechanics is an *intrinsically statistical* theory. This means that one is able to make in general only statistical statements about physical systems. But more than this, the statistical nature is not due to a distribution of well defined properties of individual elements in an ensemble as in the classical statistical mechanics of molecules in a gas. Rather, it is a limitation in principle of the knowledge of the state of these elements. This knowledge cannot be exceeded by any known experiment. There are no "hidden" degrees of freedom which, when averaged, yield quantum mechanics.*

The predictions of quantum mechanics are, then, in the nature of averages. These are obtained in the following way. (1) One associates with each observable an *operator*. (2) One associates with each state i of a system a *state vector* ψ_i. (For example, in classical physics the states of the system sun-planet are characterized by the various conic sections, their eccentricities, etc., which describe the possible planetary orbits.) These state vectors form a space which is mathematically a Hilbert space, $\mathcal{H} = \{\psi_i; i = 1, 2, \ldots\}$. The operators O act on $\mathcal{H}$. (3) When the system is in the state ψ_i, the measurement of O will most likely yield $\bar{O}_i \equiv (\psi_i, O\psi_i)$ which is the inner product of ψ_i and $O\psi_i$ in $\mathcal{H}$.

Of the various representations of $\mathcal{H}$, the configuration-space representation is of special interest (wave mechanics). In this representation the state vector ψ_i is given by the *wave function* $\psi_i(\mathbf{r}, t)$. The operators then become multiplicative and differential operators.

The quantum mechanical operators of momentum and energy are in the wave mechanics formulation

$$\mathbf{p}_{op} = \frac{\hbar}{i} \nabla, \qquad E_{op} = i\hbar \frac{\partial}{\partial t}. \tag{8–42}$$

The Schrödinger equation (8–38) can therefore be obtained heuristically from $p^2/2m = E$ by letting the corresponding operator equation act on the wave function $\psi(\mathbf{r}, t)$.

The mathematical relation between quantum and classical mechanics is formally very similar to the relation between wave optics and ray optics.

---

* More precisely, while such hidden degrees of freedom may possibly be amenable to mathematical description, they are neither accessible to experiment nor necessary for a logically consistent formulation of the theory.

As in wave optics, the quantum mechanical particle is not well localized; $\psi(\mathbf{r}, t)$ can be written as a "wave packet," a coherent superposition of waves of various wave lengths. Consider first a plane wave,

$$\psi = e^{i\mathbf{k}\cdot\mathbf{r} - i\omega t}.$$

Application of the operators (8–42) yields

$$\mathbf{p}_{op}\psi = \hbar\mathbf{k}\psi, \qquad E_{op}\psi = \hbar\omega\psi. \tag{8-43}$$

From (8–39) it follows that the average values are

$$\overline{\mathbf{p}_{op}} = \int \psi^* \mathbf{p}_{op}\psi \, d^3x = \hbar\mathbf{k}, \qquad \overline{E}_{op} = \int \psi^* E_{op}\psi \, d^3x = \hbar\omega. \tag{8-43}$$

The above plane wave therefore corresponds to a particle of momentum $\hbar\mathbf{k}$ and energy $\hbar\omega$.

Generalizing this simple example, we have

$$\psi(\mathbf{r}, t) = A e^{(i/\hbar)S(\mathbf{r}, t)}, \tag{8-45}$$

with $S(\mathbf{r}, t)/\hbar$ a generalization of the phase function $\mathbf{k}\cdot\mathbf{r} - \omega t$. Substitution into the Schrödinger equation (8–38) yields

$$\frac{1}{2m}(\nabla S)^2 - \frac{i\hbar}{2m}\nabla^2 S = -\frac{\partial S}{\partial t}. \tag{8-46}$$

In general, S will be a function of $\hbar$. In order to apply the approximation (8–37), it is therefore necessary to expand in powers of $\hbar$:

$$S = \sum_{n=0}^{\infty} \left(\frac{\hbar}{i}\right)^n S_n = S_0 + \frac{\hbar}{i} S_1 + \cdots \tag{8-47}$$

Equation (8–46) now becomes a system of coupled equations, one for each power of $\hbar$. The zeroth order equation is*

$$\frac{1}{2m}(\nabla S_0)^2 = -\frac{\partial S_0}{\partial t}. \tag{8-48}$$

This equation is indeed exactly the equation for a free particle in classical mechanics, but written in Hamilton-Jacobi formulation. Indeed, the Hamiltonian of a free particle is $H = \mathbf{p}^2/(2m)$, and the Hamilton-Jacobi equation

$$H = -\frac{\partial S}{\partial t} \tag{8-49}$$

is obtained by the transformation $\mathbf{p} = \nabla S$. This yields exactly (8–48). This equation therefore has the same contents as $m\ddot{\mathbf{r}} = 0$.

---

* It is instructive to compare this derivation with the derivation of the eikonal equation of geometrical optics from Maxwell's equations.

The reader will recall* that a system of n degrees of freedom can be expressed in terms of n generalized coordinates q_k ($k = 1, 2, \ldots, n$) and n generalized momenta p_k, canonically conjugate to one another. The Hamiltonian is a function of these $2n$ variables and the time, $H(q_1 \ldots, q_n, p_1 \ldots, p_n, t)$. It can be obtained from the Lagrangian $L(q_1 \ldots q_n, \dot{q}_1 \ldots \dot{q}_n, t)$ by a Legendre transformation,

$$H = \sum_{k=1}^{n} p_k \dot{q}_k - L.$$

The $\dot{q}_k$ are expressed in terms of the p_k and q_k by means of the n equations

$$p_k = \frac{\partial L}{\partial \dot{q}_k}, \qquad (k = 1, 2, \ldots, n).$$

The Euler-Lagrange equations (3–51′) then become the Hamilton equations of motion

$$\dot{q}_k = \frac{\partial H}{\partial p_k}, \qquad \dot{p}_k = -\frac{\partial H}{\partial q_k}.$$

A canonical transformation (i.e. one which leaves these equations, as well as the Poisson brackets, invariant) from the p_k, q_k to P_k, Q_k yields a new Hamiltonian K,

$$K = H + \frac{\partial F}{\partial t},$$

where F is the generating function of this transformation. If the $2n$ P_k and Q_k are chosen to be the $2n$ initial values of the p_k and q_k, i.e. constants, then $\dot{P}_k = 0$ and $\dot{Q}_k = 0$. This makes $K = 0$ and yields

$$H + \frac{\partial F}{\partial t} = 0.$$

Furthermore, if the generating function F is chosen to be a function of the q_k and the P_k, $F = S(q_1 \ldots, q_n, P_1 \ldots, P_n, t)$, the transformation equations become

$$p_k = \frac{\partial S}{\partial q_k}, \qquad Q_k = \frac{\partial S}{\partial P_k}.$$

The above equation then becomes the Hamilton-Jacobi equation,

$$H\left(q_1 \ldots, q_n, \frac{\partial S}{\partial q_1}, \ldots, \frac{\partial S}{\partial q_n}, t\right) + \frac{\partial S}{\partial t} = 0. \qquad (8\text{–}49)$$

Known as Hamilton's principle function, S differs from the action integral $I = \int L\, dt$ of Hamilton's principle, at most, by a constant, because

$$\frac{dS}{dt} = \sum_k \frac{\partial S}{\partial q_k} \dot{q}_k + \sum_k \frac{\partial S}{\partial P_k} \dot{P}_k + \frac{\partial S}{\partial t} = \sum_k p_k \dot{q}_k + 0 + \frac{\partial S}{\partial t} = \sum_k p_k \dot{q}_k - H = L.$$

* For example, see H. Goldstein, *Classical Mechanics*, Addison-Wesley Publishing Co., Reading, Mass., 1950.

PROBLEM 8-6

Show that the classical approximation to the Schrödinger equation

$$\left[-\frac{\hbar^2}{2m}\nabla^2 + V(\mathbf{r})\right]\psi(\mathbf{r}, t) = i\hbar\,\frac{\partial\psi(\mathbf{r}, t)}{\partial t}$$

is the Hamilton-Jacobi form of

$$m\ddot{\mathbf{r}} = \mathbf{F} \quad \text{with} \quad \mathbf{F} = -\nabla V(\mathbf{r}).$$

At this point we should inquire into the generalization of the above free-particle case to the system of a charged particle in interaction with an electromagnetic field. However, as we have seen in Section 8-1, the nonrelativistic theory of this system is not so meaningful and complete as the relativistic one. We shall therefore turn now to the relativistic treatment.*

8-6 RELATIVISTIC QUANTUM MECHANICS

(a) Particles of zero spin. A relativistic generalization of the Schrödinger equation (8-38) is suggested by the same heuristic procedure that was indicated in the nonrelativistic case. The energy of a free particle is given by

$$\mathbf{p}^2 + m^2 = E^2.$$

Turning this into an operator relation which acts on a wave function and using (8-42), we find

$$(\square - \kappa^2)\phi(x) = 0, \tag{8-50}$$

where $\kappa \equiv mc/\hbar$ and $x = (t, \mathbf{r})$. This equation is known as the *Klein-Gordon* equation.†

A probability density can again be defined by

$$\rho = \phi^*\phi, \tag{8-51}$$

which is to be normalized to 1. But there is also a current density *four*-vector,

$$s^\mu(x) = \frac{\hbar}{2im}(\phi^*\partial^\mu\phi - \phi\,\partial^\mu\phi^*), \tag{8-52}$$

---

* The classical approximation of the (nonrelativistic) Schrödinger equation for a particle in interaction with an electromagnetic field is treated, for example, by W. Pauli in *Encyclopedia of Physics*, Vol. V/1, Chapter 2, § 12, Springer, 1958.

† For the sake of historical accuracy, it must be mentioned that this equation was obtained by Schrödinger before the corresponding nonrelativistic approximation which bears his name. He concentrated on the latter because it showed better agreement with the known hydrogen spectrum. Only much later was it discovered that the relativistic and spin effects nearly cancel each other.

whose fourth component is not the above expression* ρ. This four-vector will later be generalized to the electromagnetic current density. It satisfies the conservation law

$$\partial_\mu s^\mu(x) = 0 \qquad (8\text{–}53)$$

in view of (8–50).

PROBLEM 8–7

Show that the Klein-Gordon equation reduces to the Schrödinger equation (8–38) in the nonrelativistic approximation. *Hint:* Split a factor $\exp(i\kappa ct)$ off the relativistic wave function, since the zero point on the relativistic energy scale is mc^2 higher than on the nonrelativistic scale.

The covariant equation (8–50) for the scalar function ϕ can be treated by the same substitution (8–45) as in the nonrelativistic case,

$$\phi(x) = Ae^{(i/\hbar)S(x)}, \qquad (8\text{–}45')$$

with S a scalar function. One finds easily

$$\partial_\mu S \, \partial^\mu S - i\hbar \Box S + (mc)^2 = 0. \qquad (8\text{–}54)$$

The expansion (8–47) of S in powers of $\hbar$ yields the zeroth-order equation

$$\partial_\mu S_0 \, \partial^\mu S_0 + (mc)^2 = 0 \qquad (8\text{–}55)$$

and the first-order equation

$$2\partial_\mu S_0 \, \partial^\mu S_1 + \Box S_0 = 0. \qquad (8\text{–}56)$$

Equation (8–55) is the relativistic Hamilton-Jacobi equation (a proof is given below); the relation $p^\mu = \partial^\mu S_0$ yields $p^\mu p_\mu + (mc)^2 = 0$, which is equivalent to the free-particle equations.

The first-order equation (8–56) expresses the conservation law (8–53); by means of (8–45'), the current density (8–52) gives to lowest order,

$$s^\mu = \frac{\hbar}{2im} \, 2 \, \frac{i}{\hbar} \, \partial^\mu S_0 \, \phi^* \phi = \frac{e^{2S_1}}{m} \, \partial^\mu S_0.$$

The divergence of this vector establishes (8–56) as equivalent to (8–53).

We are now ready to consider the interaction of the particle described by the Klein-Gordon equation (8–50) with an electromagnetic field. The interaction terms which yield *minimal coupling* are uniquely determined by gauge invariance. "Minimal coupling" means that the interaction is restricted to

---

* Since ϕ and therefore ρ are invariants, $\int \rho \, d^3x$ is *not* invariant because of the volume contraction. On the other hand, $\int s^\mu \, d\sigma_\mu$ is a constant because of (8–53).

terms which depend explicitly on the potentials and do not violate gauge invariance. This coupling is obtained by making the replacement

$$\partial_\mu \to \partial_\mu - \frac{ie}{\hbar c} A_\mu \tag{8-57}$$

in (8-50) yielding

$$\left[\left(\partial_\mu - \frac{ie}{\hbar c} A_\mu\right)\left(\partial^\mu - \frac{ie}{\hbar c} A^\mu\right) - \kappa^2\right]\phi(x) = 0. \tag{8-58}$$

PROBLEM 8-8

Prove that the simultaneous application of the gauge transformations of the first kind,

$$\phi \to \phi' = e^{(ie/\hbar c)\Lambda(x)}\phi,$$

and of the gauge transformations of the second kind,

$$A^\mu \to A'^\mu = A^\mu + \partial^\mu\Lambda,$$

leave (8-58) form invariant. The transformations of the first kind are unobservable just like those of the second kind, because the phase of ϕ is not an observable quantity.

The classical approximation of (8-58) is again found by means of the form (8-45') of the wave function. The zeroth-order equation is

$$\left(\partial_\mu S_0 - \frac{e}{c} A_\mu\right)\left(\partial^\mu S_0 - \frac{e}{c} A^\mu\right) - (mc)^2 = 0, \tag{8-59}$$

while the first-order equation becomes

$$2\partial_\mu S_0 \, \partial^\mu S_1 - \frac{2e}{c} A_\mu \partial^\mu S_1 + \Box S_0 = 0, \tag{8-60}$$

as can easily be verified.

We want to show that (8-59) is the Hamilton-Jacobi form of the particle equation

$$ma^\mu = \frac{e}{c} F^{\mu\nu} v_\nu, \tag{6-102}$$

while (8-60) is equivalent to charge conservation. The electromagnetic current density $j_\mu = es_\mu$ satisfies

$$\partial^\mu j_\mu = 0, \qquad j_\mu \equiv \frac{e\hbar}{2im}(\phi^* \partial_\mu \phi - \phi \partial_\mu \phi^*) - \frac{e^2}{mc} A_\mu \phi^* \phi, \tag{8-61}$$

as a consequence of (8-58).

PROBLEM 8-9

Verify that Eq. (8-58) implies the conservation law (8-61).

PROBLEM 8-10

The current density j_μ of (8–61) is in first approximation

$$j_\mu = \frac{e}{m} e^{2S_1} \left(\partial_\mu S_0 - \frac{e}{c} A_\mu \right) . \tag{8-62}$$

Prove that its divergence vanishes if and only if (8–60) holds.

In order to establish the equivalence of (8–59) and (6–102), it is necessary to cast the latter equation into its Hamilton-Jacobi form. To this end, we recall that this equation is the Euler-Lagrange equation of the variational principle

$$I = \int L \, d\lambda,$$

where

$$L = -mc^2 \sqrt{-\dot{z}^\mu \dot{z}_\mu} + eA_\mu \dot{z}^\mu. \tag{6-101}$$

The dot indicates the derivative with respect to λ. The canonical "momentum" is by definition*

$$\Pi^\mu \equiv \frac{\partial L}{\partial \dot{z}_\mu} = \frac{mc^2 \dot{z}^\mu}{\sqrt{-\dot{z}^\alpha \dot{z}_\alpha}} + eA^\mu. \tag{8-63}$$

The Hamiltonian is defined by

$$H = \Pi^\mu \dot{z}_\mu - L(z, \dot{z}). \tag{8-64}$$

By means of (8–63) it can be given the form

$$H(\lambda) = (\sqrt{-\dot{z}^\alpha \dot{z}_\alpha}/mc^2)[(\Pi^\mu - eA^\mu)(\Pi_\mu - eA_\mu) + (mc^2)^2]. \tag{8-65}$$

This Hamiltonian is associated with the arbitrary parameter λ. If we define the proper time by

$$c \, d\tau = \sqrt{-\dot{z}^\alpha \dot{z}_\alpha} \, d\lambda \tag{6-97}$$

and $H(\tau)$ by

$$H(\lambda) \, d\lambda \equiv H(\tau) c \, d\tau,$$

we can introduce $p^\mu \equiv \Pi^\mu/c$ and obtain

$$H(\tau) = \frac{1}{m} \left[\left(p^\mu - \frac{e}{c} A^\mu \right) \left(p_\mu - \frac{e}{c} A_\mu \right) + (mc)^2 \right] . \tag{8-66}$$

Equation (8–63) can now be written as

$$p^\mu = mv^\mu + \frac{e}{c} A^\mu, \tag{8-67}$$

---

* Π^μ has the dimensions of an energy. This is a consequence of our choice of the dimensions of I as energy times length. The correct physical dimensions of momentum are obtained below in the transition from λ to τ as the independent parameter.

if v^μ is defined by $dz^\mu/d\tau$. This shows immediately that H vanishes as a consequence of $v_\mu v^\mu + c^2 = 0$;

$$\left(p^\mu - \frac{e}{c} A^\mu\right)\left(p_\mu - \frac{e}{c} A_\mu\right) + (mc)^2 = 0. \tag{8-68}$$

The Hamilton-Jacobi equation is obtained by a canonical transformation from the old variables p and z to the new variables P and Z. The generating function S is chosen to depend on z and P, so that

$$p^\mu = \frac{\partial S}{\partial z_\mu}, \qquad Z^\mu = \frac{\partial S}{\partial P_\mu} \tag{8-69}$$

defines the transformation. Since this transformation is canonical, it must leave L invariant within a total derivative $d\Omega/d\tau$,

$$p^\mu \frac{dz_\mu}{d\tau} - H = P^\mu \frac{dZ_\mu}{d\tau} - K + \frac{d\Omega}{d\tau}.$$

The new Hamiltonian is K. If Ω is chosen to be $S - P^\mu Z_\mu$ and (8–69) is taken into account, a simple calculation yields

$$K = H + \frac{\partial S}{\partial \tau}. \tag{8-70}$$

The new variables P and Z can be chosen to be the (constant) initial values of p and z. Hamilton's equations

$$\frac{dP^\mu}{d\tau} = -\frac{\partial K}{\partial Z_\mu}, \qquad \frac{dZ^\mu}{d\tau} = \frac{\partial K}{\partial P_\mu}, \tag{8-71}$$

then require that K be constant. It can be chosen to vanish. Since $H = 0$ also, as was seen above, Eq. (8–70) reduces to $K = 0$, $\partial S/\partial \tau = 0$, and

$$H\left(\frac{\partial S}{\partial z}, z, \tau\right) = 0. \tag{8-72}$$

This is the relativistic Hamilton-Jacobi equation. In our case, it is just Eq. (8–68) with p_μ replaced by $\partial_\mu S$,

$$\left(\frac{\partial S}{\partial z_\mu} - \frac{e}{c} A^\mu\right)\left(\frac{\partial S}{\partial z^\mu} - \frac{e}{c} A_\mu\right) + (mc)^2 = 0. \tag{8-73}$$

Since this equation was derived from the same Lagrangian as (6–102), it is equivalent to it. This completes the proof of the equivalence of (8–59) and (6–102).

A special case of this result is the equivalence of (8–55) with the free-particle equation $ma^\mu = 0$.

What can we now conclude concerning the comparison of the descriptions of a charged particle in relativistic quantum mechanics and in relativistic classical mechanics? The particle equation (6–51) is obtained as the classical approximation of quantum mechanics only when the potentials in the Klein-Gordon equation (8–58) are the potentials $\bar{A}_\mu = A_\mu^{\text{in}} + A_\mu^-$, i.e. solutions of the homogeneous equations, or are external fields, A_μ^{ext}. In the usual formulation of quantum mechanics only external fields are used and the radiation from the charged particles is ignored.

This radiation is included, however, in the *semiclassical theory of radiation* where the particles are treated quantum-mechanically while the electromagnetic fields and potentials are the classical ones. The Maxwell-Lorentz equations are therefore the classical ones with the quantum-mechanical current as source,

$$\partial_\mu F^{\mu\nu} = -\frac{4\pi}{c} j^\nu;$$

j^ν is given by (8–61) for a relativistic charged particle.

The classical set of equations for a single charge in an electromagnetic field, (6–51), (4–51), and (6–106) of p. 162, including external fields, have their analogue in relativistic quantum mechanics as follows:

$$\left\{ \left[\partial_\mu - \frac{ie}{\hbar c} \left(A_\mu^{\text{ext}} + \bar{A}_\mu \right) \right] \left[\partial^\mu - \frac{ie}{\hbar c} \left(A_{\text{ext}}^\mu + \bar{A}^\mu \right) \right] - \kappa^2 \right\} \phi(x) = 0, \qquad (8\text{–}74)$$

$$\partial_\mu F_+^{\mu\nu} = -\frac{4\pi}{c} j^\nu, \qquad (8\text{–}75)$$

$$\partial_\mu \bar{F}^{\mu\nu} = 0, \qquad (6\text{–}106)$$

with j^ν given in terms of $\phi(x)$ and $A_\mu^{\text{ext}} + \bar{A}_\mu$ by (8–61). The potential A_μ^+ cannot appear in the current, as is apparent from the action principle associated with these equations. In analogy to (6–105), the action integral for the relativistic semiclassical theory of radiation becomes

$$I = \int \mathcal{L}(\phi, \phi^*, \bar{A}_\mu, A_\mu^+) \, d^4x, \qquad (8\text{–}76)$$

$$\mathcal{L} = -\frac{\hbar^2}{2m} (D_\mu^* \phi^*)(D^\mu \phi) - \tfrac{1}{2} mc^2 \phi^* \phi - \frac{1}{16\pi} \bar{F}_{\mu\nu} \bar{F}^{\mu\nu} - \frac{1}{8\pi} \bar{F}_{\mu\nu} F_+^{\mu\nu}, \qquad (8\text{–}77)$$

$$D_\mu \equiv \partial_\mu - \frac{ie}{\hbar c} (A_\mu^{\text{ext}} + \bar{A}_\mu). \qquad (8\text{–}78)$$

Variation of ϕ^* and $\bar{A}^\mu$ yield Eqs. (8–74) and (8–75); variation of A_+^μ yields (6–106). Since these equations yield Eqs. (6–51), (4–51), and (6–106) of the classical theory in the classical approximation, as was proven above, we have established that the relativistic semiclassical radiation theory is indeed a covering theory of the classical theory of a charged particle.

The generalization to *systems* of charged particles will not be presented here.

(b) Particles of spin one-half. The above theory describes a particle without spin, like a charged pi-meson, for example. The correct quantum-mechanical description of an electron is offered only by the theory for particles of spin $\tfrac{1}{2}\hbar$.

This theory we owe to Dirac.* It is an alternative relativistic generalization of the Schrödinger theory [Eq. (8–38)].

In this theory the wave function is not a scalar function, but a four-component spinor indicating two degrees of freedom of spin ($+\frac{1}{2}\hbar$ and $-\frac{1}{2}\hbar$) and two degrees of freedom of charge, describing both negative and positive electrons. But this latter degree of freedom is not correctly described in relativistic quantum mechanics and finds its satisfactory formulation only in quantum field theory (cf. the following section). Here we shall be concerned only with the description of the electron spin.

For particles of spin $\frac{1}{2}\hbar$ the Dirac equation takes the place of the Klein-Gordon equation. For a free particle, it has the form

$$(\gamma^\mu \partial_\mu + \kappa)\psi(x) = 0. \tag{8–79}$$

The four quantities γ^μ ($\mu = 0, 1, 2, 3$) are numerical matrices of four rows and four columns which satisfy

$$\gamma^\mu\gamma^\nu + \gamma^\nu\gamma^\mu = 2\eta^{\mu\nu}, \tag{8–80}$$

and $\kappa = mc/\hbar$ as before. The wave function $\psi(x)$ has four components, $\psi^{(\alpha)}(x)$ ($\alpha = 1, 2, 3, 4$); i.e. it is a column matrix which, in (8–79), is multiplied by the matrices γ^μ from the left. The property (8–80) ensures that multiplication of (8–79) from the left by $\gamma^\nu\partial_\nu - \kappa$ yields

$$(\square - \kappa^2)\psi(x) = 0.$$

The Dirac wave function therefore also satisfies the Klein-Gordon equation.

The classical limit of the Dirac equation is obtained as in (8–45′) by substitution of

$$\psi(x) = A(x)e^{(i/\hbar)S(x)}. \tag{8–45″}$$

However, it is now necessary that A be a spinor. In the resulting equation,

$$[\gamma^\mu(\partial_\mu S) - imc]A(x)e^{iS/\hbar} = i\hbar\gamma^\mu(\partial_\mu A)e^{iS/\hbar},$$

one can again expand in powers of $\hbar/i$. Instead of expanding S, it is more convenient here to expand A,

$$A = \sum_{n=0}^{\infty} \left(\frac{\hbar}{i}\right)^n A_n = A_0 + \frac{\hbar}{i} A_1 + \cdots \tag{8–81}$$

The result obtained by equating to zero the coefficient of each power in $\hbar$ is an infinite system of coupled equations,

$$[\gamma^\mu(\partial_\mu S) - imc]A_0 = 0, \tag{8–82}$$

$$[\gamma^\mu(\partial_\mu S) - imc]A_n = -\gamma^\mu\partial_\mu A_{n-1} \quad (n = 1, 2, \ldots). \tag{8–83}$$

---

* P. A. M. Dirac, *Proc. Roy. Soc. (London)* A **117**, 610 and **118**, 341 (1928).

The first equation is a homogeneous linear equation for the four-component spinor A_0; i.e. it is itself a set of four simultaneous equations for the four components $A_0^{(\alpha)}(\alpha = 1, 2, 3, 4)$ of A_0. This set can have solutions only when the determinant vanishes,

$$\det [\gamma^\mu(\partial_\mu S) - imc1] = 0. \tag{8-84}$$

The unit matrix is indicated by 1. The determinantal equation is found to be

$$\partial_\mu S \, \partial^\mu S + (mc)^2 = 0,$$

that is exactly the classical equation (8–55) of a particle *without spin*. The classical approximation of a Dirac particle is therefore a particle without spin.

Since the value of the determinant is independent of the representation, we can choose the following convenient representation of the γ_μ satisfying (8–80):

$$\gamma^0 = \begin{pmatrix} 1 & 0 \\ 0 & -1 \end{pmatrix}, \quad \gamma_k = \begin{pmatrix} 0 & -i\sigma_k \\ +i\sigma_k & 0 \end{pmatrix}.$$

The matrix elements here are all 2×2 matrices:

$$1 = \begin{pmatrix} 1 & 0 \\ 0 & 1 \end{pmatrix}, \quad \sigma_1 = \begin{pmatrix} 0 & 1 \\ 1 & 0 \end{pmatrix}, \quad \sigma_2 = \begin{pmatrix} 0 & -i \\ i & 0 \end{pmatrix}, \quad \sigma_3 = \begin{pmatrix} 1 & 0 \\ 0 & -1 \end{pmatrix}.$$

The above results then easily follow from (8–84).

This conclusion can also be drawn in a different way. The interaction of a Dirac particle with an electromagnetic field in minimal coupling is given by the replacement (8–57), that is,

$$\left[\gamma^\mu \left(\partial_\mu - \frac{ie}{\hbar c} A_\mu \right) + \kappa \right] \psi(x) = 0. \tag{8-85}$$

This equation corresponds to Eq. (8–58) in the spinless case. If we multiply on the left by the operator $\gamma^\mu(\partial_\mu - (ie/\hbar c) A_\mu) - \kappa$, we obtain, after an easy calculation by means of (8–80) and

$$\sigma_{\mu\nu} \equiv \frac{1}{2i} (\gamma_\mu\gamma_\nu - \gamma_\nu\gamma_\mu), \tag{8-86}$$

the result

$$\left[\hbar^2 \left(\partial_\mu - \frac{ie}{c} A_\mu \right) \left(\partial^\mu - \frac{ie}{c} A^\mu \right) - (mc)^2 + \frac{\hbar}{2} \frac{e}{c} \sigma_{\mu\nu} F^{\mu\nu} \right] \psi(x) = 0. \tag{8-87}$$

A comparison with Eq. (8-58) shows that the only difference lies in the last term which can be interpreted as the interaction of the dipole moment [due to the spin angular momentum (cf. Section 7-4)] with the electromagnetic field. The factor $\hbar$ thereby indicates the order of magnitude of this effect compared with the other terms ($\hbar\partial_\mu \sim p_\mu$ is *not* a small quantity!). Thus, we find again the result that in the classical limit there is no electron spin. Since the quantum number $\frac{1}{2}$ associated with it cannot take on larger values, the spin can never be a macroscopic quantity, in accordance with Eq. (8-37).

For this reason we have paid relatively little attention to the classical description of charged particles with (macroscopic) spin.

PROBLEM 8-11

Find the variational principle for the semiclassical theory of radiation for a Dirac particle, in analogy to (8-76) through (8-78) for a spinless particle. Take the electromagnetic current density as $j_\mu = e\psi^+\gamma_0\gamma_\mu\psi$, where ψ^+ is the row spinor which is complex conjugate to the column spinor ψ.

8-7 QUANTUM ELECTRODYNAMICS

While a full description of quantum electrodynamics would exceed the present scope, a brief discussion of its relation to classical electrodynamics is important here to emphasize the logical coherence of these theories. We shall therefore limit ourselves to those features of the theory which are relevant in connection with its classical approximation. The semiclassical theory of radiation is severely limited by the fact that it is unable to account for the spontaneous emission of radiation from atoms. Such a process requires a description of the radiation in terms of photons. Furthermore, relativistic quantum mechanics is beset by difficulties. Characteristic of these is the appearance of the squared energy-momentum relation which implies that $E = \pm c\sqrt{\mathbf{p}^2 + (mc)^2}$. There is a need to restrict the physical solutions to *positive* energies in a theory which is completely symmetrical in positive and negative energies. The associated difficulties cannot be resolved in terms of the wave mechanics of a single particle, but require a many-particle theory with an unlimited number of particles. This is accomplished in quantum field theory, where the symmetry of relativistic quantum mechanics in positive and negative energy is replaced by a symmetry with respect to particles and antiparticles.

Quantum electrodynamics, the quantum field theory of the electromagnetic interaction of charged particles, is a covering theory of the semiclassical theory of radiation. In this higher-level theory, both fields, the classical electromagnetic field $F_{\mu\nu}(x)$ and the wave function field $\psi(x)$, are replaced by *operator* fields. For example, the current density $j^\mu(x)$ of (8-61) becomes an operator. As a consequence, the Maxwell-Lorentz equations, as well as the Klein-Gordon and the Dirac equations, are field equations for field *operators*. This operator character of $F_{\mu\nu}$ (or A_μ) and $\psi(x)$ has the following physical meaning. The field operators are combinations of operators which create or annihilate the *particles* associated with these fields, e.g. the photons and the electrons (positive and negative ones), respectively. When they act on state vectors that describe the state of a system, they will produce new state vectors describing a state with, in general, a different number of particles.

The following three points are important for an appreciation of the problem involved in discussing the classical limit of quantum electrodynamics.

The first is an observation concerning the Lagrangian or Hamiltonian formulation and the role played by the action principle. These formulations are of great importance in the classical theory and, in fact, we have built the theory on the action principle (Section 6–9). General relativity can also be based on a variational principle (cf. Appendix 2, especially Section A2–5). The situation is quite different in quantum field theory. The operator character of the dependent variables $A_\mu(x)$ and $\psi(x)$ does not admit a variational principle; there does not exist a variation calculus for such operators. At best, this principle can be used in a heuristic way to deduce the basic equations of the theory in analogy to the classical or semiclassical case. Quantum electrodynamics therefore starts with a set of equations (field equations) rather than with a variational principle. Consequently, Noether's theorem cannot be used to deduce the conservation laws. Instead, these laws follow from the group structure of the underlying symmetry.

The second point is the dominant role played by the *scattering matrix*. The importance of the asymptotic conditions in the classical theory is evident. These conditions acquire additional importance because the physically interesting dynamical initial-value problem is an asymptotic one. The spacelike plane $\tau =$ const. on which the initial data are specified is the plane $\tau = -\infty$. This is necessarily so in the interaction of fundamental particles where observations are always made in a space-time domain which is asymptotic to the domain of interaction. The specification of the initial state at $\tau = -\infty$, the "in-state," is accompanied by a specification of the final state at $\tau = +\infty$, the "out-state." The situation is best compared to the formulation (6–81) of the classical equations of motion in terms of p_{in}^μ and p_{out}^μ of a particle.* The transition-probability amplitude for given in- and out-states is one element of the scattering matrix. All such elements determine the whole matrix. The complete knowledge of this matrix is expected to give all the relevant physical information. In the classical approximation of the theory, the information in the scattering matrix must therefore be the same as that in the equations of motion (6–81).

The last remark concerns the difference between the older and the newer formulations of quantum electrodynamics. The older formulation is based on

---

* Characteristically, a free particle in quantum mechanics is specified by its momentum, angular momentum, spin, mass, charge, and other intrinsic properties. The position does not play an important role.

field equations for the operator fields A_μ and ψ. The knowledge of these fields then permits the computation of the scattering matrix. This formulation, while highly successful in its agreement with experiments, is mathematically unsatisfactory, because it involves divergent integrals which have to be removed by a regrouping of terms in the scattering matrix (known as *renormalization*).*

The newer formulation is free of these divergence difficulties and does not involve renormalization, while it gives exactly the same physical results as before. This formulation, here simply called *asymptotic quantum field theory*, contains the earlier form of quantum field theory only for the in- and out-fields A_μ^{in}, A_μ^{out}, and ψ^{in}, ψ^{out}. The operator fields A_μ and ψ, which are of central importance in the older formulation and which "interpolate" between the in-fields and the out-fields (the asymptotic condition enters here), play a minor role and can actually be completely eliminated from the scattering matrix. The field equations are then replaced by equations that relate the different scattering matrix elements to one another. In this formulation, the various physical processes therefore determine one another to a large extent.

One of the main difficulties still remaining in the theory is the need for solving the basic equations in perturbation expansion. Only in very special limiting cases are nonperturbative solutions known. Since the convergence of the perturbation expansion is in doubt, one has no assurance of mathematical consistency. Unfortunately, this convergence has so far not been established even in the classical limit, as we saw in Section 6–8.

It is clear from these remarks that a mathematical demonstration of the assertion that quantum electrodynamics is a covering theory of the classical theory of charged particles would go far beyond the scope of this exposition. The mathematically satisfactory, more recent formulation of quantum electrodynamics is also much more remote from the classical equations than the older, renormalization theory. The following very brief qualitative statement, however, may give the reader at least some orientation concerning the classical approximation.

Physically, the most characteristic features of quantum electrodynamics absent in the semiclassical radiation theory of the previous section are the existence of electron (or meson) pairs and the existence of photons.

Pair production and annihilation is possible because the interaction permits transitions between states of no electrons to states of two electrons (a negaton-positon pair), or vice versa. In going to the semiclassical theory, such transitions become less and less probable and, in fact, do not occur at all in the semiclassical approximation where the electron is described by the Dirac equation and its positive energy solutions only.

---

* To be sure, this theory needs to be renormalized in any case, irrespective of the divergence of the integrals involved. Similarly, mass renormalization in the classical theory must be carried out whenever a bare mass is introduced. (See p. 137.)

Similarly, the presence of a large number of photons will in this approximation be observable only as a frequency spectrum of radiation. No individual quanta exist. The classical measurements are too inaccurate to notice the "graininess" of the electromagnetic field.

When these and other similar physical concepts are expressed mathematically, quantum electrodynamics reduces to the semiclassical theory of radiation, where only a fixed number of charged particles is involved. Since the latter is known to lead to the classical theory as formulated in Chapter 6, the logical coherence of the hierarchy of theories of electromagnetic interactions is hereby indicated.

CHAPTER 9

The Theory's Structure and Place in Physics

This final chapter begins with a review of electromagnetic fields (Section 9-1). It is followed by an exposition of the symmetry properties of the theory under space and time reversal (Secion 9-2). Much confusion exists in the literature concerning the latter in view of the point particle limit in which the equations of motion are derived and used. Time reversal also touches on the philosophically famous problem of the 'arrow of time' of electromagnetic radiation (Section 9-3). The last two sections are devoted to an overview of the structure of the theory (Section 9-4), and to its standing in the larger context of scientific theories (Section 9-5).

9-1 RADIATION, COULOMB, AND OTHER KINDS OF ELECTROMAGNETIC FIELDS

The electromagnetic field appears in a number of different forms: in terms of potentials, field strengths, field energy, momentum, and angular momentum, and in the forces that act on charged particles. Mathematically, the most fundamental of these forms are the potentials because the other quantities can be constructed from them. They are also sometimes convenient for casting field equations and their solutions into simpler expressions.

Nevertheless, potentials have no direct physical significance in classical physics. They cannot be obtained uniquely from measurements in classical physics. They are undetermined within a gauge transformation. Only field strengths are measurable. And, finally, only *retarded* potentials are physically meaningful; advanced potential (and the corresponding advanced field strengths) exist mathematically but not physically.

All physical properties of a field can be expressed in terms of the field strengths and no reference to potentials needs to be made, although it is often convenient to do so.* Fields can enter a system as incident free fields. They can be produced by

---

*In some cases, a *kinematic* variable can be expressed in terms of potentials rather than field strengths (Problem 6-5). But these expressions are consequences of the field equations. The measurability of fields is due to their action on matter (particle dynamics) and requires the equations of motion which involve only the field strengths.

a source within the system. Only retarded fields, $F^{\mu\nu}_{\text{ret}}$, exist and can be measured (see Section 7-3). Similarly, the force of one charged particle (e, m) on another $(e',$ $m')$ is always the retarded force,

$$F^{\mu}_{\text{ret}} = \frac{e'}{c} F^{\mu\alpha}_{\text{ret}}(z')v'_{\alpha} . \tag{9-1}$$

Therefore, only retarded field strengths are of physical interest. Advanced fields may be used in calculations (see for example, p.138) but only in an auxiliary way. They can never occur in the predictions of observable phenomena.

Field strengths separate in a Lorentz invariant way into *velocity fields* and *acceleration fields* [Eq. (4-98)]. The former are independent of the acceleration and are proportional to $1/\rho^2$; the latter are linear and homogeneous in the acceleration and proportional to $1/\rho$ They are often referred to as (generalized) Coulomb and radiation fields, respectively.

A separation of the electromagnetic field into *solenoidal* and *irrotational* fields was carried out in Section 4-4. The former is a transverse field and asymptotically becomes the radiation field; the latter is the Coulomb field. The separation is covariant in closed systems with a timelike unit vector n^{μ}. Both fields satisfy inhomogeneous equations.

Since the field equations are linear, every solution permits splitting off a solution of the homogeneous equation (expressing free fields). The advanced field, though physically meaningless by itself, can be added and subtracted from the retarded field yielding a separation into a bound field, F_+, and a free field, F_- (see Eqs. (6-46)),

$$F^{\mu\nu}_{\text{ret}} = F^{\mu\nu}_{-} + F^{\mu\nu}_{+} . \tag{9-2}$$

Here, $F^{\mu\nu}_{-}$ is a source-free field,

$$\partial_{\mu}F^{\mu\nu}_{-} = 0 \tag{9-3}$$

while $F^{\mu\nu}_{+}$ is produced by the source,

$$\partial_{\mu}F^{\mu\nu}_{+} = \partial_{\mu}F^{\mu\nu}_{\text{ret}} = -\frac{4\pi}{c}j^{\nu} . \tag{9-4}$$

The irrotational Coulomb field seems to be static and instantaneous. However, one can show that it is a retarded field that propagates with a finite speed. This seeming contradiction can be explained by a curious cancellation between the retardation of the Coulomb field and the nonlocality of its source, $j_{\parallel}$.[*]

Free fields and bound fields are obviously very different mathematically since they satisfy different equations. But they are also very different physically: free fields have an autonomous nature; they exist in some sense irrespective of the presence of sources. Bound fields are the products

---

[*]F. Rohrlich, American Journal of Physics **70** (April 2002) 411-414.

of their sources and cannot be completely divorced from them. Thus, $F^{\mu\nu}_{ret}$ consists of a velocity field, which always stays with the source, and an acceleration field, part of which can escape as a radiation field. We conclude that *not all* of $F^{\mu\nu}_{ret}$ can leave the source.

The free-field nature of $F^{\mu\nu}_-$ seems to cause some conceptual difficulties. According to (9-2) it is the difference of two bound fields produced by the same charge at two different instances. Nevertheless, it satisfies the free-field equation everywhere. This can best be understood on the basis of the expression

$$A^\mu_-(x) = \frac{2\pi}{c} \int D(x - x')\, j^\mu(x')\, d^4x' \qquad (9\text{-}5)$$

according to (4-72), (4-79), and (4-80). The field $F^{\mu\nu}_-$ vanishes when the current density j^μ vanishes. Nevertheless, j^μ does not enter as the source of $F^{\mu\nu}_-$ since this field satisfies the source-free Maxwell-Lorentz equations. The possibility of such a situation is apparently closely related to the linearity and time symmetry of the theory (see Section 9-2).

The autonomous character of the radiation field finds its full expression in the quantum nature of radiation, i.e. in the "grainy" structure of its momentum, energy, and angular momentum corresponding to a collection of photons. There are no photons associated with the generalized Coulomb field (velocity field) and the latter remains a classical field also in quantum physics. The appearance of so-called timelike photons in quantum electrodynamics which could be interpreted as Coulomb quanta is a purely formal calculational device, devoid of physical meaning and spurious in the sense that these photons occur only in the intermediate stages of the calculations and not in the final results which describe observable phenomena.

In contrast to the autonomous radiation field, the permanently source-dependent generalized Coulomb field plays an auxiliary role only; it does not enter the fundamental action integral of the theory: only the direct Coulomb interaction between charges occurs. Indeed, this field, $F^{\mu\nu}_+$, is introduced only artificially in order to give the equations a simpler appearance, the expression (7-11) being simpler than (7-10). The free fields, $F^{\mu\nu}_-$, do enter the action integral in an essential way and can be eliminated from it only by certain nontrivial and nonobvious assumptions (see action at a distance, Section 7-2).

The velocity fields-acceleration fields dichotomy is based on the *kinematical* properties of the source at the instant of the production of the fields. The free field-bound field dichotomy is *dynamical*, being based on the nature of the field equations: homogeneous versus inhomogeneous equations.*

The Liénard-Wiechert expressions for the fields permit us to eliminate the fields produced by the particles and to replace them by particle variables. Conversely, in some instances, it is possible to express the particle variables

---

* The dynamics of the field is characterized by its source; that of a particle, by the force acting on it.

by field variables. These eliminations can give rise to various apparent contradictions and conceptual difficulties. For example, when the fields $F_{-}^{\mu\nu}$ in the equation

$$ma^{\mu} = F_{\text{ext}}^{\mu} + \frac{e}{c} F_{-}^{\mu\nu} v_{\nu} \tag{6-53}$$

are expressed in terms of the kinematical properties of the source, one obtains the Lorentz-Dirac equation

$$ma^{\mu} = F_{\text{ext}}^{\mu} + \frac{2}{3}\frac{e^{2}}{c^{3}}(\dot{a}^{\mu} - a^{\lambda}a_{\lambda}v^{\mu}), \tag{6-57}$$

which exhibits preacceleration.

Another example is the occurrence of the $\overline{F}^{\mu\nu}$ and $F_{+}^{\mu\nu}$ rather than the $F_{\text{in}}^{\mu\nu}$ and $F_{\text{ret}}^{\mu\nu}$ in the action principle: despite this fact, the particle equations (7–13) contain $F_{\text{ret}}^{\mu\nu}$ for the interaction of one charge with another. Part of $F_{\text{ret}}^{\mu\nu}$, namely $F_{+}^{\mu\nu}$, is implicit in the direct interaction terms of the action integral.

9-2 SYMMETRY PROPERTIES OF THE THEORY

In the past chapters we have emphasized repeatedly the importance of Noether's theorem for establishing conservation laws on the basis of certain symmetry properties of the theory. In particular, the invariance of the action integral under the continuous group of inhomogeneous Lorentz transformations was essential. We have also discussed the invariance of the theory under gauge transformations. In this section we want to study other symmetry properties which are expressed as invariance under transformation groups that are discrete rather than continuous.

(a) Space inversion. The transformation of space inversion, P, is defined by

$$P: \begin{array}{l} \mathbf{r} \to \mathbf{r}' = -\mathbf{r} \\ t \to t' = t. \end{array} \tag{9-6}$$

Since the repetition of this operation yields the original quantities, we are dealing with a group of two elements only.

The transformation of all the other kinematic quantities $\mathbf{v}$, $\mathbf{a}$, etc., follows from this definition:

$$\mathbf{v} = \frac{d\mathbf{r}}{dt} \to \mathbf{v}'(t') = \frac{d\mathbf{r}'}{dt'} = -\frac{d\mathbf{r}}{dt} = -\mathbf{v}(t),$$

$$\mathbf{a}(t) \to \mathbf{a}'(t') = -\mathbf{a}(t). \tag{9-7}$$

The proper time τ, given by

$$d\tau = \sqrt{1 - (d\mathbf{r}/dt)^{2}/c^{2}}\, dt, \tag{A1-35}$$

evidently remains invariant, $\tau' = \tau$.

If the inverted quantity differs from the original one only by a factor $+1$ (or -1), we say that this quantity has even (or odd) *space parity*; τ has even, $\mathbf{v}$ has odd space parity.

In order to facilitate the notation it is convenient to make use of our Cartesian Minkowski metric $\eta_{\mu\nu}$:

$$x^\mu = (t, \mathbf{r}), \qquad x_\mu = \eta_{\mu\nu}x^\nu = (-t, \mathbf{r}).$$

It permits us to write (9-6) in the form

$$P: \quad x^\mu \to x'^\mu = -x_\mu. \tag{9-6'}$$

Correspondingly,

$$v^\mu \to v'^\mu(\tau') = -v_\mu(\tau) \quad \text{and} \quad a^\mu \to a'^\mu(\tau') = -a_\mu(\tau). \tag{9-7'}$$

The transformation of v^μ implies for the current density,

$$j^\mu(x) = ec\int_{-\infty}^{\infty} \delta(x - z)\, v^\mu(\tau)\, d\tau, \tag{4-82}$$

that

$$j^\mu(x) \to j'^\mu(x') = -j_\mu(x).$$

The four-vector potential must therefore satisfy

$$A^\mu(x) \to A'^\mu(x') = -\, A_\mu(x); \tag{9-8}$$

this follows from

$$A^\mu_{\substack{\text{ret}\\\text{adv}}}(x) = \frac{4\pi}{c}\int_A D_R(x - x')\, j^\mu(x')\, d^4x' \tag{4-73}$$

and the invariance of the Green functions,

$$D_R(-\mathbf{r}, t) = D_R(\mathbf{r}, t),$$
$$\quad A \qquad\qquad A$$

which is evident from (4-79) and (4-80). Equations (9-6') and (9-8) imply

$$F'^{\mu\nu}(x') = F_{\mu\nu}(x) \tag{9-9}$$

or

$$\mathbf{E}'(x') = -\mathbf{E}(x), \qquad \mathbf{B}'(x') = +\mathbf{B}(x). \tag{9-10'}$$

In three-vector language, this means that $\mathbf{E}$ is a *polar* vector, while $\mathbf{B}$ is an *axial* vector, corresponding to odd and even parity, respectively.

If "$\sim$" indicates "transforms like," we can conclude from the above that

$$\mathbf{r} \sim \mathbf{v} \sim \mathbf{a} \sim \mathbf{E}, \qquad x^\mu \sim v^\mu \sim a^\mu \sim j^\mu \sim A^\mu, \qquad F^{\mu\nu} \sim x^\mu x^\nu. \tag{9-10}$$

The use of the metric tensor in expressing the inversion properties of tensors is obviously very convenient for a covariant formulation of the theory. For this reason it appears appropriate to define the concept of *covariant parity*.

Let $T^{\mu\cdots}(x)$ be the contravariant components of a tensor and let the mapping

$$I: \quad T^{\mu\cdots}(x) \to T'^{\mu\cdots}(x')$$

be an inversion transformation; let $s(\eta) \equiv \frac{1}{2}\sum_{\alpha=0}^{3}\eta_{\alpha\alpha}$ be the "sign" of the Minkowski metric. Then the covariant parity of $T(x)$ under I, $C_I(T)$, is defined by

$$T'^{\mu\cdots}(x') = C_I(T)\, s(\eta)\, T_{\mu\ldots}(x) \tag{9-11}$$

with

$$C_I(T) = \pm 1.$$

If T does not satisfy an equation of the form (9–11), covariant parity is not defined for it under this inversion. As examples we see that the vectors x, v, a, j, and A have odd covariant space parity ($C_P = -1$), while the tensor $F^{\mu\nu}$ has even covariant space parity ($C_P = +1$). Our metric has $s(\eta) = +1$.

It is now easily seen that the basic equations of the theory are invariant under space inversion. For the Maxwell-Lorentz field equations this is trivial, because it is implied in the above derivation of the transformation properties of A^{μ} from (4–73). For the Lorentz-Dirac equation of the closed one-particle system,

$$ma^{\mu} = F^{\mu}, \tag{6-51}$$

we have

$$F^{\mu} = \frac{e}{c}F^{\mu\nu}v_{\nu} \to \frac{e}{c}F'^{\mu\nu}v_{\nu}' = -\frac{e}{c}F_{\mu\nu}v^{\nu} = -F_{\mu} \tag{9-12}$$

so that, with (9–7'), Eq. (6–51) remains invariant. Clearly, as in (9–12),

$$F'^{\mu}_{\text{in}}(z') = -F^{\text{in}}_{\mu}(z). \tag{9-13}$$

But, since $\bar{F}^{\mu} = F^{\mu}_{\text{in}} + \Gamma^{\mu}$, this implies that

$$\Gamma'^{\mu}(z') = -\Gamma_{\mu}(z). \tag{9-14}$$

This can also be verified directly by means of the transformation (9–7') on

$$\Gamma^{\mu} = \frac{2}{3}\frac{e^2}{c^3}\left(\dot{a}^{\mu} - \frac{1}{c^2}a^{\lambda}a_{\lambda}v^{\mu}\right). \tag{6-55}$$

The equations of motion are also invariant, because the asymptotic conditions do not violate this invariance.

The invariance under space inversion of the basic equations of the theory can be proven more elegantly by means of the action integral. One shows that this integral is invariant under space inversion and that none of the operations involved in deriving the basic equations and the conservation laws violate this invariance. The invariance of I in (6–105) is obvious from (9–7') and (9–9).

PROBLEM 9–1

Show that the closed system of n charges in interaction with radiation (Section 7–1) is space-inversion invariant.

(b) **Time reversal.** Minkowski space treats time much like any space dimension, except for a suitable minus sign in the metric. Cognitively, however, time plays a very different role. The *direction of time* is of special importance physically. One cannot even speak of the fundamental notion of *causality* without first specifying the time direction. Only after the time direction is specified, can one begin to describe physical change.

What matters in time reversal is not the instant of time as measured relative to the origin of a time coordinate. Rather, it is the *direction* in time that is reversed. Therefore, time reversal is not a coordinate mapping but a reversal of motion. Time reversal is *motion reversal*; it is therefore conceptually quite different from space reversal that is a mapping (see Section **9-2 (a)** above).

The speed of light enters relativistic classical dynamics long before electromagnetic phenomena are being considered. The velocity fourvector of a moving mass point must lie within the future light cone whose origin is that mass. Thus, one must be able to distinguish between the future and past in order to know how the particle traverses its world line in Minkowski space. Again, the *direction* of time is what matters. The point of time relative to some origin of time is irrelevant. Once the motion of a point charge is reversed, the corresponding reversal of the electromagnetic fields follows as a consequence.

Electromagnetic radiation is caused by the acceleration (or deceleration) of a charge. Radiation is emitted along the *future* light cone originating on the charge. Only *retarded* radiation exists. This fact is therefore closely tied to the notion of causality and the direction of time. Emission of radiation follows acceleration. The direction in time is determined by the light cone that contains the four-velocity of the source. Radiation is always retarded no matter which time direction the charge follows: just as radiation was emitted relative to the old time direction of its source, after time reversal, *retarded radiation remains retarded* relative to the motion of its source. This invariance is a consequence of motion reversal: one cannot time reverse radiation without first time reversing the motion that causes radiation.

After these qualitative introductory remarks, the mathematical description of time reversal becomes very easy to follow. It is defined by...

$$dt \rightarrow dt' = -dt, \quad dr \rightarrow dr' = dr, \qquad (9\text{-}15)$$

or,

$$T: \qquad dx^\mu \rightarrow dx'^\mu = dx_\mu \qquad (9\text{-}15')$$

where the metric was used to write the transformation in a compact way. The proper time transformation follows from (9-15),

$$d\tau \rightarrow d\tau' = -d\tau.$$

(9-16)

From this and (9-15) follows as a matter of consistency that (see (A1–35)),

$$dt/d\tau = \gamma \rightarrow \gamma' = \gamma.$$

(9-17)

The other kinematic quantities then follow easily,

$$T: \qquad v'^{\mu}(\tau') = -v_{\mu}(-\tau), \qquad a'^{\mu}(\tau') = a_{\mu}(-\tau), \quad \text{etc.}$$

(9-18)

Here, the arguments $-\tau$ are symbolic: they are meant to indicate that at the time point τ the world line is to be traversed in the negative τ direction. The relation of v^{μ} to the current density, (4-82), yields

$$j'^{\mu}(x') = ec \int_{-\infty}^{\infty} \delta(x' - z') v'^{\mu}(\tau') d\tau' = ec \int_{-\infty}^{\infty} \delta(x - z(-\tau))[-v_{\mu}(-\tau)] d\tau = -j_{\mu}(x).$$

(9-19)

The transformation properties of the electromagnetic fields follow from those of the charged particle because these fields are a consequence of the existence of charges and their motion. This can best be seen by the vector potential A^{μ}. That potential is related to the motion of the charge that produces it by the Lienard-Wiechert equations (4-91),

$$A^{\mu}(x) = \frac{e}{c} \frac{v^{\mu}}{\rho(x, z)}.$$

(4-91)

It is important that $\rho_{\text{ret}} = \rho_{adv}$; these spacelike distances are distinguished only by the location of x relative to z. It can be either on the future or on the past light cone centered at z. But which one it is plays no role in the value of $\rho(x, z)$. The transformation properties of A^{μ} are therefore the same as those of v^{μ}. Under T,

$$A^{\mu}(x) \rightarrow A'^{\mu}(x') = -A_{\mu}(x).$$

(9-20)

Thus, retarded potentials transform into retarded ones, and advanced ones into advanced ones. This is how moving charges generate electromagnetic fields. The time direction of the particle motion determines the nature of the radiated fields. All fields are emitted into the future: only retarded fields are allowed by causality. This can be indicated as follows:

$$T: \qquad F_{\text{ret}}'^{\mu\nu}(x') = -F_{\mu\nu}^{\text{ret}}(x).$$

(9-21)

The same holds for advanced fields. This equation implies for the three-vector fields,

$$T: \qquad \mathbf{E}'(x') = \mathbf{E}(x), \qquad \mathbf{B}'(x') = -\mathbf{B}(x). \qquad (9\text{-}22)$$

Combining equations (9-21) and (9-19) shows that *Maxwell's equations remain invariant under time reversal.*

The important question is now whether the equations of motion are also time reversal invariant. Consider the Lorentz-Dirac equation first and the Landau-Lifshitz approximations after that (see the supplement).

From (9-21) and (9-18) one sees that the electromagnetic force transforms under T as

$$T: \qquad F'^{\mu} = F'^{\mu\alpha}v'_{\alpha} = F_{\mu\alpha}v^{\alpha} = F_{\mu}. \qquad (9\text{-}23)$$

All forces and fields in this equation are retarded, and remain retarded under T. One can assume that non-electromagnetic forces, F^{μ}_{ext}, transform the same way.

The Abraham-Laue four-vector, F^{μ}, was derived by Dirac by a limiting procedure. First, he added and subtracted the advanced field thereby splitting the total retarded self-field into two parts, (6-46),

$$F^{\mu\nu}_{\text{ret}} = F^{\mu\nu}_{+} + F^{\mu\nu}_{-}.$$

This separation does not attach an observable meaning to the advanced field since both fields, $F^{\mu\nu}_{+}$ and $F^{\mu\nu}_{-}$, interact simultaneously, although quite differently. The self-action due to the field $F^{\mu\nu}_{+}$ produces the Coulomb self-energy that is absorbed as electromagnetic mass into the observed mass, m (see equations (6-47) and (6-40')). The self-interaction due to $F^{\mu\nu}_{-}$ yields the vector

$$F^{\mu}(x, z(\tau)) = \frac{e}{c}F^{\mu\alpha}_{-}(x)v_{\alpha}(\tau) \qquad (9\text{-}24)$$

where $F^{\mu\nu}_{-}$ is given by (6-46)$_{-}$ and (4-101). This vector still transforms like a force as in (9-23). When the Lorentz-Dirac equation is written in the form

$$m\dot{v}^{\mu} = F^{\mu} + [\Gamma^{\mu}(x, z)]_{\rho \to 0}$$

where the limit $\rho \to 0$ is meant as in equation (6-63), *the equation remains invariant under T.*

However, when the limit is carried out first, resulting in the Lorentz-Dirac equation (6-57) with Γ^{μ} given by (6-55), it is *not* time reversal invariant. The Abraham fourvector (6-55) changes sign under T, $\Gamma^{\mu} \to -\Gamma_{\mu}$. Indeed, mathematically, the limit $\rho \to 0$ is prohibited because of the Coulomb self-energy divergence. It is inconsistent to treat the two parts of $F^{\mu\nu}_{\text{ret}}$ differently; they are physically inseparable and always occur together.

9-3   THE RADIATION ARROW OF TIME

The preceding Section on symmetries of the theory is appropriately followed by an important asymmetry. It is the following qualitative claim:

Given a finite volume containing interacting charged particles that may be subject to external forces. These particles are capable of emitting and absorbing radiation. The radiation arrow of time claims that *some radiation will always escape* (along a future light cone) while the rest will be absorbed by other particles.

Before justifying this claim, it must be noted that such asymmetries have been well known in other branches of physics as well as in other sciences. An example is the irreversibility of energy lost due to friction. Much more general is the law of entropy: in any closed physical system, the entropy never decreases. As a special case, it is common knowledge that when two bodies of different temperatures are able to exchange heat, the hotter body will give heat to the cooler one until a uniform temperature is reached. Arrows of time are also claimed in the cosmological expansion of the universe and in biological evolution.

We know that causality ensures radiation emission along the future light cones from a given source: radiation is always retarded and never advanced. In addition, radiation is emitted into all spatial directions (with various angular distributions depending on the source). Absorption of radiation by a charge within the system involves only very narrow cones of radiation from other sources. Some of the absorbed radiation converts into kinetic energy; some converts into internal energy (in case the absorber has internal structure) and only a (usually small) fraction of it is reradiated. All this provides for the "arrow of time": an overall asymmetry in time due to radiation emission into the future rather than the past. The latter preference is due to the lack of existence of advanced fields. That, in turn, is a consequence of causality.

The radiation arrow of time is of course fully compatible with the time reversal invariance of the theory for finite size charges.*

9-4   THE STRUCTURE OF THE THEORY

The classical theory of charged particles is the application of the special theory of relativity to the description of the interaction of a finite number of structureless charged particles (point charges) with electromagnetic fields. Non-electromagnetic external forces are also allowed. If the interacting system is closed, the following statements apply.

(A1) *Invariance.* The theory is invariant under the group of orthochronous inhomogeneous Lorentz transformation with determinant -1, as well as under space reversal.

(A2) *Time reversal.* Invariance under time reversal holds for finite size charge

---

*F. Rohrlich, *American Journal of Physics* **74** (April 2006) 313-315.

distributions; it is lost in the point charge limit.

(B) *Action integral.* The theory is based on an action integral (see (7-6)) for a system of one or more point charges in interaction with one another and with an external force that may include an external electromagnetic field.

The last postulate suffices to deduce the Maxwell-Lorentz equations and the Lorentz-Dirac equation for a single charge as well as for their generalization to a system of charges. However, the Lorentz-Dirac equation needs to be further restricted because it permits unphysical solutions. For many years, an asymptotic condition was used to this effect (see Section 6-6). That condition eliminates run-away solutions but does not eliminate preacceleration. Only at the turn of the century, a satisfactory solution for the restriction of the Lorentz-Dirac equation to physical solutions has been found (see the Supplement).

It is ensured by the following requirement.

(C) *The Landau-Lifshitz approximation.* (See the Supplement to this third edition). While this is an approximation mathematically, it is not an approximation physically: it reaches to the end of the validity domain of classical physics. The characteristic time interval, τ_0, is so small that further corrections would lead into the domain of quantum mechanics. The resulting equations of motion are in this sense accurate and no further corrections are believed to be necessary within a classical description.

The theory begins with the action integral (6-105) that involves the physical mass, m. From it, the Maxwell-Lorentz field equations are derived as well as the Lorentz-Dirac equations for point charges. The conservation laws follow from Noether's theorem. The equations of motion for point charges are obtained from the Lorentz-Dirac equations by restricting these equations to small characteristic times, τ_0, by means of the Landau-Lifshitz approximation (see the Supplement).

9-5 THE THEORY IN ITS LARGER CONTEXT

Physicists are usually not interested in the relationship between different branches of physics. They know intuitively whether, say, quantum mechanics or classical mechanics is appropriate for solving a particular problem. 'Philosophy of science' is regarded with suspicion. The German word 'Wissenschaftstheorie' (theory of science) is less offensive to scientists than 'philosophy of science'. In the following, I shall deduce some useful and interesting information from the 'theory of physics'. It will bear directly on the physics of classical charged particles.

At about the time when the first edition of the present book was published, there appeared a book by Thomas Kuhn, *The Structure of Scientific Revolutions.** It caused lots of attention. The claims in that book seem to oppose an old view in the theory of science known as logical empiricism that dates from a much earlier part of the twentieth century, and that has been embraced by many physical sci-

---

*Thomas Kuhn, *The Structure of Scientific Revolutions*, University of Chicago Press 1962.

entists. They used it often without being aware of that term. For example, Kuhn
characterized scientific revolutions as paradigm shifts while most scientists believed
in reductionism. He saw radical changes of fundamental concepts that cannot be
related to the old concepts, while physicists saw the old theory deducible from the
new one by certain approximations and limiting processes. The term 'theory reduction' means simply that a new and more sophisticated theory can be simplified in
some limit of suitable parameters so that the old and well-known theory emerges.
Kuhn's ideas received wide attention from people within the theory of science as
well as from outside. Intensive discussions on this subject abated only near the end
of the last century. I shall outline some of the results as I see them especially as
they affect the classical theory of charged particles.*

Looking at the large number of branches in the physical sciences, we find them
to be specialized in various ways. Astronomy, geology and atomic physics clearly
differ in the scale of the phenomena under study. Call these domains of scale
'levels'. Starting with the level of classical physics, one can go deeper into the
atomic level (quantum mechanics), and even deeper into the level of fundamental
particles (quantum field theory). How are the theories on different levels related?

This raises the question of the validity domain of a given physical theory. A
theory must satisfy certain basic requirements such as internal consistency and
completeness; and when its predictions are compared with empirical data, they
must agree with the measurements within their accuracy. But the accuracy of a
theory's predictions has no limit *unless* another theory restricts it.

Consider the example of nonrelativistic classical mechanics. We did not know
any of its validity limits until relativity theory was developed (1905). That theory
tells us that when $(v/c)^2$ is larger than the error of the measurement, nonrelativistic
mechanics will no longer predict the correct result. Thus, the requirement that
terms of order $(v/c)^2$ are negligible, *i.e.* $(v/c)^2 \ll 1$, establishes a limit for the
validity of nonrelativistic mechanics. A validity limit of the old theory is imposed
by the new theory.

But this is not the only such limitation. Since the 1920s, we know that for atomic
scales, a new theory, quantum mechanic, provides the correct description. This
leads to another validity limit for nonrelativistic mechanics; it is due to quantum
mechanics. Let us characterize quantum mechanics by the Compton wavelength of
a particle of mass m, $\lambda = h/(mc)$. When a length, l, becomes comparable to λ,
the classical prediction for it will be incorrect unless $\lambda/l \ll 1$. The validity domain,
D_T, of a theory T may need to be characterized by more than one limit. There
is a different limit for each 'adjacent' theory that is unknown until that theory
becomes known. Each limit provides another dimensionless number that must be
small compared to *1*. When all neighboring theories and the corresponding validity
limits are known, the domain of a theory is complete.

*F. Rohrlich, *Realism Despite Cognitive Antireductionism*, International Studies in Philosophy
of Science **18** (March 2004) 73-88.

The theory is then called 'established'.* Classical electrodynamics including the dynamics of charged particles is an established theory.

Returning to our classical relativistic theory of charged particles, we see its validity domain restricted by relativistic quantum electrodynamics. The restriction is similar to the nonrelativistic case: it is again $\lambda/l \ll 1$ where l is a length characteristic of the classical system. But there is no longer a restriction on the speed as in the nonrelativistic case. The limitation in size may be accompanied by a limitation in time intervals. Given $\lambda/(c\Delta t) \ll 1$ and $\lambda \sim \tau_0 c/\alpha = 137\tau_0 c$ where $1/137$ is the value of the fine structure constant, α, it follows that $\Delta t \gg \lambda/c \sim 137\tau_0 \gg \tau_0$. The constant τ_0 is our well-known constant of equation (6-70). This result implies that derivatives higher than the first, *i.e.* $(\tau_0 d/d\tau)^n$ with $n > 1$, are expected to be outside the validity domain of classical physics. Recalling that the largest value of τ_0 is that of the electron, 0.62×10^{-23}, and that for classical objects it is much, much smaller, *the Landau-Lifshitz approximation is actually exact within classical physics.*

Having elaborated on the validity limits of a theory, the next question is the relationship of the superseding theory, S, to the superseded theory, T. Does the new, deeper level theory reduce to the old one? This requires the equations of the new theory to become those of the old theory in a suitable limit. That this is indeed the case is well-known for relativistic mechanics: it does reduce to non-relativistic mechanics in the limit $(v/c)^2 \ll 1$. For other theories, similar reductions hold. Example: the classical limit of the equations of quantum mechanics yields indeed the equations of classical mechanics. That is shown in many texts on quantum mechanics.

But theories in physics do not only involve equations, they also involve the *meaning of the symbols* in the equations. In special relativity, we know that the mass, m, of an object is equivalent to an amount of energy, mc^2. No such relation exists in nonrelativistic theory. But in this case, the concepts of mass and energy occur in both theories. In other cases, the theory on the deeper level, S, contains *incommensurable* concepts that are qualitatively different from the concepts in the superseded theory, T.

Here is one striking example: gravitation. According to Einstein's gravitation theory (general relativity), gravitation is described by a *curved* four-dimensional space-time but no force; in Newtonian theory, gravitation is described by a *flat* four-dimensional spacetime in which there is a force acting. In the mathematical reduction of Einstein's theory to Newton's, a symbol related to the curvature of spacetime must be interpreted as symbol for gravitational force. This bridges the incommensurability of the two theories.**

One comes to the conclusion that the more sophisticated, deeper theory, S, may have equations that reduce *mathematically* to those of T, but that their meaning

---

*F. Rohrlich and L. Hardin, "Established Theories", *Philosophy of Science* **50** (1983), 603.

F. Rohrlich, *Foundations of Physics* **19, 1151-1170 (1989), for physicists, and *Philosophy of Science* **68** (June 2001) 185-202, for philosophers of science.

does not: the meaning of the symbols may be entirely different; they may require reinterpretation. Thus, while the mathematics of the theory can be reduced, the *meaning of the symbols* changes. One cannot reduce one meaning to another. That's the essence of the incommensurability claim. Each level of physical description has at least some concepts that are not shared with other levels. But within its validity domain a theory is 'good forever'; it is established.

The lesson to be learned is that even if there were a 'theory of everything', and even if its mathematics could be reduced step by step down to the theory of the common (everyday) level of reality, *the basic concepts of the 'theory of everything' cannot be reduced.* There exist incommensurabilities that prevent such a reduction. That is an insight that is to a large extent due to Kuhn.

After this diversion into physical theory in general, the obvious question arises: can relativistic quantum electrodynamics be reduced to relativistic classical dynamics of charged particles? The answer must be positive, but so far, nobody has yet provided such a mathematical reduction. This open question is at present the biggest deficiency of the theory. Nevertheless, the classical relativistic theory of charged particles is now in an excellent state: not only the field equations are well established; the dynamical equations for the charged particles are also (see the following Supplement). At the end of a century of inadequacy, the equations of motion are free of unphysical solutions and they are valid throughout the classical domain.

Supplement

The Physically Correct Dynamics

A. THEORY

The equation of motion derived by Lorentz, Abraham, and Dirac has been discussed at length in Chapter 6. The classical dynamics of charged particles implied by that equation has dominated the field during the entire twentieth century. However, that equation has been well known to be defective. It violates Newton's first law of motion (Galileo's law of inertia) that holds for neutral particles both relativistically and non-relativistically. For electrically charged particles, the Lorentz-Abraham-Dirac equation contains a term that expresses the effect of the generated radiation. That term violates Newton's first law; it allows for self-accelerating solutions, and it permits acceleration in advance of the onset of a force (preacceleration). These solutions were discussed in Section C of Chapter 6.

An extensive literature has developed dealing with attempts at solving this problem. But no solution seemed satisfactory. That sad state of affairs was not resolved until the beginning of the current century. The present supplement is devoted to this resolution. Here, I cannot give a full historical account of this development (it would be much too long) but I shall present a physical argument that leads to this solution.

The offending term in a charged particle's equation of motion is entirely due to its charge. That term is responsible for the emission of electromagnetic radiation whenever the particle is accelerated. In order to account for this radiation emission, Abraham and Lorentz deduced that additional term from Maxwell's equations; it was relativistically correct even though they worked before 1905 when Einstein published his fundamental paper on special relativity. The term was later given its manifestly covariant form by von Laue (equation (2-30)).

The key to the resolution of the difficulties of the Lorentz-Abraham-Dirac equation lies in the observation that this additional term is *small*. What exactly is meant by 'small' will be specified below. Remarkably, this smallness was first observed already half a century ago but nobody had realized its significance. In the 1951 English edition of *The Classical Theory of Fields*, Landau and Lifshitz[*] derived an

---

[*]L. Landau and E. Lifshitz, *The Classical Theory of Fields*, Addison-Wesley Press, Cambridge, Mass., 1951.

approximation to the Lorentz-Dirac equation that took account of that smallness.
I shall call their method the Landau-Lifshitz approximation.

In order to exhibit this approximation in a quantitative and also transparent
way, it is desirable to write the equations in as simple a notation as possible. This
will permit an easier identification of the physical meaning of the terms involved.

Start with the Lorentz-Abraham-Dirac (LAD) equation (6-57). The two forces
acting on the charged particle can be combined,

$$F^\mu = F_{in}^\mu + F_{ext}^\mu .$$ (S-1)

The Abraham-Laue four-vector, Γ^μ, can be written in the compact form

$$\Gamma^\mu = P^{\mu\nu}\tau_0 m \frac{da_\nu}{d\tau}$$ (S-2)

where τ_0 is given by (6-70), and

$$P^{\mu\nu} = \eta^{\mu\nu} + v^\mu v^\nu$$ (S-3)

is the projection tensor into the spacelike hyperplane whose normal is the velocity
v^μ. In these equations, we chose the units of time such that the speed of light $c = 1$.
(For conventional units, the correct powers of c can easily be restored by checking
the dimensions.) The dimensionless tensor $P^{\mu\nu}$ in (S-3) satisfies

$$P^{\mu\alpha}P_\alpha^\nu = P^{\mu\nu} = P^{\nu\mu} \quad \text{and} \quad P^{\mu\alpha}v_\alpha = 0 .$$ (S-4)

With these preliminaries, the LAD equation can be written compactly as

$$ma^\mu = F^\mu + P^{\mu\nu}\tau_0 \frac{d(ma_\nu)}{d\tau} .$$ (S-5)

Landau and Lifshitz (L-L), in their text on classical fields (see above), suggested
an approximation of this equation. Assuming the second term on the right is negligible, we have just the relativistic equation for neutral particles, $ma^\mu = F^\mu$. This
might be called the zeroth approximation. When τ_0 is finite but small, one can use
that zeroth approximation for ma_ν on the right hand side, replacing ma_ν by F_ν.
This results in

$$ma^\mu = F^\mu + P^{\mu\nu}\tau_0 \frac{dF_\nu}{d\tau} = F^\mu + \tau_0 \left(\frac{dF^\mu}{d\tau} + v^\mu v^\alpha \frac{dF_\alpha}{d\tau} \right) .$$ (S-6)

That equation was suggested by Ford and O'Connell* who, at the time, were un-
aware of the work by L-L. Very recently, Medina derived (S-6) directly for an extended spherical charge in contradistinction to Dirac's point particle approximation.**

The smallness of the second term in (S-5) is strongly supported empirically:
$c\tau_0 = \lambda/137$ where λ is the Compton wave length. The largest value of τ_0 for any
particle occurs for the electron: $\tau_0 = 0.62 \times 10^{-23}$ sec. Terms of order τ_0^2 are no
longer in the classical domain.

*G.W. Ford and R.F. O'Connell, *Physics Letters* A **174** (1993) 182.
R. Medina, *Journal of Physics* A, **39 (2006) 3801-3816.

Equation (S-6) is a natural consequence of the L-L method. But one must be aware of the limitation of that equation. It is restricted by the assumption made in its derivation,

$$|P^{\mu\nu}\tau_0\frac{dF_\nu}{d\tau}| \ll |F^\mu|. \tag{S-7}$$

The meaning of the τ_0 term in the Abraham-Laue fourvector (the last term on the right in (S-6)) is the same as in the LAD equation. The first part, $\tau_0 dF^\mu/d\tau$, is the Schott term, the second part is the radiation reaction fourvector,

$$\frac{dP^\mu_{rad}}{d\tau} = -v^\mu \mathscr{R}'. \tag{S-8}$$

The invariant radiation rate is

it dot

$$\mathscr{R}' = \tau_0 \dot{v}^\alpha \frac{dF_\alpha}{d\tau} = \tau_0 \frac{1}{m}\frac{dF^\alpha}{d\tau}\frac{dF_\alpha}{d\tau} = \mathscr{R}. \tag{S-9}$$

In this L-L approximation, all three of these equations hold and $\mathscr{R}$ is given by the relativistic Larmor formula (5-14).

Half a century after L-L came the work by Spohn. As a mathematical physicist, he attacked the problem in a very different way. He studied the space of solutions of the Lorentz-Dirac equation, and proved the existence of a critical subspace (manifold). Only those solutions that are inside that subspace are physical solutions.[*] That, of course, recalls the approximation made by L-L. *They treated the LAD equation as a small deviation from the equation for neutral particles.* This suggests that only the physical solutions are kept in this approximation. The importance of the L-L approximation, therefore, became clear to me through Spohn's work.[**]

In the LAD equation, the acceleration and its derivative on the right hand side might be blamed for the unphysical solutions. To avoid such occurrence in (S-6) in the case when $F^\mu = eF^{\mu\alpha}v_\alpha$, L-L made additional approximations.

This L-L approximation

$$dF^\mu/d\tau = ev_\alpha \partial^\alpha F^{\mu\nu}v_\nu + \frac{e}{m}F^{\mu\nu}eF_{\nu\alpha}v^\alpha$$

when substituted into (S-6) leads to

$$m\dot{v}^\mu = eF^{\mu\alpha}v_\alpha + \tau_0 e\left[v^\alpha \partial_\alpha F^{\mu\nu}v_\nu + \frac{e}{m}(F^{\mu\alpha}F_{\alpha\beta}v^\beta + v^\mu v_\alpha F^{\alpha\beta}F_{\beta\nu}v^\nu\right]. \tag{S-10}$$

This is the equation actually given by Landau and Lifshitz. They did not consider non-electromagnetic forces. But the approximation method in their work can also be applied to external forces that do not depend on the velocity as was done above. The elimination of the velocity dependence of the electromagnetic forces obviously complicates the equation; it replaces (S-6) by (S-10). The validity condition (S-7) is also more complicated. Written symbolically, it becomes

$$|P^{\mu\nu}\tau_0(ev_\alpha \partial^\alpha F^{\mu\nu}v_\nu + \frac{e}{m}F^{\mu\nu}eF_{\nu\alpha}v^\alpha)| \ll |F^\mu|. \tag{S-11}$$

---

[*]Spohn, *Europhysics Letters* **50** (2000) 287. See also H. Spohn, *Dynamics of Charged Particles and their Radiation Field*, Cambridge University Press, Cambridge (2004)

[**]F. Rohrlich, *Physics Letters A* **283** (2001) 276, and **303** (2002) 307.

The set of equations, (S-10) and (S-11), as well as the preceding set, (S-6) and (S-7), each provide a physically acceptable dynamics because *each set satisfies the law of inertia*: each set leads to uniform motion when the external force and its time derivative vanish. Thus, the first set is also valid for velocity dependent forces such as the electromagnetic ones. The set (S-10) and (S-11) involves further approximations, is unnecessary and is also more complex.

B. APPLICATIONS

The claim to be proven in this section is that the unphysical features that occur in solutions of the LAD equation are *all* absent in the L-L approximations (S-6) and (S-7) valid for velocity independent forces as well as for electromagnetic forces. But the inequalities (S-7) must be satisfied.

Runaway solutions. A comparison of equations (S-5) and (S-6) shows that the Landau-Lifshitz equation *satisfies Newton's first law* (the law of inertia) while the LAD equation does not. When the force vanishes during a finite time interval, (*F and its derivative, dF/dt* vanish), the acceleration vanishes also; therefore the particle moves uniformly. It follows that for the free particle, there are *no run-away (self-accelerating) solutions* in the L-L approximation while they do exist for the LAD equation (see Section (6 – 10)).

Uniform acceleration. From Section 6-11 we see that $ma^\mu = F^\mu$ holds by definition so that $P^{\mu\nu}dF_\nu/d\tau = mP^{\mu\nu}da_\nu/d\tau = 0$. This latter equation is exactly the definition of uniform acceleration, (5-23) or (5-23'). For this motion, the results of the equation (S-6) are identical to those of the LAD equation. Both lead to hyperbolic motion. Only a higher order difference, (S-9), appears in the radiation rate. The difference between the Abraham-Laue fourvector and the radiation reaction fourvector is now very transparent: the former vanishes, and the latter does not. The radiation reaction four-vector is the negative of the fourvector of the radiation emission rate in accordance with Newton's third law. In uniform acceleration, the radiation loss is compensated exactly by the Schott fourvector. The latter is just the time rate of change of the external force. If the acceleration is not uniform, the fourvector of the radiation emission rate is only partly compensated by the Schott term. The rest comes from the rate of work done by the external force.

Preacceleration. A well-known defect of the LAD equation is preacceleration, the acceleration of a charge before the onset of a force. This defect is also absent in the L-L equation. To show this, we return to Section 6-13. Using the convenient coordinates (6-131) and notation (6-135), the equation (S-6) for one-dimensional motion simplifies to

$$\dot{w} = f + \tau_0 \dot{f}. \tag{S-12}$$

We are now dealing with an ordinary first order differential equation rather than an integro-differential equation as in (6-140). It is clear that no preacceleration will

occur. But consider the problem in more detail. Equation (S-12) is accompanied
by the condition (S-7) that, in terms of the present variables, reads

$$|\tau_0 \dot{f}| \ll |f|. \tag{S-13}$$

A discontinuity in the force, f, (that describes a pulse in Section (6-13)) is therefore *excluded*: f must be a continuous function of time. With that restriction and
(S-13), no preacceleration arises. Similarly, one can show that no post-acceleration
exists. Consider a continuous step function of finite slope,

$$f(\tau) = \frac{A}{2}\left(\frac{\tau}{\varepsilon} + 1\right) \text{ for } /\tau/ < \varepsilon, \quad f(\tau) = A \text{ for } \tau > \varepsilon, \quad f = 0 \text{ for } \tau < -\varepsilon. \tag{S-14}$$

It is applicable provided the condition (S-13) holds, *i.e.* $\varepsilon \gg \tau_0/2$ for $\tau > 0$. This
condition provides the limit to which the discontinuity that occurs in (6-147) can
be approximated by a slope. It follows that the applied force must be continuous
and its slope restricted. With this limitation, the force (S-14) is applicable to the
equation (S-12) and no preacceleration arises. For $\tau < -\varepsilon$, $f = 0$ and the restriction
(S-13) becomes meaningless.

One can approximate a step function by a smooth function that contains a
parameter such that a step occurs only in a limit:

$$f(\tau) = \frac{A}{2}\left[\tanh\left(\frac{\tau}{\varepsilon}\right) + 1\right]. \tag{S-15}$$

For large $\tau > 0$ this function approaches A, while for large $\tau < 0$, it approaches
0; for $\tau = 0$ it is $A/2$. Finding the slope, we see that when the parameter A/ε is very
large, $f(\tau)$ increasingly approximates a step function. But the validity condition
(S-13) limits this approximation just as for the force (S-14).

PROBLEM S-1

Solve the equation (S-6) for the force (S-15) and find the conditions on the
parameters that satisfy (S-13).

For many applications of the equation of motion, it is convenient to express it
in terms of the w variables (6-131). First, one casts (S-6) into the form

$$\dot{v}^\mu = f^\mu + \tau_0(\dot{f}^\mu + v^\mu v \cdot \dot{f}). \tag{S-16}$$

The Frenet equations (6-130) generalized to Minkowski space permit this equation
to be written as

$$\dot{w} = f_v + \tau_0(\dot{f}_v - \kappa f_n \cosh^2 w)$$

$$\kappa \tanh w = f_n + \tau_0(f_n \dot{w} \tanh w + \dot{f}_n + \kappa f_v - \kappa' f_b) \tag{S-17}$$

$$0 = f_b + \tau_0(f_b \dot{w} \tanh w + \dot{f}_b + \kappa' f_n).$$

In contrast to (6-136) for the LAD equations, these equations are of first order
rather than of second order in the variable w. They are therefore easier to solve. A

comparison between (6-136) and (S-16) provides little insight except for the obvious replacement of $\ddot{w}$ by $\dot{f}_v$. Note that equations (S-17) just as (6-136) involve only three equations that replace the corresponding fourvector equations because they describe the motion in the three-dimensional spacelike hyperplane orthogonal to the velocity fourvector.

The validity condition (S-7) must also be written in terms of the variables w and the intrinsic variables κ and κ' of the Frenet equations (6-130). Thus, (S-17) is restricted by

$$\tau_0 |\dot{f}_v - \kappa f_n \cosh^2 w| \ll |f_v|$$

$$\tau_0 |\dot{f}_n + f_n \dot{w} \tanh w + \kappa f_v - \kappa' f_b| \ll |f_n| \tag{S-18}$$

$$\tau_0 |\dot{f}_b + f_b \dot{w} \tanh w + \kappa' f_n| \ll |f_b| \,.$$

Recall that the equations (S-6) and (S-7) and correspondingly (S-17) and (S-18) can be used for *all* forces including electromagnetic ones. As stated earlier, the Landau-Lifshitz equations (S-10) and (S-11) are an unnecessary complication.

Annotated References

The books and papers listed below provide deeper insight into selected topics treated in the present text.

Bargmann, V.L., Michel, L. and V.L. Telegdi, *Physical Review Letters* **2** (1959) 435.

> This classic paper provided the name 'BMT equation' to the equation for the Pauli-Lubanski fourvector (7-51) that describes the relativistic dynamics of the spin.

Jackson, J.D., *Classical Electrodynamics*, 3rd edition, 1999, John Wiley and Sons, Minneola, New York.

> This is the most comprehensive and authoritative text on this subject presently available. It contains more than can be reasonably included in a year's course. Nevertheless, its treatment of charged particle dynamics is considerably briefer than the present specialized book that can therefore be used supplementary to this general text.

Jammer, M., *Concepts of Mass*, Dover Publications, Minneola, New York 1997.
Jammer, M., *Concepts of Force*, Dover Publications, Minneola, New York 1999.

> These two important books, originally published in 1961 and 1957, respectively, describe the difficulties in the history of physics that had to be overcome before their present understanding emerged.

Rohrlich, F., "The Electron: Development of the First Elementary Particle Theory," pp.331-369 in J. Mehra (ed.) *The Physicist's Conception of Nature*, D. Reidel, Dordrecht-Holland, 1973.

This paper is based on my lecture in honor of Dirac's seventieth birthday. It includes additional material on the history of the electron and on the dynamics of its spin and magnetic moment.

Spohn, H., *Dynamics of Charged Particles and their Radiation Fields*, Cambridge University Press, Cambridge, 2004.

A mathematically rigorous treatment of electrodynamics not easily found elsewhere. In particular, it derives the Landau-Lifshitz approximation from singular perturbation theory, and it includes several of its applications.

Teitelboim, C. *Physical Review* D **1** (1970) 1572, **3** (1971) 297, **4** (1971) 345.

This series of papers prove that the Lorentz-Dirac equation can also be derived without Dirac's trick of separating the retarded fields using the advanced fields as auxiliaries (see equation (6-46)). Teitelboim's derivation is based on the energy-momentum-stress tensor.

Teitelboim C., D. Villarroel, and Ch. Van Weert, *Revista Nuovo Cimento* **3**, (1980) 1.

An excellent review of the dynamics of point charges with and without magnetic dipole moments.

Yaghjian, A. D., *Relativistic Dynamics of a Charged Sphere*, 2nd ed. Springer Science 2006.

This book presents a detailed derivation of the equations of motion first obtained by Lorentz and Abraham. For this purpose, the author uses non-covariant notation as in the original papers that were written before special relativity (1905) and before covariant notation became known in physics (1908). He also discusses the unphysical solutions of the Lorentz-Dirac equation and suggests ways to remove them.

Appendices

The Space-Time of Special Relativity

The special theory of relativity could be defined as that branch of physics which is governed by the space-time symmetry imposed by the Lorentz group of coordinate transformations.

One is led to this symmetry by a combination of experimental facts and axiomatic reasoning. The reasoning that leads to the requirement of invariance of the laws of physics under the 7-parameter Euclidean group (four space-time translations, three space rotations) was described in Chapter 3. The homogeneous Lorentz transformations which form the essential part of the group of Lorentz transformations can be derived as follows.

A1-1 DERIVATION OF THE LORENTZ TRANSFORMATION

Let x, y, z and x', y', z' be the rectilinear orthogonal Cartesian coordinates of two reference systems S and S'. Let t and t' be their respective time coordinates. Assume that S and S' are in relative motion with constant velocity **v**. We are seeking the transformation from S to S' subject to the following restrictions.

(1) If a particle is in uniform motion relative to S, it must be in uniform motion also relative to S' for all values of the coordinates. (Equivalence of inertial systems in uniform relative motion.)

(2) A light wave in a vacuum expanding spherically with velocity c relative to S must expand spherically with velocity c also relative to S'. (Independence of light velocity from source and observer motion.)

The first restriction requires the transformation to be linear.* The homogeneity of space-time, already implied in translation invariance, permits us to restrict ourselves to the *homogeneous* transformations. Thus, we can require that at $t = 0$ we have $t' = 0$ and the origins of S and S' coincide: $x = y = z = 0$ implies $x' = y' = z' = 0$. Furthermore, the isotropy of space, already implied in rotation invariance, permits us to *orient* the space coordinates so

---

* Note that the words "for all values of the coordinates" are essential in order to exclude from the transformation singularities which would otherwise be present. See, for example, V. Fock, *The Theory of Space, Time, and Gravitation*, Pergamon Press, New York, 1959, Appendix A.

that at $t = 0$ the axes of x and x', y and y', z and z' are collinear and pointing in the same directions, respectively. This ensures that for $\mathbf{v} = 0$ the transformation $S \to S'$ is the identity transformation. We can also choose the x-axis such that the origin of S' as seen by S is moving with velocity $\mathbf{v}$ along the positive x-axis.

Consider now a point in the ($z = 0$)-plane moving uniformly relative to S, and parallel to the x-axis at a distance y. That distance will remain constant relative to S. This same point will move relative to S' parallel to the x'-axis at a distance y' and in the ($z' = 0$)-plane, because its z'-position is independent of t' and is $z' = 0$ at time $t' = 0$. Therefore, y will transform according to

$$y' = ay. \tag{A1-1}$$

The factor a can depend on the velocity but, by isotropy of space, cannot depend on its direction. Thus, $a = a(v^2)$. If the argument leading to (A1-1) is repeated for a transformation from S' to S'' which is moving with velocity $\mathbf{v}$ along the *negative* x'-axis as seen from S', one obtains $y'' = ay'$. But S'' must be identical with S. Therefore, $a^2 = 1$ and $a = +1$. The solution $a = -1$ is ruled out because for $v = 0$ we must have the identity transformation. Thus,

$$y' = y \quad \text{and} \quad z' = z. \tag{A1-2}$$

The latter equation follows by interchanging y, y' with z, z' in the above argument.

One is left with the transformation

$$ct' = \alpha x + \beta ct,$$
$$x' = \gamma x + \delta ct. \tag{A1-3}$$

The coefficients α, β, γ, δ are functions of $\mathbf{v}$. Since we require the identity transformation for $\mathbf{v} = 0$,

$$\beta(0) = \gamma(0) = 1, \quad \alpha(0) = \delta(0) = 0. \tag{A1-4}$$

Furthermore, the origin of S (velocity $dx/dt = 0$) has velocity $dx'/dt' = -v$ relative to S'. Differentiating (A1-3) with respect to t, we find*

$$(dx'/dt)/(dt'/dt) = c\delta/\beta = -v. \tag{A1-5}$$

Conversely, the origin of S' (velocity $dx'/dt' = 0$) has velocity $dx/dt = v$ relative to S, yielding

$$-(dx/dt')/(dt/dt') = c\delta/\gamma = -v. \tag{A1-6}$$

These two equations give $\beta = \gamma$.

_____

* Clearly dt'/dt can never vanish.

So far, only the principle of relativity was used, i.e. the equivalence of inertial observers.

In the Newtonian case one assumes, in addition, the absolute nature of time (except for changes of origin), so that $\alpha = 0$, $\beta = 1$. Combined with (A1-5) and (A1-6), this yields for the transformation (A1-3),

$$x' = x - vt, \qquad t' = t, \tag{A1-7}$$

the *Galilean* transformation.

In special relativity, the principle (2) of the constancy of the velocity of light is added to the principle of relativity. It requires that

$$x^2 + y^2 + z^2 - c^2t^2 = 0$$

should imply

$$x'^2 + y'^2 + z'^2 - c^2t^2 = 0. \tag{A1-8}$$

The transformation (A1-3) must therefore yield

$$x^2 + y^2 + z^2 - c^2t^2 = a(x'^2 + y'^2 + z'^2 - c^2t'^2). \tag{A1-9}$$

But $a = 1$ because of (A1-2) and the fact that (A1-3) is homogeneous. Thus,

$$x^2 - c^2t^2 = x'^2 - c^2t^2 \tag{A1-10}$$

is the restriction on the coefficients of (A1-3) by the principle (2). It yields

$$\beta^2 - \delta^2 = 1, \qquad \gamma^2 - \alpha^2 = 1, \qquad \alpha\beta = \gamma\delta. \tag{A1-11}$$

Combined with (A1-4) and (A1-5), the coefficients are thereby uniquely determined as

$$\beta = \gamma, \qquad \alpha = \delta = -\frac{v\gamma}{c}, \qquad \gamma = \frac{1}{\sqrt{1 - v^2/c^2}}. \tag{A1-12}$$

The desired transformation (A1-3) finally results as

$$t' = \gamma(t - vx/c^2),$$
$$x' = \gamma(x - vt). \tag{A1-13}$$

Equations (A1-2) and (A1-13) constitute the *Lorentz* transformation in the x-direction from S to S'. Its inverse is easily seen to be

$$t = \gamma(t' + vx'/c^2),$$
$$x = \gamma(x' + vt'), \tag{A1-14}$$

and therefore differs from (A1-13) only by the replacement of v by $-v$, as was to be expected.

The relationship between the Galilean transformations (A1–7) and the Lorentz transformations (A1–13) is now evident: if $v/c \ll 1$ so that terms of order $(v/c)^2$ are negligible, the Lorentz transformation reduces to the Galilean transformation.* This limiting process also expresses the relationship between Newtonian physics and special relativity, because these theories can be characterized by their invariance properties under these transformation groups (see, however, Section 8–1).

A1-2 LORENTZ TRANSFORMATIONS AND SPACE ROTATIONS

The combination of the above Lorentz transformations with the space rotations can be carried out in two steps by means of the mathematical formalism indigenous to space rotations, namely three-vectors. If each observer, S as well as S', rotates his coordinates by the same angle and about the same axis (e.g. by an angle α about the z- and z'-axes, respectively), one obtains a transformation in which v is no longer collinear with the x- and x'-axes. But the coordinate axes of S and S' are still parallel, since they were rotated by the same angle α. The corresponding Lorentz transformation is obtained from (A1–2) and (A1–13) by expressing it in terms of $\mathbf{r} = (x, y, z)$ and $\mathbf{r}' = (x', y', z')$. This yields the transformation (3–9) of Section 3–6 which is valid for arbitrary $\mathbf{v}$ as long as S and S' are parallel.

The second step consists in rotating the axes of only one of the two systems, S', say. This yields the most general homogeneous Lorentz transformation without inversion,

$$\mathbf{R} \cdot \mathbf{r}' = \gamma(\mathbf{r} - \mathbf{v}t) + (\gamma - 1)\hat{\mathbf{v}} \times (\hat{\mathbf{v}} \times \mathbf{r})$$
$$t' = \gamma(t - \mathbf{r} \cdot \mathbf{v}/c^2), \tag{A1-15}$$

in which the axes of S and S' are no longer parallel. This combination of "pure" Lorentz transformations and space rotations is essential for the group character of these transformations: only when the Lorentz transformations are generalized to (A1–15) will they form a group. When two Lorentz transformations of the type (A1–13) or (3–9) with axes parallel to each other are carried out in succession, $S \rightarrow S'$ and $S' \rightarrow S''$ say, the resulting transformations $S \rightarrow S''$ will *not* involve parallel axes but will be of type (A1–15) unless the velocities $\mathbf{v}(S \rightarrow S')$ and $\mathbf{v}(S' \rightarrow S'')$ are collinear.†

The generalization of (A1–13) to (A1–15) can be obtained in a much more elegant way as a rotation in a four-dimensional space. To this end, one notices

---

* Note that one must assume here that x and vt are of the same order of magnitude so that x/ct is of order v/c.

† This fact is responsible for the Thomas precession observable, for example, in atomic spectra.

first that (A1–13) is a rotation in a two-dimensional hyperbolic space x^1 and $x^0 = ct$,

$$x^{0'} = x^0 \cosh \theta - x^1 \sinh \theta$$
$$x^{1'} = -x^0 \sinh \theta + x^1 \cosh \theta \qquad \text{(A1–16)}$$
$$\cosh \theta = \gamma, \qquad \sinh \theta = v\gamma/c.$$

Indeed, with $\theta = i\alpha$ and $x^0 = -ix^4$ this transformation reduces to the rotations in a two-dimensional Euclidean space (x^1, x^4). The rotations in three-space $(x = x^1, y = x^2, z = x^3)$ can therefore be combined with the Lorentz transformations of type (A1–16) to yield a general rotation in a four-dimensional space x^μ $(\mu = 0, 1, 2, 3)$. This realization and the development of the corresponding formalism we owe to a large extent to Minkowski.

The starting point is Eq. (A1–9) with $a = 1$, which expresses the invariance of the square of the "length" of a vector in a four-dimensional space under the transformations in question. We define this space as the set of all points x (called *events*) which are characterized by the four *contravariant* components x^μ $(\mu = 0, 1, 2, 3)$: $x^0 = ct$, $x^1 = x$, $x^2 = y$, $x^3 = z$, which are always real. Two sets x^μ and y^μ determine a product

$$(x, y) \equiv \eta_{\mu\nu}x^\mu y^\nu \qquad \text{(summation convention*)}. \qquad \text{(A1–17)}$$

The *metric* $\eta_{\mu\nu}$ is so defined that

$$(x, x) = x^2 + y^2 + z^2 - c^2t^2. \qquad \text{(A1–18)}$$

This determines $\eta_{\mu\nu}$ as

$$\eta_{\mu\nu} = \begin{array}{c|cccc} \mu\backslash\nu & 0 & 1 & 2 & 3 \\ \hline 0 \\ 1 \\ 2 \\ 3 \end{array} \left(\begin{array}{cccc} -1 & 0 & 0 & 0 \\ 0 & 1 & 0 & 0 \\ 0 & 0 & 1 & 0 \\ 0 & 0 & 0 & 1 \end{array} \right). \qquad \text{(A1–19)}$$

It is convenient to define for each x also four *covariant* components x_μ which are given in terms of the contravariant ones by

$$x_\mu \equiv \eta_{\mu\nu}x^\nu. \qquad \text{(A1–20)}$$

The product (A1–17) can therefore also be written as

$$(x, y) = x_\mu y^\mu = x^\mu y_\mu. \qquad \text{(A1–21)}$$

Since the "square" of x can take on both positive and negative values, as is evident from (A1–18), the metric (A1–19) is called *indefinite*. This four-

---

* All Greek indices which occur twice, once as subscript and once as superscript, are to be summed over from 0 to 3.

dimensional space is therefore called *pseudo-Euclidean*, a Euclidean space being positive definite. It is referred to as *Minkowski space.*

The lowering of indices by means of $\eta_{\mu\nu}$ can be supplemented by a raising of indices by means of $\eta^{\mu\nu}$, defined as the reciprocal of $\eta_{\mu\nu}$,

$$\eta^{\mu\alpha}\eta_{\nu\alpha} = \delta^\mu_\nu, \tag{A1-22}$$

where δ^μ_ν is the Kronecker delta ($\delta^\mu_\nu = 1$ for $\mu = \nu$, 0 otherwise). Thus, $\eta^{\mu\nu}$ is numerically identical with $\eta_{\mu\nu}$ [Eq. (A1-19)]. The raising of indices now proceeds according to $x^\mu = \eta^{\mu\nu}x_\nu$.

We now seek the most general homogeneous linear transformation on x,

$$x'^\mu = \alpha^\mu{}_\rho\, x^\rho, \tag{A1-23}$$

which will preserve the square,

$$\eta_{\mu\nu}x'^\mu x'^\nu = \eta_{\mu\nu}\alpha^\mu{}_\rho\alpha^\nu{}_\sigma x^\rho x^\sigma = \eta_{\rho\sigma}x^\rho x^\sigma.$$

Thus,*

$$\eta_{\mu\nu}\,\alpha^\mu{}_\rho\,\alpha^\nu{}_\sigma = \eta_{\rho\sigma}. \tag{A1-24}$$

Multiplication of (A1-23) by $\alpha_{\mu\sigma} \equiv \eta_{\mu\nu}\alpha^\nu{}_\sigma$ yields, because of the restriction (A1-24),

$$x_\sigma = \alpha_{\mu\sigma}x'^\mu \quad \text{or} \quad x^\rho = \alpha_\mu{}^\rho\, x'^\mu. \tag{A1-25}$$

The last step requires index raising by multiplication with $\eta^{\rho\sigma}$.

This equation is the inverse of (A1-23). The invariance of the square therefore yields a restriction† on $\alpha_\mu{}^\rho$ analogous to (A1-24),

$$\eta_{\rho\sigma}\,\alpha_\mu{}^\rho\,\alpha_\nu{}^\sigma = \eta_{\mu\nu}. \tag{A1-26}$$

In this way the transformation (A1-22) and its inverse (A1-24) for contravariant components imply the transformation of covariant components according to

$$x_\mu{}' = \alpha_\mu{}^\rho x_\rho, \tag{A1-27}$$

$$x_\rho = \alpha^\mu{}_\rho x'_\mu. \tag{A1-28}$$

The transformation $x \to x'$ in terms of covariant components is exactly that of $x' \to x$ in terms of contravariant components, and vice versa.

The identification of x^μ with ct, x, y, z is more than formal. The metric (A1-19) ensures an identification of both the contravariant and the covariant

---

* This condition will also ensure the invariance of the product (A1-21) of any two vectors.

† Note that (A1-26) is not a *new* restriction but is equivalent to (A1-24). They can both be written as $\alpha^\mu{}_\nu\,\alpha_\nu{}^\rho = \delta^\mu_\nu$, (Eq. 3-17), by shifting the indices up and down appropriately. The beginner should pay special attention to the order of indices since $\alpha^\mu{}_\nu \neq \alpha_\nu{}^\mu$ in general. This order can be ignored only for symmetric quantities such as $\delta_\mu{}^\nu = \delta^\nu{}_\mu = \delta^\nu_\mu$.

components of x with rectilinear orthogonal coordinates of a point. Correspondingly, x can be regarded as a vector in this four-dimensional space. More precisely, a *four-vector* A is a set of four quantities which transform according to either (A1-23) (contravariant components A^μ) or (A1-27) (covariant components A_μ). The vector is called *spacelike*, *timelike*, or *null* when $A_\mu A^\mu$ is positive, negative, or zero.

The definition of four-vectors given above can be generalized to tensors. $T_{\mu\nu\dots}$ are the covariant components of a tensor T, provided they transform like $x_\mu x_\nu \dots$. Analogous definitions hold for $T^{\mu\nu\dots}$, the contravariant components of T, and $T^{\mu\dots}_{\nu\dots}$, its mixed components.*

The graphical representation of the Lorentz transformation in Minkowski space corresponds to the generalization (A1-16) of rotations in a Euclidean plane. Specifically, the transformation (A1-13) yields an axis system, S', which appears oblique when viewed from S. The parallel projections of a vector V in the $x'_1 - x'_0$ plane, V'_1 and V'_0, can be related to the components of V in S, V_1 and V_0, via the orthogonal projections OA and OB (see Fig. A1-1).

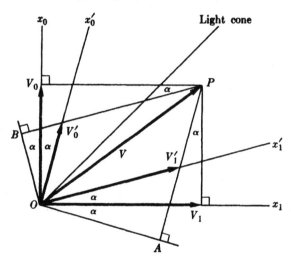

Fig. A1-1. Graphical representation of a Lorentz transformation in Minkowski space. The orthogonality of the two coordinates x'_1 and x'_0 in the moving frame appears graphically as follows: the light cone from O bisects the angle between x'_1 and x'_0.

Let $\alpha < \pi/4$ be the angle between the x_1 and x'_1 axes as well as between the x_0 and x'_0 axes, such that the x'_1 and x'_0 axes make an angle of $\pi/2 - 2\alpha$. The orthogonal projections of V are

$$OA = V_1 \cos \alpha - V_0 \sin \alpha,$$
$$OB = V_0 \cos \alpha - V_1 \sin \alpha.$$

---

* Since the transformations under consideration are orthogonal ones, these tensors and the four-vectors are more precisely known as Cartesian tensors and vectors.

The parallel projections are given by

$$V'_1 = \frac{OA}{\cos 2\alpha} = V_1 \frac{\cos \alpha}{\cos 2\alpha} - V_0 \frac{\sin \alpha}{\cos 2\alpha},$$

$$V'_0 = \frac{OB}{\cos 2\alpha} = V_0 \frac{\cos \alpha}{\cos 2\alpha} - V_1 \frac{\sin \alpha}{\cos 2\alpha}.$$

Therefore, the equalities

$$\gamma = \frac{\cos \alpha}{\cos 2\alpha} > 1, \qquad \frac{v}{c} = \tan \alpha < 1$$

identify this transformation with the Lorentz transformation (A1–13). The restriction $\alpha < \pi/4$, which is equivalent to $v/c < 1$, ensures $\gamma > 1$.

As a result of the representation in Fig. A1–1, the orthogonality of space and time in S' is lost when viewed from S. Correspondingly, the normal to the plane $t' = \text{const.}$ is as indicated, for example, in Fig. A1–3 of Section A1–5.

A1-3 THE LORENTZ GROUPS

The 16 real parameters $\alpha^\mu{}_\nu$ of (A1–23) which determine the transformations are restricted by the 10 conditions (A1–24) and therefore specify (A1–23) as a 6-parameter family of transformations. Three parameters, $\alpha^i{}_k = \alpha_{ik} = -\alpha_{ki}$, involve only "space"-indices* ($i = 1, 2, 3; k = 1, 2, 3$) and describe the space rotations (3–2). The remaining three linearly independent parameters can be given by $\alpha^i{}_0$ or $\alpha^0{}_i$ and characterize the Lorentz transformations proper. Their knowledge is equivalent to the knowledge of the relative velocity three-vector $\mathbf{v}$. It can easily be verified that (A1–23) reduces to (A1–16) for the case

$$\alpha^0{}_0 = \alpha^1{}_1 = \gamma, \qquad \alpha^0{}_1 = \alpha^1{}_0 = -v\gamma,$$
$$\alpha^0{}_k = \alpha^k{}_0 = \alpha^1{}_k = \alpha^k{}_1 = 0, \qquad \alpha^k{}_l = \delta^k_l, \qquad (k = 2, 3; l = 2, 3).$$

In this way the equivalence of (A1–23), restricted by (A1–24), with the transformations (A1–15) is established, provided that no coordinate inversions are permitted and that the identity transformation can always be reached in a suitable limit, $\alpha^\mu{}_\nu \to \delta^\mu{}_\nu$. This can be ensured by specifying the determinant

$$\det \alpha^\mu{}_\nu = +1 \qquad (A1–29)$$

and the sign of $\alpha^0{}_0$,

$$\alpha^0{}_0 > 0. \qquad (A1–30)$$

---

* Latin indices will always be assumed to range from 1 to 3 only.

With these restrictions, (A1-23) characterizes the group* of *proper ortho-chronous homogeneous* Lorentz transformations $L_+^\uparrow$ [corresponding to (A1-15)].

This group can be extended to include improper transformations (det $\alpha^\mu{}_\nu = -1$), yielding the orthochronous homogeneous group $L^\uparrow$; a further generalization to $\alpha^0{}_0 \gtrless 0$ yields the (extended) homogeneous group L.

The 10-parameter *inhomogeneous Lorentz groups* are obtained by pairing these groups with the translation group (3-13), yielding $\mathcal{L}_+^\uparrow$, $\mathcal{L}^\uparrow$, and $\mathcal{L}$, respectively.

A1-4 PROPER TIME AND KINEMATICS

The specification of Lorentz transformations as *orthogonal* linear transformations is basic. The metric tensor $\eta_{\mu\nu}$, (A1-19), is correspondingly diagonal. Of course, it is possible to introduce arbitrary curvilinear coordinates in three-space x, y, z; but this is a trivial generalization as far as Lorentz transformations are concerned: the essential feature is the absence of space-time crossterms in the product of two vectors. This *time orthogonality* of the transformations is conditioned by the underlying physics. The argument is as follows.[†]

Assume that one wants to study the feasibility of differential geometry in four-dimensional Minkowski space. To this end, one first expresses the invariant (A1-18) in terms of infinitesimal elements,

$$(dx, dx) \equiv (dx)^2 + (dy)^2 + (dz)^2 - c^2(dt)^2. \qquad (A1-31)$$

The fundamental assumption which permits one to relate Minkowski space to physics is that the natural time rate, i.e. the rate of an atomic clock at rest, is the same for any inertial system and is to be identified with $d\tau > 0$, where

$$(d\tau)^2 \equiv -(dx, dx)/c^2 \qquad (A1-32)$$

whenever $(dx, dx)^2 < 0$ (dx a timelike vector). This invariant time is called *proper* time. When $v/c \ll 1$, this time must be identical with the Newtonian time t associated with the coordinate x^0. Thus, if (A1-31) were generalized to

$$c^2 (d\tau)^2 = c^2 (dt)^2 + 2ac\, dt\, dx - (dx)^2, \qquad (A1-33)$$

assuming $dy = dz = 0$, the relation between the proper time τ and the coordinate time $t = x^0/c$ would become

$$\tau_2 - \tau_1 = \int_1^2 dt\sqrt{1 + 2av/c - (v/c)^2}$$

---

* While the term *Lorentz group* is generally used, the designation *Poincaré group* would be more appropriate, since Poincaré was the first to point out the group property of the Lorentz transformations.

† A. S. Eddington, *The Mathematical Theory of Relativity*, Cambridge University Press, Cambridge, 1924.

with $v = dx/dt$. When $(v/c)^2 \ll 1$,

$$\tau_2 - \tau_1 = t_2 - t_1 + a(x_2 - x_1)$$

and the relation to Newtonian physics is recovered only for $a = 0$. This result requires a restriction of the metric $\eta_{\mu\nu}$ in $\eta_{\mu\nu}\, d\xi^\mu\, d\xi^\nu$, where ξ may be curvilinear coordinates, to the time-orthogonal case $\eta_{k0} = 0$ which does not allow (A1-33). This means that the use of curvilinear coordinates is restricted to the subspace of Minkowski space which is ordinary three-space. In general relativity, where the presence of a gravitational field limits the choice of $g_{\mu\nu}$, this argument is no longer valid unless it is restricted to the local tangent space (see Appendix 2).

Since each point in Minkowski space specifies a position at a particular time (*event*), the motion of a mass point will be described by a line (*world line*) $x^\mu(\tau)$ which has a timelike unit tangent vector, $n^\mu = v^\mu/c$ [cf. (A1-37) below] at every point, because no particle can attain the velocity of light in vacuo.

The specification of proper time permits one to generalize the basic kinematical concepts of velocity, acceleration, momentum, etc., from the three-vector formalism characteristic for the description of invariance under the group of rotations in three-dimensional Euclidean space to the four-vector formalism characteristic for the description of invariance under the group of rotations in four-dimensional Minkowski space. From (A1-12), (A1-13), and (A1-32) one obtains for the motion of a point mass m

$$\mathbf{v} = \left(\frac{dx}{dt}, \frac{dy}{dt}, \frac{dz}{dt}\right), \tag{A1-34}$$

$$\frac{d\tau}{dt} = \sqrt{1 - v^2/c^2} = 1/\gamma, \tag{A1-35}$$

$$v^\mu \equiv \frac{dx^\mu}{d\tau} = \left(c\frac{dt}{d\tau}; \frac{dx}{d\tau}, \frac{dy}{d\tau}, \frac{dz}{d\tau}\right)$$

$$= (\gamma c, \gamma \mathbf{v}). \tag{A1-36}$$

This four-velocity satisfies the identity

$$v_\mu v^\mu = -c^2. \tag{A1-37}$$

Similarly, one defines the four-acceleration

$$a^\mu \equiv \frac{dv^\mu}{d\tau} = \frac{d^2 x^\mu}{d\tau^2} \equiv \ddot{x}^\mu, \tag{A1-38}$$

which can be expressed in terms of $\mathbf{a} \equiv d\mathbf{v}/dt$ [Eq. (5-26)].

The four-momentum is defined by*

$$p^\mu = mv^\mu \equiv (E/c, \mathbf{p}),\qquad\qquad\text{(A1–39)}$$

$$\mathbf{p} = m\gamma\mathbf{v} = \gamma\mathbf{p}_N,\qquad\qquad\text{(A1–40)}$$

in terms of the Newtonian momentum $\mathbf{p}_N$ and the *total* energy

$$E \equiv \gamma mc^2.\qquad\qquad\text{(A1–41)}$$

The mass m is the rest mass of the particle, i.e. the mass measured when it is at rest relative to the measuring inertial observer. The identification of $m\gamma$ with a "transverse mass" and $m\gamma^3$ with a "longitudinal mass," as is the case in the older literature, is due to the useless and uncalled-for attempt to bring $d\mathbf{p}/dt$ into the Newtonian form of mass times acceleration three-vector.

The kinetic energy of a free particle is obtained by omitting the rest energy mc^2 from (A1–41),

$$E_{\text{kin}} \equiv (\gamma - 1)mc^2.\qquad\text{(A1–42)}$$

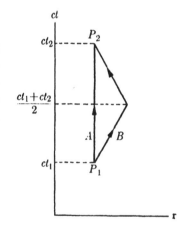

It follows from (A1–36) that a particle moving unaccelerated, i.e. with constant velocity v^μ pursues a straight line in Minkowski space. Such a straight line is also a *geodesic* in this space. This follows from the observation that the variational problem

$$\int \sqrt{|dx^\mu\, dx_\mu|} = \text{extremum}\qquad\text{(A1–43)}$$

for timelike dx, that is

$$0 = \delta\int\sqrt{-dx^\mu\, dx_\mu} = \delta\int\sqrt{-\dot{x}_\mu\dot{x}^\mu}\, d\tau\qquad\text{(A1–44)}$$

Fig. A1–2. Minkowski diagram for the clock paradox.

has the solution $\ddot{x}^\mu = 0$, according to the Euler-Lagrange equations (3–51′).

The proper time, i.e. the time measured by the comoving observer, is exactly given by the length of the geodesic,

$$\int_1^2 \sqrt{-dx_\mu\, dx^\mu} = \int_1^2 \sqrt{-v_\mu v^\mu}\, d\tau = c(\tau_2 - \tau_1),\qquad\text{(A1–45)}$$

using (A1–37). As an example, consider the following two world lines: (A) the world line of a particle at rest during the time interval $t_2 - t_1$ in S, (B) the world line of a particle moving with velocity $\mathbf{v}$ during the time interval between

---

* The collection of the three space-components of p^μ into a three-vector $\mathbf{p}$ which differs from the Newtonian $\mathbf{p}_N$ is of great convenience in relativistic calculations.

t_1 and $t_1 + \frac{1}{2}(t_2 - t_1)$ followed by a motion with velocity $-\mathbf{v}$ during the interval from $t_1 + \frac{1}{2}(t_2 - t_1)$ to t_2. These world lines are shown in Fig. A1-2. When both particles start at the same point, they will meet again at that point. The proper time elapsed for A, who is at rest in S, is

$$\Delta\tau_A = \tau_2 - \tau_1 = t_2 - t_1. \tag{A1-46}$$

According to (A1-35), the proper time elapsed for B, who is at rest in a frame S' during the first half, and in S'' during the second half of his trip, is

$$\Delta\tau_B = (t_2 - t_1)/\gamma = \Delta\tau_A/\gamma. \tag{A1-47}$$

The length of the world line of B is therefore shorter than that of A by a factor $\gamma^{-1} < 1$. Less time is elapsed for B than for A. This result is obvious geometrically. Just as the sum of two sides of a triangle in a Euclidean plane is always larger than the third one, so is that sum *smaller* than the third one in a hyperbolic plane [such as the $(x - ct)$-plane of Minkowski space]. Since the lengths of the world lines are a measure of the proper time elapsed, the result follows: *the longest proper time between two points P_1 and P_2 timelike with respect to each other is that elapsed along a geodesic.*

This result has been repeatedly stated in the form of an apparent contradiction known as the *twin paradox* or *clock paradox*. (Since the author has great difficulty constructing paradoxes from clear mathematical facts, he will not attempt to do so.)

At this point a remark concerning the description of nonuniform motion in special relativity is necessary. Such a motion can be related to an inertial observer at any point of its world line. To this end, one defines the *instantaneous Lorentz frame* as that Lorentz frame in which the particle is at rest at a particular instant. Nonuniform motion can therefore be characterized by the limit $\Delta\mathbf{v} \to 0$ during $\Delta t \to 0$ ($0 < \Delta\mathbf{v}/\Delta t < \infty$) of a discontinuous change $\Delta\mathbf{v}$ from one Lorentz transformation to another, $S \to S'$ with velocity $\mathbf{v}$ followed by $S' \to S''$ with velocity $\mathbf{v} + \Delta\mathbf{v}$. It is assumed that the proper time is thereby not affected; a comoving standard clock will, by assumption, at all times go at a rate equal to that of the clock in the instantaneous Lorentz frame. Consequently, the length of a world line which is not a straight line can still be given correctly by

$$\int_1^2 \sqrt{-dx_\mu\,dx^\mu} = c\int_1^2 d\tau.$$

A repeatedly encountered problem is that of constructing a four-vector in Minkowski space which is orthogonal to a given vector and which satisfies other conditions. Consider, for example, the velocity

$$v^\mu = (\gamma c, \gamma\mathbf{v}) \tag{A1-36}$$

and construct a vector F^μ which is orthogonal to v^μ and which has a given space part, $\gamma \mathbf{F}$. Let

$$F^\mu = (\gamma F^0, \gamma \mathbf{F}). \tag{A1–48}$$

The orthogonality requires that

$$v_\mu F^\mu = \gamma^2 \mathbf{v} \cdot \mathbf{F} - \gamma^2 c F^0 = 0$$

or

$$F^0 = \frac{1}{c} \mathbf{v} \cdot \mathbf{F}.$$

The desired vector is therefore

$$F^\mu = \left(\frac{1}{c} \gamma \mathbf{v} \cdot \mathbf{F}, \gamma \mathbf{F} \right). \tag{A1–49}$$

In the rest system

$$F^\mu_{(0)} = (0, \mathbf{F}_{(0)}). \tag{A1–50}$$

An application of this result is the relativistic form of Newton's second law,

$$m a^\mu = F^\mu. \tag{3–12}$$

Since differentiation of (A1–37) leads to

$$v_\mu a^\mu = 0, \tag{A1–51}$$

Eq. (3–12) will be consistent only when the force four-vector is orthogonal to v^μ. Thus, it must be of the form (A1–49). According to (A1–50), $\mathbf{F}_{(0)}$ can then be identified with the Newtonian (nonrelativistic) force.

A1–5 SPACELIKE PLANES AND GAUSS'S INTEGRAL THEOREM

A spacelike surface in Minkowski space is defined as a three-dimensional *surface* such that the distance between any two points x and y on it is spacelike,

$$(x_\mu - y_\mu)(x^\mu - y^\mu) > 0. \tag{A1–52}$$

As explained in the previous section (time-orthogonality), an instant of an inertial observer is characterized by a spacelike *plane*. It is given geometrically by a unit normal vector n which is necessarily timelike,

$$n_\mu n^\mu = -1, \tag{A1–53}$$

and which can be chosen to point into the future,

$$n^0 > 0. \tag{A1–54}$$

The equation of the spacelike plane σ is

$$n_\mu x^\mu + c\tau = 0. \tag{A1–55}$$

The invariant τ can be identified with the proper time τ defined in (A1-32). To this end, one notes that when $n^\mu = (1; 0, 0, 0)$, $\tau = t$, and the spacelike plane σ describes three-space at the instant t (*now-plane*). When $n^\mu = v^\mu/c$, where v^μ is the four-velocity with which S' is moving relative to S, the plane is "tilted" in S, while a Lorentz transformation to S' transforms σ to the plane $\tau = t'$. Thus, $n^\mu = v^\mu/c$ in (A1-55) describes the three-space $t' = \tau$ in S', as seen by S. The world line of a particle determines uniquely its now-plane at every instant and as seen by any inertial observer.

The surface element on such a plane is given by the vector

$$d\sigma^\mu = n^\mu \, d\sigma, \tag{A1-56}$$

where the invariant area element is most easily determined by its value in the frame S_0 in which $n^\mu = (1; 0, 0, 0)$,

$$d\sigma = -n_\mu \, d\sigma^\mu = dx^{(0)} \, dy^{(0)} \, dz^{(0)}.$$

Since $d\sigma^\mu$ is a vector whenever n^μ is specified accordingly (e.g. as v^μ/c), surface integrals of the type

$$S = \int_\sigma f_\mu(x) \, d\sigma^\mu \tag{A1-57}_S$$

are Lorentz invariant (scalars), while those of the type

$$V_\mu = \int_\sigma f_{\mu\nu}(x) \, d\sigma^\nu \tag{A1-57}_V$$

are vectors, provided, of course, that f_μ is a covariant vector (i.e. the covariant components of a vector) and $f_{\mu\nu}$ a covariant tensor. On the other hand, if n^μ is not specified in such a way that a Lorentz transformation on x induces one on n (i.e. so that n is specified in a Lorentz covariant way), S will not be a scalar and V will not be a vector *unless these integrals are independent of σ.*

Whether integrals of the type (A1-57) are independent of σ will clearly depend on the integrand. The necessary and sufficient condition is

$$\partial^\mu f_\mu = 0 \quad \text{and} \quad \partial^\mu f_{\mu\nu} = 0, \tag{A1-58}$$

respectively, for the above two cases. Here

$$\partial^\mu = \eta^{\mu\nu}\partial_\nu, \quad \partial_\mu \equiv \frac{\partial}{\partial x^\mu} = \left(\frac{1}{c}\frac{\partial}{\partial t}; \frac{\partial}{\partial x}, \frac{\partial}{\partial y}, \frac{\partial}{\partial z}\right) \tag{A1-59}$$

is the four-vector operator of differentiation; the expressions in (A1-58) are correspondingly denoted as four-divergences.

The remainder of this section will be devoted to the proof of the conditions (A1-58) for the surface-independence of the above integrals. We note that Gauss's integral theorem is valid not only in three-dimensional space, where it is so often used, but also in any space of n dimensions. It relates an n-dimensional volume integral to an $(n-1)$-dimensional surface integral, the surface being the boundary of the volume. For $n = 4$ and Minkowski space, we picture a space-time volume V_4 bounded by two spacelike planes σ_1 and σ_2 and a timelike surface Σ (whose normal is spacelike) at large distance. These constitute the three-dimensional boundary surface S_3 of V_4. Gauss's theorem now states that

$$\int_{V_4} \partial^\mu f_{\mu\nu\dots}(x)\, d^4x = \int_{S_3} \epsilon_\sigma f_{\mu\nu\dots}(x)\, d\sigma^\mu. \qquad (A1\text{-}60)$$

Here, $d^4x = dx^0\, dx^1\, dx^2\, dx^3$, $f_{\mu\nu\dots}(x)$ is a tensor function of one or more indices, and the surface integral is understood to be taken with the surface normal pointing *out of* V_4. The sign function ϵ_σ is $+1$ or -1 when n^μ is spacelike or timelike, respectively. If $f_{\mu\nu\dots}(x)$ vanishes sufficiently fast in spacelike infinity so that the surface integral over Σ vanishes,

$$\int_{V_4} \partial^\mu f_{\mu\nu\dots}(x)\, d^4x = -\int_{\sigma_1} f_{\mu\nu\dots}(x)\, d\sigma^\mu + \int_{\sigma_2} f_{\mu\nu\dots}(x)\, d\sigma^\mu, \qquad (A1\text{-}61)$$

where both planes are now taken with their normals pointing into the future $(n^0_{(1)} > 0,\ n^0_{(2)} > 0)$, in accordance with (A1-54). Figure A1-3 shows that the volume integral gives contributions of opposite signs on the two sides of the intersection of σ_1 with σ_2.

The desired result now follows from (A1-61). If the integrals (A1-57)$_S$ or (A1-57)$_V$ are independent of σ, the right-hand side of (A1-61) vanishes for any two planes σ_1 and σ_2. This is possible only when the integrand on the left-hand side vanishes, i.e. when (A1-58) is satisfied.

If, in particular, the two planes σ_1 and σ_2 are parallel, i.e. if both are associated with the same normal vector n,

$$\sigma_1: \quad n^\mu x_\mu + c\tau_1 = 0 \qquad x\epsilon\sigma_1,$$
$$\sigma_2: \quad n^\mu x_\mu + c\tau_2 = 0 \qquad x\epsilon\sigma_2,$$

then, in the limit $\tau_1 = \tau_2 + d\tau$, (A1-61) implies that

$$\int_\sigma \partial^\mu f_{\mu\nu\dots}(x)\, d\sigma = -\frac{d}{c\,d\tau} \int_\sigma f_{\mu\nu\dots}(x)\, d\sigma^\mu \qquad (A1\text{-}62)$$

because $d^4x = c\, d\sigma\, d\tau$. In this equation τ is given by Eq. (A1-55) for σ.

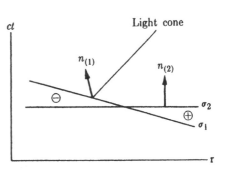

FIG. A1-3. Spacelike planes and their normals. The volume between the planes changes sign beyond their intersection.

The proof of (A1-58) can alternatively be given by means of (A1-62). Firstly, the surface integrals (A1-57) must be invariant under translations of σ parallel to itself. Equation (A1-62) then requires its left-hand side to vanish. Secondly, if this invariance is to hold for an arbitrary choice of n, the conditions (A1-58) follow.

A1–6 THE LIGHT–CONE SURFACE ELEMENT*

Consider an arbitrary event z^μ as the center of a pseudosphere (sphere whose radius is a timelike vector) in its future light cone,

$$R^\mu R_\mu = -\lambda^2, \tag{A1–63}$$

with $R^\mu \equiv x^\mu - z^\mu$, x^μ being an event on the (three-dimensional) surface of that pseudosphere; $\lambda > 0$ is its "radius." Its normal vector

$$N^\mu = R^\mu/\lambda \tag{A1–64}$$

is timelike according to (A1–63). (Compare with Fig. A1–4.) The surface element of this pseudosphere will be denoted by $d\sigma_p$,

$$d\sigma^\mu{}_p = N^\mu \, d\sigma_p = R^\mu \, d\sigma_p/\lambda. \tag{A1–65}$$

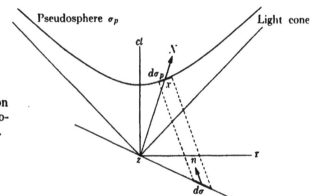

FIG. A1–4. The projection of an element of a pseudo-sphere on a spacelike plane.

We want to compare $d\sigma_p$ with the surface element $d\sigma$ of a spacelike hyper-plane with normal n. It is given by

$$d\sigma^\mu = n^\mu \, d\sigma$$

according to (A1–56). By the projection formula*

$$d\sigma = |N^\mu n_\mu| \, d\sigma_p = |R^\mu n_\mu| \, d\sigma_p/\lambda$$

so that

$$d\omega \equiv \frac{d\sigma_p}{\lambda} = \frac{d\sigma}{|R_n|} \qquad (R_n \equiv R^\mu n_\mu). \tag{A1–66}$$

The essential point here is that the right side of this equation is independent

---

* The following discussion is based on J. L. Synge, *The Special Theory of Relativity,* North-Holland Publishing Company, Amsterdam, 1956, Appendix D.

of n since the left side clearly does not depend on it; it is therefore an invariant. Thus, with (A1–65),

$$d\sigma^\mu{}_p = R^\mu \, d\omega = R^\mu \, d\sigma/|R_n|. \tag{A1–67}$$

In order to ensure that the surface normal points into the future, a minus sign has to be inserted in (A1–64) and (A1–67) in case the pseudosphere lies in the past light cone of z.

This result for the surface element of a pseudosphere can be extended to hold also for the surface element of the light cone. In the limit $\lambda \to 0$ the pseudosphere (A1–63) goes over into the light cone

$$R^\mu R_\mu = 0. \tag{A1–68}$$

At the same time, the two-dimensional differential volume element $d\omega$ is still meaningfully represented by the arbitrarily chosen spacelike plane σ by means of (A1–66). Thus, for the light cone,

$$d\sigma^\mu{}_l = R^\mu \, d\omega = R^\mu \, d\sigma/|R_n| \tag{A1–69}$$

with R being the null vector (A1–68).

The Space-Time of General Relativity

As is evident from Chapters 1, 3, and 8, a full appreciation of a theory based on special relativity requires at least some notion of its relation to general relativity. The basic mathematical tool of the theory is a four-dimensional Riemannian space with indefinite metric. This appendix presents very briefly the most important concepts, definitions, and theorems which *lead to the adoption of such a space*. It will enable the reader to follow the corresponding discussions in the text without previous knowledge of general relativity.

A2-1 GENERALIZING MINKOWSKI SPACE

We wish to generalize the four-dimensional space-time continuum characterized by orthogonal rectilinear coordinates x^μ and by the inner product of Eqs. (A1–17) through (A1–19) of Appendix 1. This generalization should proceed from a flat space to a curved space in analogy to the generalization of Cartesian coordinates in a plane to those on a wiggly surface.

As a first step in this direction, one must work with curvilinear rather than rectilinear coordinates. Let x^μ ($\mu = 0, 1, 2, 3$) be such coordinates of a given point and let

$$x^\mu \rightarrow x'^\mu = f^\mu(x) \tag{A2-1}$$

be a transformation to another set of curvilinear coordinates of that same point; let a neighboring point have coordinates $x^\mu + dx^\mu$. Then dx^μ transforms according to*

$$dx^\mu \rightarrow dx'^\mu = \frac{\partial f^\mu}{\partial x^\nu}\, dx^\nu = \frac{\partial x'^\mu}{\partial x^\nu}\, dx^\nu. \tag{A2-2}$$

A *contravariant vector* is a set of four quantities A^μ ($\mu = 0, 1, 2, 3$) which transform in the same way as dx^μ. These quantities may themselves be functions of x:

$$A^\mu(x) \rightarrow A'^\mu(x') = \frac{\partial x'^\mu}{\partial x^\nu}\, A^\nu(x). \tag{A2-3}$$

---

* Note the implicit assumption of differentiability of (A2–1). We use the summation convention as in Appendix 1.

Any coordinate transformation (A2–1) is assumed to have an inverse

$$x'^\mu \to x^\mu = g^\mu(x') = g^\mu(f(x)). \tag{A2-4}$$

The inverse to (A1–2) is therefore

$$dx'^\mu \to dx^\mu = \frac{\partial g^\mu(x')}{\partial x'^\nu} dx'^\nu = \frac{\partial x^\mu}{\partial x'^\nu} dx'^\nu. \tag{A2-5}$$

It is clear that the existence of the inverse requires that the Jacobian, i.e. the determinant of the coefficient in (A2–2), does not vanish,*

$$\det\left(\frac{\partial x'^\mu}{\partial x^\nu}\right) \neq 0. \tag{A2-6}$$

A *scalar* or invariant function $S(x)$ is defined by

$$S'(x') = S(x) \tag{A2-7}$$

under a coordinate transformation. The derivative of a scalar, i.e. the set

$$\partial_\mu S(x) \equiv \frac{\partial S(x)}{\partial x^\mu} \qquad (\mu = 0, 1, 2, 3) \tag{A2-8}$$

does not transform like a contravariant vector. Rather,

$$\partial'_\mu S'(x') = \partial'_\mu S(x) = \frac{\partial x^\nu}{\partial x'^\mu} \partial_\nu S(x).$$

Any set $B_\mu(x)$ which transform in this way,

$$B_\mu(x) \to B'_\mu(x') = \frac{\partial x^\nu}{\partial x'^\mu} B_\nu(x), \tag{A2-9}$$

is called a *covariant vector*. The transformation coefficients in (A2–9) are exactly those of the inverse transformation (A2–5), but the numerator is summed rather than the denominator.

Equations (A2–3) and (A2–9) are the generalizations of (A1–23) and (A1–27). The generalization of a *tensor* in Minkowski space is then clearly given by the characterizing transformation

$$T'^{\mu\nu\cdots}{}_{\alpha\beta\ldots}(x') = \partial_\kappa x'^\mu \partial_\lambda x'^\nu \cdots \partial'_\alpha x^\rho \partial'_\beta x^\sigma \cdots T^{\kappa\lambda\cdots}{}_{\rho\sigma\ldots}(x). \tag{A2-10}$$

Upper indices transform according to the contravariant law (A2–3); lower ones, according to the covariant law (A2–9), *provided* the quantity is a tensor.

---

* A coordinate transformation can be defined by (A2–2) and (A2–6) and need not be integrable into (A2–1).

The definitions of covariant and contravariant tensors is such that the product of a covariant and a contravariant vector (tensor of first rank) satisfies

$$A^\mu B_\mu = A^\mu(x) B_\mu(x) = A'^\nu(x') B_\nu(x') \tag{A2–11}$$

and is an invariant. This can easily be verified by application of (A2–3) and (A2–9).

The notion of distance is introduced by means of a symmetric, covariant tensor function of second rank, $g_{\mu\nu}(x)$, called the *metric tensor*. The infinitesimal distance ds of two neighboring points is then defined by

$$ds = +\sqrt{(ds)^2}$$

and*

$$(ds)^2 \equiv \epsilon g_{\mu\nu}(x)\, dx^\mu\, dx^\nu, \qquad g_{\mu\nu} = g_{\nu\mu}. \tag{A2–12}$$

Here, $\epsilon = \pm 1$ and is so chosen that $(ds)^2 \geq 0$; ds is clearly an invariant. In the special case $g_{\mu\nu}(x) = \eta_{\mu\nu}$ with $\eta_{\mu\nu}$ given by (A1–19), Eq. (A2–12) reduces to the Minkowski line element (A1–31). We can then again speak of spacelike and timelike separation of the two points when ϵ must be chosen $+1$ and -1, respectively, in order to yield $(ds)^2 > 0$. If $(ds)^2 = 0$, the distance is called null.

In order to ensure that Minkowski space be recovered at least locally from our space (Section A2–5), it is necessary to restrict $g_{\mu\nu}$ by the requirements

$$g_{00} < 0, \qquad \sum_{i,j=1}^{3} g_{ij}(x)\xi^i\xi^j > 0. \tag{A2–13}$$

The ξ^i are the three space-components of an arbitrary contravariant spacelike vector ξ^μ.

The metric tensor can be used to *define* a covariant vector when a contravariant one is given,

$$A_\mu(x) \equiv g_{\mu\nu}(x) A^\nu(x). \tag{A2–14}$$

Since the determinant

$$g(x) \equiv \det g_{\mu\nu}(x) \tag{A2–15}$$

is assumed to be different from zero (because of A2–13 it is, in fact, negative, $g < 0$), it is possible to define an inverse, $g^{\mu\nu}(x)$, to $g_{\mu\nu}(x)$ by means of

$$g^{\mu\alpha}g_{\alpha\nu} = \delta^\mu_\nu. \tag{A2–16}$$

This *contravariant* metric tensor permits one to define contravariant vectors from covariant ones,

$$B^\mu(x) \equiv g^{\mu\nu}(x) B_\nu(x). \tag{A2–17}$$

---

* The question of nonsymmetric metric tensors will not be considered in this short treatment.

Equations (A2-14) and (A2-17) can be verbally stated, "the metric tensor permits one to move indices up and down."

A2-2 THE COVARIANT DERIVATIVE

In order to develop a calculus of tensors which, in turn, produces quantities of well-defined and general transformation properties, we want the derivative of a vector to transform like a tensor. This is not trivial, because $\partial_\mu A^\nu(x)$ is *not* a tensor. In fact, according to (A2-3)*

$$\partial'_\mu A'^\nu(x') = \partial'_\mu \big(\partial_\beta x'^\nu A^\beta(x)\big) = \partial'_\mu \partial_\beta x'^\nu A^\beta(x) + \partial_\beta x'^\nu \partial'_\mu x^\alpha \, \partial_\alpha A^\beta(x). \tag{A2-18}$$

Therefore, $T_\mu{}^\nu(x) \equiv \partial_\mu A^\nu(x)$ is a tensor, i.e. satisfies the transformation law (A2-10) only when the first term on the right vanishes. But this is in general not the case; using (A2-1), we have

$$\partial'_\mu \partial_\beta x'^\nu = \partial'_\mu x^\alpha \partial_\alpha \partial_\beta f^\nu(x) \neq 0,$$

unless $f^\nu(x)$ is a *linear* function of x. This is exactly fulfilled for Lorentz transformations in Minkowski space. Thus, the problem of defining the derivative of a vector so that it is a tensor does not arise there. It arises when curvilinear coordinates are used.

A *covariant* derivative ∇_μ can be defined by means of a *linear connection* $L^\nu_{\mu\alpha}(x)$. The latter is a set of functions which is so chosen that

$$\nabla_\mu A^\nu(x) \equiv \partial_\mu A^\nu(x) + L^\nu_{\mu\alpha}(x) A^\alpha(x) \tag{A2-19}$$

is a tensor. The connection $L^\nu_{\mu\alpha}$ is called *linear*, because the additional term is linear in $A(x)$.

The requirement that $\nabla_\mu A^\nu(x)$ be a tensor yields a transformation equation for $L^\nu_{\mu\alpha}(x)$. This equation must be independent of $A(x)$ in order to permit this definition of ∇_μ to be characteristic of the space. Subtraction of (A2-18) from

$$\nabla'_\mu A'^\nu(x') = \partial'_\mu x^\alpha \, \partial_\beta x'^\nu \nabla_\alpha A^\beta(x),$$

which must be true by definition, and substitution of (A2-19) yield

$$L'^\nu_{\mu\alpha} A'^\alpha(x') = -\partial'_\mu \partial_\beta x'^\nu A^\beta(x) + \partial'_\mu x^\alpha \, \partial_\beta x'^\nu L^\beta_{\alpha\gamma} A^\gamma(x).$$

One can express $A'^\alpha(x')$ in terms of the unprimed components by means of (A2-3). After relabeling dummy indices, one finds that the coefficients of the components of $A(x)$ must satisfy

$$L'^\nu_{\mu\lambda} \partial_\beta x'^\lambda = -\partial'_\mu \partial_\beta x'^\nu + \partial'_\mu x^\alpha \, \partial_\gamma x'^\nu L^\gamma_{\alpha\beta}, \tag{A2-20}$$

---

* We recall our convention that a derivative such as ∂_α acts only on the immediately following quantity unless indicated otherwise.

so that (A2–19) will be valid for arbitrary vectors $A^\mu(x)$. This equation for $L^\gamma_{\alpha\beta}$ depends only on the coefficients of the covariant and contravariant coordinate transformations, $\partial_\alpha x'^\beta$ and $\partial'_\beta x^\alpha$. Thus, we can define a linear connection as a set of functions $L^\gamma_{\alpha\beta}$ which transform according to (A2–20) under a coordinate transformation. Note that $L^\gamma_{\alpha\beta}$ is *not a tensor*.

Equation (A2–20) can be written more conveniently, by making use of

$$\partial_\rho x'^\mu \, \partial'_\mu x^\alpha = \delta^\alpha_\rho,$$

as

$$L''^\nu_{\mu\lambda} \, \partial_\beta x'^\lambda \, \partial_\rho x'^\mu = -\partial_\rho \, \partial_\beta x'^\nu + \partial_\gamma x'^\nu L^\gamma_{\rho\beta}. \tag{A2–20'}$$

This equation shows that the antisymmetric part of $L^\gamma_{\rho\beta}$,

$$\Omega^\gamma_{\rho\beta} \equiv \tfrac{1}{2}(L^\gamma_{\rho\beta} - L^\gamma_{\beta\rho}), \tag{A2–21}$$

is a tensor, while only the symmetric part,

$$\Gamma^\gamma_{\rho\beta} \equiv \tfrac{1}{2}(L^\gamma_{\rho\beta} + L^\gamma_{\beta\rho}), \tag{A2–22}$$

is not a tensor.

PROBLEM A2–1

Express $\partial_\sigma \, \partial_\rho \, \partial_\beta x'^\nu = \partial_\rho \, \partial_\sigma \, \partial_\beta x'^\nu$ in terms of the connection, using (A2–20'), and establish that the integrability condition of (A2–20) requires the quantity

$$R^\mu{}_{\nu\alpha\beta} \equiv \partial_\beta L^\mu_{\alpha\nu} - \partial_\alpha L^\mu_{\beta\nu} + L^\mu_{\beta\lambda} \, L^\lambda_{\alpha\nu} - L^\mu_{\alpha\lambda} \, L^\lambda_{\beta\nu} \tag{A2–23}$$

to be a tensor.

The covariant derivative of a *contra*variant vector, (A2–19), can be extended to a *co*variant vector by the requirement that ∇_μ satisfy the usual law of differentiation of products (Leibnitz's rule),

$$\nabla_\mu(AB) = (\nabla_\mu A)B + A(\nabla_\mu B),$$

and that it be identical with the usual derivative for scalar functions,

$$\nabla_\mu S(x) = \partial_\mu S(x).$$

When these requirements are applied to the scalar product of a vector with itself, $g_{\mu\nu}A^\mu A^\nu = A^\mu A_\mu$, one finds easily that the covariant derivative of a covariant vector must be

$$\nabla_\mu B_\nu(x) = \partial_\mu B_\nu(x) - L^\alpha_{\mu\nu} B_\alpha(x). \tag{A2–24}$$

Assume we are given a curve in our space $x^\mu = x^\mu(\tau)$, where τ is a suitable scalar parameter. Then

$$\frac{d}{d\tau} \, A^\mu\big(x(\tau)\big) = \frac{dx^\alpha}{d\tau} \, \partial_\alpha A^\mu\big(x(\tau)\big).$$

This quantity will *not* be a tensor, because only the factor $dx^\alpha/d\tau$ is a tensor (a vector is a tensor of first rank). One must therefore define the *covariant* derivative with respect to τ:

$$\frac{\delta}{\delta\tau}\, A^\mu(x(\tau)) \equiv \frac{dx^\alpha}{d\tau}\, \nabla_\alpha A^\mu(x(\tau)) = \frac{dA^\mu}{d\tau} + L_{\alpha\beta}^\mu A^\beta \frac{dx^\alpha}{d\tau}. \qquad \text{(A2-25)}$$

The covariant derivative can also be extended to the differentiation of higher-rank tensors. For example,

$$\nabla_\mu g_{\alpha\beta} = \partial_\mu g_{\alpha\beta} - L_{\mu\alpha}^\rho g_{\rho\beta} - L_{\mu\beta}^\rho g_{\alpha\rho}.$$

A2-3 PARALLELISM AND CURVATURE TENSOR

The study of differentiation in the previous section permits one to develop a complete covariant calculus. However, the geometrical implications are not established until one defines a parallel displacement.

The *ordinary* differential of a vector function $A^\mu(x)$ is clearly

$$dA^\mu(x) = \partial_\alpha A^\mu(x)\, dx^\alpha. \qquad \text{(A2-26)}$$

However, Eq. (A2-25) enables us to define also a *covariant differential*,

$$\delta A^\mu(x) \equiv \nabla_\alpha A^\mu(x)\, dx^\alpha = dA^\mu(x) + L_{\alpha\beta}^\mu(x) A^\beta(x)\, dx^\alpha. \qquad \text{(A2-27)}$$

The latter is a vector; the former is not a vector.

Let P and Q be two neighboring points with coordinates x^μ and $x^\mu + dx^\mu$. Then

$$A^\mu(Q) = A^\mu(x + dx) = A^\mu(x) + \partial_\alpha A^\mu(x)\, dx^\alpha;$$

this means that

$$dA^\mu(x) = A^\mu(Q) - A^\mu(P) = \partial_\alpha A^\mu(x)\, dx^\alpha$$

is the ordinary differential. However, if we displace the vector A^μ from the point P to the point Q *parallel* to itself, we obtain the differential

$$d_{||}A^\mu(x) = A_{||}^\mu(Q) - A^\mu(P),$$

which is in general different from $dA^\mu(x)$. The value depends on the properties of the space. We now *define parallel displacement* by

$$d_{||}A^\mu(x) = -L_{\alpha\beta}^\mu(x) A^\beta(x)\, dx^\alpha. \qquad \text{(A2-28)}$$

This definition implies that the difference between the ordinary displacement and the parallel displacement, $dA^\mu - d_{||}A^\mu$, or between the vector at Q and the parallel displaced vector, $A^\mu(Q) - A_{||}^\mu(Q)$, is the covariant differential (A2-27),

$$dA^\mu - d_{||}A^\mu = [A^\mu(Q) - A^\mu(P)] - [A_{||}^\mu(Q) - A^\mu(P)] = \delta A^\mu. \qquad \text{(A2-29)}$$

It follows that if, in a given vector field $A^\mu(x)$, the covariant derivative vanishes at a given point P, the vectors in the neighborhood of P are parallel to the vector at P,

$$A^\mu(x + dx) \parallel A^\mu(x) \quad \text{if and only if} \quad \delta A^\mu(x) = 0. \qquad \text{(A2–30)}$$

The parallel displacement of a covariant vector is defined analogously,

$$d_{\parallel} B_\mu(x) = +L^\beta_{\alpha\mu}(x) B_\beta(x)\, dx^\alpha. \qquad \text{(A2–31)}$$

The length of a vector, $l^2_A \equiv g_{\mu\nu} A^\mu A^\nu = A^\mu A_\mu$, is invariant under parallel displacement,

$$\delta(A^\mu A_\mu) = dx^\alpha\, \nabla_\alpha (A^\mu A_\mu) = dx^\alpha\, \partial_\alpha (A^\mu A_\mu), \quad \text{so that} \quad d_{\parallel}(A^\mu A_\mu) = 0 \qquad \text{(A2–32)}$$

according to (A2–29).

A linear connection is called an *affine connection* when it determines parallel displacements as in (A2–28) and (A2–31).

Let C be a closed, simply connected curve.* Then the path integral along C of the vector $A^\mu(x)$, displaced parallel with respect to C, is

$$\oint d_{\parallel} A^\mu = -\oint L^\mu_{\alpha\nu} A^\nu\, dx^\alpha.$$

This line integral can be converted into a two-dimensional surface integral extended over the surface enclosed by C. To this end, one needs the generalization of Stokes' integral theorem,

$$\oint B_\alpha\, dx^\alpha = \tfrac{1}{2} \iint (\partial_\alpha B_\beta - \partial_\beta B_\alpha)\, d\sigma^{\alpha\beta},$$

where $d\sigma^{\alpha\beta} = -d\sigma^{\beta\alpha}$ is the two-dimensional surface element. Thus,

$$\oint d_{\parallel} A^\mu = -\tfrac{1}{2} \iint [\partial_\alpha(L^\mu_{\beta\nu} A^\nu) - \partial_\beta(L^\mu_{\alpha\nu} A^\nu)]\, d\sigma^{\alpha\beta}$$

$$= -\tfrac{1}{2} \iint [\partial_\alpha L^\mu_{\beta\nu} A^\nu + L^\mu_{\beta\nu}(-L^\nu_{\alpha\lambda} A^\lambda) - \partial_\beta L^\mu_{\alpha\nu} A^\nu - L^\mu_{\alpha\nu}(-L^\nu_{\beta\lambda} A^\lambda)]\, d\sigma^{\alpha\beta}.$$

The derivatives $\partial_\alpha A^\nu$ and $\partial_\beta A^\nu$ were here eliminated by means of (A2–28), since we are dealing with parallel displacements: $\partial_\alpha A^\nu = -L^\nu_{\alpha\lambda} A^\lambda$. The integral thus becomes

$$\oint d_{\parallel} A^\mu = +\tfrac{1}{2} \iint R^\mu_{\nu\alpha\beta} A^\nu\, d\sigma^{\alpha\beta}, \qquad \text{(A2–33)}$$

where $R^\mu_{\nu\alpha\beta}$ is the tensor (A2–23) first encountered in Problem A2–1. In the

----

* More precisely, we also assume that the enclosed surface is orientable and that C can be shrunk to zero in a continuous fashion.

limit, as C shrinks to zero,

$$\lim_{C \to 0} \frac{\oint d_{\parallel} A^{\mu}}{\iint d\sigma^{\alpha\beta}} = \tfrac{1}{2} R^{\mu}{}_{\nu\alpha\beta} A^{\nu}. \tag{A2-34}$$

When the space is flat, a vector transported along a closed curve in parallel displacement must return to its starting position, pointing in exactly the original direction. This is not the case when the space is curved, as can be studied easily on two-dimensional spaces such as spheres. The tensor $R^{\mu}{}_{\nu\alpha\beta}$ is called the *curvature tensor* or Riemann-Christoffel tensor.

It follows that a space is flat if and only if $R^{\mu}{}_{\nu\alpha\beta} = 0$ everywhere.

A2–4   GEODESICS

A *geodesic* is a curve with the following property: when the tangent vector at P, $t^{\mu}(P)$, is parallel displaced to the neighboring point Q on the curve, $t^{\mu}_{\parallel}(Q)$, it has the same direction as $t^{\mu}(Q)$,

$$t^{\mu}(Q) - t^{\mu}_{\parallel}(Q) = c(\lambda)\, d\lambda\, t^{\mu}, \qquad t^{\mu} = \frac{dx^{\mu}}{d\lambda},$$

$$\frac{\delta}{\delta\lambda} \frac{dx^{\mu}}{d\lambda} = c(\lambda) \frac{dx^{\mu}}{d\lambda}.$$

From (A2–27),

$$\frac{d^2 x^{\mu}}{d\lambda^2} + L^{\mu}_{\alpha\beta} \frac{dx^{\beta}}{d\lambda} \frac{dx^{\alpha}}{d\lambda} = c(\lambda) \frac{dx^{\mu}}{d\lambda}.$$

Introducing the new parameter $\tau = \tau(\lambda)$, this equation becomes

$$\frac{d^2 x^{\mu}}{d\tau^2} + L^{\mu}_{\alpha\beta} \frac{dx^{\alpha}}{d\tau} \frac{dx^{\beta}}{d\tau} = - \left(\frac{d\tau}{d\lambda}\right)^{-2} \frac{dx^{\mu}}{d\tau} \left(\frac{d^2\tau}{d\lambda^2} - c(\lambda) \frac{d\tau}{d\lambda}\right).$$

We now choose τ such that the right-hand side vanishes,

$$\frac{d^2\tau}{d\lambda^2} = c(\lambda) \frac{d\tau}{d\lambda}. \tag{A2-35}$$

If this choice is possible, τ is called an *affine parameter* and the equation which characterizes the curve $x^{\mu}(\tau)$ as a geodesic becomes

$$\frac{\delta^2 x^{\mu}}{\delta\tau^2} = \frac{d^2 x}{d\tau^2} + L^{\mu}_{\alpha\beta} \frac{dx^{\alpha}}{d\tau} \frac{dx^{\beta}}{d\tau} = 0. \tag{A2-36}$$

This equation is sometimes called the *geodesic equation*. But it should be noted that the existence of an affine parameter is not required for the existence of geodesics in a general space.

Under what conditions is a geodesic a curve of extremum length between two given points? Consider the line integral between two points P and Q

and require it to be an extremum,

$$\delta \int_P^Q ds = \delta \int_P^Q \sqrt{\epsilon g_{\mu\nu}(dx^\mu/d\lambda)(dx^\nu/d\lambda)}\, d\lambda = 0, \qquad \text{(A2–37)}$$

according to (A2–12). This variational problem has the well-known Euler-Lagrange equations as a consequence (cf. 3–51'),

$$\left(\frac{d}{d\lambda}\frac{\partial}{\partial(dx^\lambda/d\lambda)} - \frac{\partial}{\partial x^\lambda}\right)\sqrt{\epsilon g_{\mu\nu}\frac{dx^\mu}{d\lambda}\frac{dx^\nu}{d\lambda}} = 0.$$

We carry out the differentiations and then substitute s for λ:

$$ds = \sqrt{\epsilon g_{\mu\nu}(dx^\mu/d\lambda)(dx^\nu/d\lambda)}\, d\lambda. \qquad \text{(A2–38)}$$

The result is

$$\frac{d}{ds}\left(g_{\lambda\nu}\frac{dx^\nu}{ds}\right) - \frac{1}{2}\frac{dx^\alpha}{ds}\frac{dx^\beta}{ds}\frac{\partial g_{\alpha\beta}}{\partial x^\lambda} = 0.$$

Multiplication by $g^{\mu\lambda}$ and summation over λ yield

$$\frac{d^2 x^\mu}{ds^2} + \Gamma^\mu_{\alpha\beta}\frac{dx^\alpha}{ds}\frac{dx^\beta}{ds} = 0, \qquad \text{(A2–39)}$$

where the *Christoffel symbol* is defined by

$$\Gamma^\mu_{\alpha\beta} \equiv \tfrac{1}{2}g^{\mu\lambda}(\partial_\alpha g_{\lambda\beta} + \partial_\beta g_{\lambda\alpha} - \partial_\lambda g_{\alpha\beta}). \qquad \text{(A2–40)}$$

Comparison of this result with (A2–36) tells us that *a geodesic is an extremum curve whenever the affine connection is given by (A2–40)*, i.e. whenever

$$L^\mu_{\alpha\beta} = \Gamma^\mu_{\alpha\beta}. \qquad \text{(A2–41)}$$

The length is thereby an affine parameter. A metric space characterized by (A2–12) and (A2–41) is called a *Riemann space*. It has a symmetric affine connection which is determined completely by the metric tensor and its first derivatives.* Thus, a Riemann space is clearly a very special kind of space, Eq. (A2–41) being a very strong limitation.

PROBLEM A2–2

Show that in a Riemann space $\nabla_\mu g_{\alpha\beta} = 0$. Note that this property implies that $\nabla_\mu A_\alpha = g_{\alpha\beta}\nabla_\mu A^\beta$.

---

* One verifies that (A2–40) satisfies (A2–20).

A2–5 RIEMANN SPACE AND MINKOWSKI SPACE

General relativity postulates a four-dimensional Riemann space with the restrictions (A2–13) as the geometry of the space-time continuum endowed with a gravitational field. Alternatively, one can postulate only (A2–12) and (A2–13) and require that (1) the equation of motion of a neutral test particle freely falling in a gravitational field be described by a timelike geodesic $x^\mu(\tau)$,

$$\frac{d^2 x^\mu}{d\tau^2} + L^\mu_{\alpha\beta} \frac{dx^\alpha}{d\tau} \frac{dx^\beta}{d\tau} = 0, \qquad (A2\text{–}36)$$

with the proper time

$$d\tau = \sqrt{-g_{\mu\nu}\, dx^\mu\, dx^\nu}$$

as the affine parameter and (2) these geodesics be the extremum lines. This will also lead to a Riemann space because of the implied restriction (A2–41). The two requirements can be justified, respectively, as follows: (1) by *the principle of equivalence (E)* of Section 3–8 and (2) by *Hamilton's principle of least action*, which thus continues to be at the basis of the theory.

A variational principle is also used for the fields in general relativity, as one postulates that the presence of a true gravitational field manifests itself as a curvature of the Riemann space whose integral over all space-time is an extremum,*

$$\delta \int R\sqrt{-g}\, d^4 x = 0. \qquad (A2\text{–}42)$$

Here R is the *scalar curvature* obtained from the curvature tensor by contraction,

$$R \equiv R^\mu_\mu, \qquad R_{\mu\nu} \equiv R^\alpha_{\ \mu\nu\alpha}. \qquad (A2\text{–}43)$$

The Euler-Lagrange equations of the variational problem (A2–42) are the Einstein field equations

$$R_{\mu\nu} - \tfrac{1}{2} g_{\mu\nu} R = 0. \qquad (A2\text{–}44)$$

These are the fundamental equations of the general theory of relativity.

This theory is based on a Riemann space. But it must also be the covering theory of special relativity which is based on Minkowski space. It follows therefore that there must exist a relationship of these two spaces in the sense that the latter must be an approximation of the former for sufficiently small gravitational fields. This is a necessity for the logical coherence of physical theory (Section 1–2).

The mathematical formulation of this statement is as follows. First, we can show that given an arbitrary point P in a Riemann space, we can always transform to a coordinate system in which $\Gamma^\mu_{\alpha\beta}(P) = 0$ (*local geodesic coordinate*

---

* The appearance of a factor $\sqrt{-g}$ is due to the fact that the four-dimensional volume element $d^4 x$ is not an invariant while $\sqrt{-g}\, d^4 x$ *is* an invariant; $\sqrt{-g}$ is the Jacobian resulting from the transformation of the invariant volume element to $d^4 x$.

system). In fact, this is possible in much more general spaces so long as the linear connection is symmetric. From (A2-20′) it follows that $L'^\mu_{\alpha\beta}(P) = 0$, provided $L^\mu_{\alpha\beta} = L^\mu_{\beta\alpha}$ and

$$\partial_\alpha \, \partial_\beta x'^\mu = L^\lambda_{\alpha\beta} \, \partial_\lambda x'^\mu. \qquad (A2\text{--}45)$$

In the vicinity of P, the transformation to the geodesic coordinate system x'^μ must therefore be

$$x'^\mu = x^\mu - x^\mu(P) + \tfrac{1}{2}L^\mu_{\alpha\beta}(P)(x^\alpha - x^\alpha(P))[x^\beta - x^\beta(P)], \qquad (A2\text{--}46)$$

because this transformation satisfies (A2-45). Since the transformation implies $\partial x'^\mu/\partial x^\nu = \delta^\mu_\nu$ at the point P, it leaves the components of any tensor invariant. Furthermore,

$$0 = \nabla_\mu g_{\alpha\beta}(P) = \partial_\mu g_{\alpha\beta}(P), \qquad (A2\text{--}47)$$

according to Problem A2-2, such that the coordinates can be so chosen that at P, $g_{\alpha\beta}(P) = \eta_{\alpha\beta}$, the Minkowski metric. Geometrically, this means that we have a tangent space (osculating flat space) at P whose Minkowski metric coincides with $g_{\alpha\beta}$ at P. Any neighborhood of P which is so small (linear dimension $\sim \delta x^\mu$) that $R^\mu_{\ \nu\alpha\beta} \, \delta x^\alpha \, \delta' x^\beta \sim 0$ in a certain approximation, can then be identified with the neighborhood of P in the osculating Minkowski space. Physically, it means that in this approximation the true gravitational field is negligible and that the possibly residual apparent gravitational field (flat space!) can be transformed away. This can be seen from the fact that a freely falling test particle which follows a geodesic through P will at P satisfy*

$$(d^2 x^\mu/d\tau^2) = 0, \qquad (A2\text{--}48)$$

because $L^\mu_{\alpha\beta}(P) = 0$. The local geodesic coordinate system is therefore locally (in the above sense of approximation) an inertial coordinate system.

The logical coherence of general and special relativity now requires that special relativity be valid locally in the inertial system of the geodesic coordinates. We see that *any* geometrization of gravitational interactions will satisfy this requirement *so long as the linear connection is symmetrical.*

The fulfillment of this requirement of logical coherence implies the validity of Einstein's principle of equivalence, according to which the apparent gravitational field can always be transformed away: when the space is flat (which is always true at least locally in some approximation), there always exists a Minkowski coordinate system in which (A2-48) is valid.

The more general symmetric space required by logical coherence and characterized by (A2-12), (A2-13), and $L^\mu_{\alpha\beta} = L^\mu_{\alpha\beta}$ is restricted to a Riemann

---

* A local geodesic coordinate system can be constructed under the above conditions not only at a point but also along a curve C and in particular along a geodesic. (Fermi's theorem, p. 167, T. Levi-Civita, *The Absolute Differential Calculus*, Blackie, London, 1927).

space by the requirement that all geodesics be extremum curves. This is a much stronger restriction than that imposed by Einstein's principle of equivalence and logical coherence.

We can summarize these results as follows.

(1) The principle of equivalence (E) suggests a geometrization of gravitational forces but does not provide a restriction to a particular geometry. The postulate that free particles move only on geodesics is, of course, consistent with (E).

(2) The logical coherence of physical theories requires a theory of gravitation to be a covering theory for Newtonian gravitation theory as well as for special relativity. It restricts the possible geometries to those with symmetric linear connections, so that a tangent space exists which can be identified with Minkowski space. It also restricts the metric tensor by (A2-13).

(3) Einstein's principle of equivalence is fully implied in the requirement of logical coherence. The symmetry of the linear connection ensures the existence of a local inertial system. Logical coherence requires that in this system all of special relativity be valid.

(4) The postulate of Hamilton's principle, a variational principle for the motion of test particles, imposes the strongest restriction on the geometry: it limits the linear connection to (A2-40) and the space to a Riemann space. Only in such a space can geodesics be extremum curves.

Indices

Author Index

Subject Index